PHYSICAL CHEMISTRY RESEARCH FOR ENGINEERING AND APPLIED SCIENCES

VOLUME 1

Principles and Technological Implications

Edited by

Eli M. Pearce, PhD, Bob A. Howell, PhD,
Richard A. Pethrick, PhD, DSc, and
Gennady E. Zaikov, DSc

AAP | APPLE ACADEMIC PRESS

Apple Academic Press Inc.
3333 Mistwell Crescent
Oakville, ON L6L 0A2
Canada

Apple Academic Press Inc.
9 Spinnaker Way
Waretown, NJ 08758
USA

Exclusive worldwide distribution by CRC Press, a member of Taylor & Francis Group

Printed in the United States of America on acid-free paper

International Standard Book Number-13: 978-1-77188-053-4 (Hardcover)

Library and Archives Canada Cataloguing in Publication

Physical chemistry research for engineering and applied sciences.

Includes bibliographical references and index.
Contents: Volume 1. Principles and technological implications/edited by Eli M. Pearce, PhD, Bob A. Howell, PhD, Richard A. Pethrick, PhD, DSc, and Gennady E. Zaikov, DSc
ISBN 978-1-77188-053-4 (v. 1 : bound)
1. Chemistry, Physical and theoretical. 2. Chemistry, Technical. 3. Physical biochemistry. I. Pearce, Eli M., author, editor II. Howell, B. A. (Bobby Avery), 1942-, author, editor III. Pethrick, R. A. (Richard Arthur), 1942-, author, editor IV. Zaikov, G. E. (Gennadii Efremovich), 1935-, author, editor
QD453.3.P49 2015 541 C2015-900409-8

Library of Congress Cataloging-in-Publication Data

Physical chemistry research for engineering and applied sciences / Eli M. Pearce, PhD, Bob A. Howell, PhD, Richard A. Pethrick, PhD, DSc, and Gennady E. Zaikov, DSc.

volumes cm
Includes bibliographical references and index.
Contents: volume 1. Principles and technological implications
ISBN 978-1-77188-053-4 (alk. paper)
1. Chemistry, Physical and theoretical. 2. Chemistry, Technical. 3. Physical biochemistry. I. Pearce, Eli M. II. Howell, B. A. (Bobby Avery), 1942- III. Pethrick, R. A. (Richard Arthur), 1942- IV. Zaikov, G. E. (Gennadii Efremovich), 1935-

QD453.3.P49 2015 541--dc23 2015000878

Apple Academic Press also publishes its books in a variety of electronic formats. Some content that appears in print may not be available in electronic format. For information about Apple Academic Press products, visit our website at **www.appleacademicpress.com** and the CRC Press website at **www.crcpress.com**

ABOUT THE EDITOR

Eli M. Pearce, PhD

Dr. Eli M. Pearce was the President of the American Chemical Society. He served as Dean of the Faculty of Science and Art at Brooklyn Polytechnic University in New York, as well as a Professor of Chemistry and Chemical Engineering. He was the Director of the Polymer Research Institute, also in Brooklyn. At present, he consults for the Polymer Research Institute. As a prolific author and researcher, he edited the *Journal of Polymer Science* (Chemistry Edition) for 25 years and was an active member of many professional organizations.

Bob A. Howell, PhD

Bob A. Howell, PhD, is a Professor in the Department of Chemistry at Central Michigan University in Mount Pleasant, Michigan. He received his PhD in physical organic chemistry from Ohio University in 1971. His research interests include flame-retardants for polymeric materials, new polymeric fuel-cell membranes, polymerization techniques, thermal methods of analysis, polymer-supported organoplatinum antitumor agents, barrier plastic packaging, bioplastics, and polymers from renewable sources.

Richard A. Pethrick, PhD, DSc

Professor R. A. Pethrick, PhD, DSc, is currently a Research Professor and Professor Emeritus in the Department of Pure and Applied Chemistry at the University of Strathclyde, Glasglow, Scotland. He was Burmah Professor in Physical Chemistry and has been a member of the staff there since 1969. He has published over 400 papers and edited and written several books. Recently, he has edited several publications concerned with the techniques for the characterization of the molar mass of polymers and also the study of their morphology. He currently holds a number of EPSRC grants and is involved with Knowledge Transfer Programmes involving three local companies involved in production of articles made out of polymeric materials. His current research involves AWE and has acted as a consultant for BAE Systems in the area of explosives and a company involved in the production of anticorrosive coatings.

Dr. Pethrick is on the editorial boards of several polymer and adhesion journals and was on the Royal Society of Chemistry Education Board. He is a Fellow of the Royal Society of Edinburgh, the Royal Society of Chemistry, and the Institute of Materials, Metal and Mining. Previously, he chaired the 'Review of Science Provision 16-19' in Scotland and the restructuring of the HND provision in chemistry. He was also involved in the creation of the revised regulations for accreditation by the Royal Society of Chemistry of the MSc level qualifications in chemistry. For a many years, he was the Deputy Chair of the EPSRC IGDS panel and involved in a number of reviews of the courses developed and offered under this program. He has been a member of the review panel for polymer science in Denmark and Sweden and the National Science Foundation in the USA.

Gennady E. Zaikov, DSc

Gennady E. Zaikov, DSc, is the Head of the Polymer Division at the N. M. Emanuel Institute of Biochemical Physics, Russian Academy of Sciences, Moscow, Russia, and Professor at Moscow State Academy of Fine Chemical Technology, Russia, as well as Professor at Kazan National Research Technological University, Kazan, Russia.

He is also a prolific author, researcher, and lecturer. He has received several awards for his work, including the Russian Federation Scholarship for Outstanding Scientists. He has been a member of many professional organizations and on the editorial boards of many international science journals.

Physical Chemistry Research for Engineering and Applied Sciences in 3 Volumes

Physical Chemistry Research for Engineering and Applied Sciences:
Volume 1: Principles and Technological Implications
Editors: Eli M. Pearce, PhD, Bob A. Howell, PhD,
Richard A. Pethrick, PhD, DSc, and Gennady E. Zaikov, DSc

Physical Chemistry Research for Engineering and Applied Sciences:
Volume 2: Polymeric Materials and Processing
Editors: Eli M. Pearce, PhD, Bob A. Howell, PhD,
Richard A. Pethrick, DSc, PhD, and Gennady E. Zaikov, DSc

Physical Chemistry Research for Engineering and Applied Sciences:
Volume 3: High Performance Materials and Methods
Editors: Eli M. Pearce, PhD, Bob A. Howell, PhD,
Richard A. Pethrick, DSc, PhD, and Gennady E. Zaikov, DSc

CONTENTS

LIST OF CONTRIBUTORS

Arezoo Afzali
University of Guilan, Rasht, Iran

Alaz Izer
Izmir Institute of Technology Department of Chemical Engineering Gulbahce Urla Izmir, Turkey

M. Anachkov
Institute of Catalysis, Bulgarian Academy of Sciences, Bonchev St. #11, Sofia 1113, Bulgaria

Shershnev Vladimir Andreevich
N. M. Emanuel Institute of Biochemical Physics, Russian Academy of Sciences, ul. Kosygina 4, Moscow, 119991 Russia, 499-246-4769, E-mail: shershnev@mitht.ru

M. I. Artsis
N. M. Emanuel Institute of Biochemical Physics, Russian Academy of Sciences, ul. Kosygina 4, Moscow, 119991 Russia

L. M. Baider
Emanuel Institute of Biochemical Physics, Russian Academy of Sciences, ul. Kosygina 4, Moscow, 119334 Russia

Devrim Balköse
Izmir Institute of Technology Department of Chemical Engineering Gulbahce Urla Izmir, Turkey

Mária Baňasová
Institute of Experimental Pharmacology and Toxicology, Slovak Academy of Sciences, Dúbravska cesta 9, SK-84104, Bratislava, Phone: +421259410669, E-mail address: banasova.majka@gmail.com

T. Batakliev
Institute of Catalysis, Bulgarian Academy of Sciences, Bonchev St. #11, Sofia 1113, Bulgaria

O. L. Belaya
Moscow State Medico-Stomatological University, ul. Delegatskaya 20/1, Moscow, 127473 Russia

T. N. Bogatyrenko
Institute of Chemical Physics Problems, Russian Academy of Sciences, Chernogolovka, Moscow region, 142432 Russia.

G. A. Bonartseva
A. N. Bach's Institute of Biochemistry, RAS, Leninskiy pr. 33, Moscow. 119071 RF

V. V. Chernova
Bashkir State University Russia, Republic of Bashkortostan, Ufa, 450074, ul. Zaki Validi, 32

V. Georgiev
Institute of Catalysis, Bulgarian Academy of Sciences, Bonchev St. #11, Sofia 1113, Bulgaria

M. D. Goldfein
Saratov State University named after N.G. Chernyshevsky, Russia

B. A. Howell
Central Michigan University, Chemical Faculty, Mount Pleasant, MI, USA. E-mail: bob.a.howell@cmich.edu

V. A. Ilatovsky
N.N. Semenov Institute of Chemical Physics Russian Academy of Sciences, 4 Kosygin str, Moscow 119991, Russia, E-mail: iva1947@yandex.ru

A. L. Iordanskii
N. Semenov Institute of Chemical Physics, RAS, Kosygin str. 4, Moscow. 119996 RF, A.N. Bach's Institute of Biochemistry, RAS, Leninskiy pr. 33, Moscow 119071 RF. E-mail: aljordan08@gmail.com

Ivo Juránek
Institute of Experimental Pharmacology and Toxicology, Slovak Academy of Sciences, SK-84104, Bratislava, Slovakia

Tugce Nefise Kahyaoglu
Izmir Institute of Technology Department of Chemical Engineering Gulbahce Urla Izmir Turkey

Lukáš Kerner
Department of Organic Chemistry, Faculty of Natural Sciences, Comenius University of Bratislava, SK-84215, Slovakia

G. G. Komissarov
N.N. Semenov Institute of Chemical Physics Russian Academy of Sciences, 4 Kosygin str, Moscow 119991, Russia

G. A. Korablev
Izhevsk State Agricultural Academy, Basic Research and Educational Center of Chemical Physics and Mesoscopy, URC, UrD, RAS, Russia, Izhevsk, 426000

N. V. Kozhevnikov
Saratov Chernyshevsky State University, Russia

G. V. Kozlov
FSBEI HPE "Kh.M. Berbekov Kabardino-Balkarian State University," Nal'chik - 360004, Chernyshevsky st., 173, Russian Federation

T. N. Kukhta
Scientific Research Institute BelNIIS RUE, 15b Frantsisk Skorina St., 220114, Minsk, Belarus

E. I. Kulish
Bashkir State University Russia, Republic of Bashkortostan, Ufa, 450074, ul. Zaki Validi, 32, Russia

Z. V. Kuropteva
Emanuel Institute of Biochemical Physics, Russian Academy of Sciences, ul Kosygina 4, Moscow, 119334 Russia, zvk@sky.chph.ras.ru

N. V. Kuvardin
South-West State University, 305040, Kursk, Russia

E. M. Kuvardina
South-West State University, 305040, Kursk, Russia

N. M. Livanova
Emanuel Institute of Biochemical Physics, Russian Academy of Sciences, ul. Kosygina 4, Moscow, 119991 Russia, Livanova Nadezhda Mikhaylovna, 495-939-7193, E-mail: livanova@sky.chph.ras.ru

Shima Maghsoodlou
University of Guilan, Rasht, Iran

G. M. Magomedov
FSBEI HPE Daghestan State Pedagogical University, Makhachkala, 367003, M. Yaragskii st., 57, Russian Federation

T. A. Makhina
A.N. Bach's Institute of Biochemistry, RAS, Leninskiy pr. 33, Moscow, 119071 Russian Federation

F. F. Niyazi
South-West State University, 305040, Kursk, Russia

Eli M. Pearce
Brooklyn Branch of New York University, 333 Jay Str., Six Metrotech Center, Brooklyn, NYC, NY, USA, E-mail: epearce@poly.edu

R. A. Pethrick
University of Strathclyd, 295 Cathedral Street, Glasgow, Scotland, UK, E-mail: r.a.pethrick@strath.ac.uk

V. D. Polonik
Sverdlova Str.13a, Minsk, Republic of Belarus, E-mail: v.polonik@belstu.by

A. A. Popov
Lomonosov State Academy of Fine Chemical Technology, pr. Vernadskogo 86, 117571, Moskow, Russia. Popov Anatoliy Anatol'evich, 495-939-7933, E-mail: popov@sky.chph.ras.ru

N. R. Prokopchuk
Belarusian State Technological University

P. G. Pronkin
Emanuel Institute of Biochemical Physics, Russian Academy of Sciences, ul. Kosygina 4, Moscow, 119334 Russia

G. A. Ptitsyn
N.N. Semenov Institute of Chemical Physics Russian Academy of Sciences, 4 Kosygin str, Moscow 119991, Russia

Martin Putala
Department of Organic Chemistry, Faculty of Natural Sciences, Comenius University of Bratislava, SK-84215, Slovakia

S. Rakovsky
Institute of Catalysis, Bulgarian Academy of Sciences, Bonchev St. #11, Sofia 1113, Bulgaria

E. G. Rozantsev
Saratov Chernyshevsky State University, Russia

Zh. S. Shashok
333 Jay street, Six Metrotech Centre, Brooklyn, NYC, NY, USA

V. A. Shershnev
Lomonosov State Academy of Fine Chemical Technology, pr. Vernadskogo 86, 117571, Moskow, Russia

A. S. Shurshina
Bashkir State University Russia, Republic of Bashkortostan, Ufa, 450074, ul. Zaki Validi, 32

G. V. Sinko
N.N. Semenov Institute of Chemical Physics Russian Academy of Sciences, 4 Kosygin str, Moscow 119991, Russia

E. D. Sklyanchuk
Center of Traumatology and Orthopaedy, Stavropolska str. 23 k.1, Moscow. 109386, RF.

Ladislav Šoltés
Institute of Experimental Pharmacology and Toxicology, Slovak Academy of Sciences, SK-84104, Bratislava, Slovakia

A. S. Tatikolov
Emanuel Institute of Biochemical Physics, Russian Academy of Sciences, ul. Kosygina 4, Moscow, 119334, Russia

I. F. Tuktarova
Bashkir State University Russia, Republic of Bashkortostan, Ufa, 450074, ul. Zaki Validi, 32

R. R. Usmanova
Ufa State Technical University of Aviation, Ufa, Bashkortostan, Russia

Katarína Valachová
Institute of Experimental Pharmacology and Toxicology, Slovak Academy of Sciences, SK-84104, Bratislava, Slovakia

Kh. Sh. Yakh'yaeva
FSBEI HPE Daghestan State Agrarian University, Makhachkala, 367032, M. Gadzhiev st, 180, Russian Federation

G. E. Zaikov
N. M. Emanuel Institute of Biochemical Physics, Russian Academy of Sciences, 4 Kosygin str., Moscow 119334 /119996/119991, Russia. Tel: 495-939-7191, E-mail: gezaikov@yahoo.com / chembio@sky.chph.ras.ru

LIST OF ABBREVIATIONS

AOS	Antioxidant System
ARE	Antioxidant-Response Element
BJH	Barrett-Joyner-Halenda
CaSt2	Calcium Stearate
CHD	Coronary Heart Disease
CLD	Chord-Length Distribution
COFs	Covalent Organic Frameworks
DFT	Density Functional Theory
DR	Dubinin-Radushkevich
ENB	Ethylidene Norbornene
EPR	Electron Paramagnetic Resonance
FCC	Face-Centered Cubic
GCMC	Grand Canonical Montecarlo
GMS	Gentamicin Sulfate
HCP	Hexagonal Close-Packed
ISC	Iron–Sulfur Centers
IUPAC	Union of Pure and Applied Chemists
LPO	Lipid Peroxidation
MOFs	Metal Organic Frameworks
MOPs	Microporous Organic Polymers
MP	Mercury Porosimetry
NBR	Butadiene-Nitrile Rubber
NG	Nitroglycerine
NLDFT	Nonlocal Density Functional Theory
PALS	Positron Annihilation Lifetime Spectroscopy
Phr	Per Hundred of Rubber
PIMs	Polymers of Intrinsic Microporosity
PSD	Pore Size Distribution
ROS	Reactive Oxygen Species
SANS	Small Angle Neutron Scattering
SAS	Small-Angle Scattering
SAXS	Small-Angle X-rays Scattering
SEM	Scanning Electron Microscopy
TEM	Transmission Electron Microscopy
VEGF	Vascular Endothelial Growth Factor
WAS	Wide-Angle Scattering

LIST OF SYMBOLS

Φ	angle of contact between liquid and walls
a_m	area of an adsorbate molecule
a_p	pore surface area
b, c	constant
θ	contact angle
D_e	distribution function for pore diameter
F	meniscus shape factor
$F(\vec{S})$	atomic form factor
K	porod invariant
l_p	porod length
M_v	gram molecular volume
N	Avagadro's number
n_m	monolayer capacity
P	pressure
P_0	vapor pressure of the bulk liquid, ambient pressure
R	gas constant
R_k	Kelvin radius
r_m	mean radius of curvature of the liquid/gas interface
S	surface area
S	scattering vector
S	total pore surface
S_{BET}	specific surface area
T	temperature
V	volume adsorbed per unit mass of adsorbent, pore volume
V_L	molal liquid volume
V_m	volume adsorbed at the complete monolayer point
v_p	the pore volume
V_{tot}	total pore volume
$Z(\vec{S})$	lattice factor
γ	surface tension
λ	wavelength
ρ_b	bulk density
ρ_p	particle density
σ	liquid-gas surface tension
ϕ	volume fraction of voids

PREFACE

This three-volume set covers a significant amount of new research and applications on physical chemistry for engineering and applied sciences. Physical chemistry for engineering and applied sciences shows how materials can behave and how chemical reactions occur. Physical chemistry for engineering and applied sciences can be considered as a knowledge that is relevant in nearly every area of chemistry. It covers diverse topics, from biochemistry to materials properties to the development of quantum computers.

The aim of this important book is to provide both a rigorous view and a more practical, understandable view of chemistry and biochemical physics. Physical chemistry for engineering and applied sciences is geared toward readers with both direct and lateral interest in the discipline. Physical chemistry for engineering and applied sciences applies physics and math to problems that interest chemists, biologists, and engineers. Physical chemists use theoretical constructs and mathematical computations to understand chemical properties and describe the behavior of molecular and condensed matter.

Physical chemistry for engineering and applied sciences is structured into different parts devoted to industrial chemistry and biochemical physics and their applications. In the first volume of this series, some principles and technological implications of industrial chemistry and biochemical physics are presented. This volume discusses new discoveries and realizations of the importance of key concepts and emphases are placed on the underlying fundamentals and on acquisition of a broad and comprehensive grasp of the field as a whole.

In the second volume, some fascinating phenomena associated with the remarkable features of high performance polymers are presented. This volume also provides an update on applications of modern polymers. This volume offers new research on structure–property relationships, synthesis and purification, and potential applications of high performance polymers. The collection of topics in second volume reflects the diversity of recent advances in modern polymers with a broad perspective that will be useful for scientists as well as for graduate students and engineers.

The various categories of high performance materials and their composites are discussed in the third volume. The third volume of Physical Chemistry Research for Engineering and Applied Sciences provides up-to-date synthesis details, properties, characterization, and applications for such systems in or-

der to give readers and users better information to select the required material.

Together, these three volumes highlight and present some of the most important areas of current interest in biochemical physics and chemical processes, filling the gap between theory and application. Every section of the book has been expanded, where relevant, to take account of significant new discoveries and realizations of the importance of key concepts. Furthermore, emphases are placed on the underlying fundamentals and on acquisition of a broad and comprehensive grasp of the field as a whole.

INTRODUCTION

January 7, 2015 will be the 80th birthday of Prof. G. E. Zaikov, and he has more than 60 years scientific activity. Zaikov was born in Omsk, Siberia (USSR), where he graduated from their primary, middle, and high schools. He also graduated from a musical professional school where he studied violin and pianoforte. However, his parents, Efrem and Matrena, decided that it might be better for their son to continue his education by following in the footsteps of his mother, who was a chemistry teacher in high school and at Omsk's Medical Institute (his father was a mathematician and land-surveyor). Therefore, in 1952 Gennady moved to Moscow where he entered the Moscow State University (MSU), and he graduated with a chemistry degree in December 1957. His bachelor's degree dealt with the problem of separating Li6 and Li7 isotopes. After this he joined the Institute of Chemical Physics (ICP) in Moscow in February 1958. In 1996, this institute was split into two parts: N. N. Semenov Institute of Chemical Physics (ICP) and N. M. Emanuel Institute of Biochemical Physics. At the present time Prof. G. E. Zaikov is working at the N. M. Emanuel Institute of Biochemical Physics (IBP). So, G. E. Zaikov never changed place of his job.

Gennady was originally invited to ICP by Professor Nikolai Markovich Emanuel. Under his guidance, G. E. Zaikov defended in 1963 his PhD thesis titled "Comparison of the Kinetics and Mechanism of Oxidation of the Organic Compounds in Gaseous and Liquid Phases" in 1963. These results were the foundation for industrial application. A plant floor was built in Moscow at a petrochemical plant (Kapotnya district) for production of 10,000 tons/year of acetic acid and 5000 tons/year of methylethylketone by oxidation of n-butane in liquid phase in critical conditions (50 atm, 150°C). The main contributors of this plant floor were N. M. Emanuel, E. A. Blumberg, Z. K. Maizus, M. G. Bulygin, E. B. Chizhov, and G. E. Zaikov. In 1968, Gennady defended a Doctor of Science thesis titled "The Role of Media in Radical-Chain Oxidation Reactions". In 1970, he became a full professor.

In 1966, Gennady began to become involved with polymer science. N. M. Emanuel charged Zaikov with the organization of work on problems associated with aging and stabilization of polymers, and, later, with the combustion of polymeric materials. In the 1970s, there were about 1000 scientists (about 50 research centers) in the U.S.S.R. working on these problems, including 200

scientists from ICP under Zaikov's leadership. The research was conducted on all aspects of these polymer problems, thermal degradation, oxidation, ozonolysis, photodegradation and radiation degradation, hydrolysis, biodegradation, mechanical degradation, pyrolysis, and flammability. Scientists from synthetic laboratories of this division (Prof. V. V. Ershov, E. G. Rozantsev, and K. M. Dyumaev) prepared several very important and original stabilizers for polymers and organized production of these stabilizers.

After "perestroika and degradation" of the U.S.S.R. in 1991, the new Russian government decreased the financial support of science significantly. So, G. E. Zaikov has now with him in the N. M. Emanuel Institute only 15 coworkers (instead of 200 as in 1970–1980s).

He compensated for the decrease of scientists in his institute by increasing the cooperation with other research centers in Russia and abroad.

Now G. E. Zaikov has scientific cooperation with:

- Prof. Victor Manuel de Matos Lobo and Dr. Artur Valente (Coimbra University, Coimbra, Portugal);
- Prof. Alfonso Jimenez (Alicante University, Alicante, Spain);
- Dr. Nekane Guarrotxena Arlunduaga (Institute of Polymer Science and Technology, Madrid, Spain);
- Prof. Alberto D'Amore (Second Naples University, Naples, Italy);
- Dr. Antonio Ballada (former Vice-President of Himont Co., Milan, Italy);
- Prof. Goerg Michler (Martin Luther University, Halle-Saale, Germany);
- Dr. Frank Pudel (OHMI Consulting Co., Magdeburg, Germany);
- Prof. Ryszard Kozlowski (Institute of Natural Fibers, Poznan, Poland);
- Prof. Jan Pielichowski (Cracow University of Technology, Cracow, Poland);
- Dr. Daniel Horak (Institute of Macromolecular Science, Prague, Czeck Republic);
- Prof. Slavi Kirillov Rakovsky, and Dr. Methody Anachkov (Institute of Catalysis, Sofia, Bulgaria);
- Prof. Cornelia Vasile (Polymer Research Institute, Iassi, Romania);
- Prof. Richard A. Pethrick (University of Strathclyde, Glasgow, Scotland, UK);
- Prof. Eli Pearce and Dr. Gerald Kirshenbaum (Brooklyn Polytechnic University, Brooklyn, New York, USA);
- Prof. David Schiraldi (Case Western Reserve University, Clevelend, Ohio, USA);

- Prof. Bob Howell (Central Michigan University, Mount Pleasant, Michigan, USA);
- Dr. James Summers (Former Head of Division of PolyOne Co., Cleveland, Ohio, USA);
- Dr. LinShu Liu (US Department of Agriculture, Windmoor, Pennsylvania, USA);
- Prof. Walter Focke (Pretoria University, Pretoria, South Africa);
- Prof. Hans-Joachim Radusch (Martin Luter University, Halle-Saale, Germany);
- Prof. Ryszard M. Kozlowski (ESCORENA, United Nationals, Poznan, Poland);
- Prof. Roman Jozwik (Military Institute of Chemistry and Radiometry, Warsaw, Poland);
- Dr. Raijesh Ananjiwala, Research Textile Institute, Port Elisabeth, South Africa).

He has also cooperation with CIS countries (former republics of the USSR):

- Prof. Anatolii A. Turovskii, Prof. Roman G. Makitra, and Prof Yurii G. Medvedevskikh (Pisarzhevskii Institute of Physical Chemistry, L'viv Division and Institute of Coal, L'viv, Ukraine);
- Prof. Nikolai A. Turovskii (Donetsk State University, Donetsk, Ukraine);
- Prof. Alexandr I. Burya (Dnepropetrovsk State Agriculture University, Dnepropetrovsk, Ukraine);
- Prof. Nodare G. Lekishvili and Prof. Omari Mukbaniani (I. Javakhishvili Tbilisi State University, Georgia);
- Prof. Jimsher N. Aneli (Institute of Kibernetic, Tbilisi, Georgia);
- Prof. Jenis A. Djamanbaev (Institute of Organic Chemistry, Bishkek, Kirgisia);
- Prof. Nikolai R. Prokopchuk (Belorussian State Technical University, Minsk, Belorussia);
- Prof. Norair M. Beylerian (Institute of Chemical Physics, Erevan, Armenia).

G.E. Zaikov has also cooperation with scientists from many research centers of Russia. Here are only some of these:

- Prof. A. A. Berlin, Prof. A. L. Iordanskii, and Dr. K. Z. Gumargalieva (N.N. Semenov Institute of Chemical Physics, Moscow);
- Dr. N. A. Sivov (D. I. Topchiev Institute of Pethrochemical Synthesis, Moscow);

- Dr. N. N. Komova, Dr. A. A. Ol'khov, Prof. B. Tsoi (M.V. Lomonosov Moscow State Academy Fine Chemical Technology, Moscow);
- Prof. V. S. Osipchik (D. I. Mendeleev Russian Chemical-Technical University, Moscow);
- Prof. Yu. A. Ershov (The Second Moscow State Medical University, Moscow);
- Prof. N. Ya. Yaroshenko (Institute of Pure Chemical Compounds, Moscow);
- Prof. Yu. G. Yanovsky (Institute of Applied Mathematic, Moscow);
- Dr. O. A. Legon'kova (Moscow State University of Applied Biotechnology, Moscow);
- Prof. A. K. Mikitaev (L. Ya. Karpov Physico-Chemical Institute, Moscow);
- Prof. A. M. Egorov (Oncology Center, Moscow);
- Dr. E. V. Kalugina (Plastic Company, Moscow);
- Dr. G. V. Kozlov, Prof. M. Kh. Ligidov, and Prof. N. I. Mashukov (K. Kh. Berbekov Kabardino-Balkarian State University, Nal'chik, Kabardino-Balkaria);
- Prof. Yu. B. Monakov (Institute of Organic Chemistry, Ufa, Bashkortostan);
- Prof. M. I. Abdullin, Prof. V. P. Zakharov, Prof. S. V. Kolesov, and Prof. R. Z. Biglova (Bashkirian State University, Ufa, Bashkortostan);
- Prof. S. S. Zlotsky (Ufa State Technological Oil University, Ufa, Bashkortostan);
- Prof. F. F. Niyazi (Kursk State University, Kursk);
- Prof. V. A. Babkin (Volgograd State Technical University, Volgograd);
- Prof. A. I. Rakhimov (Institute of Ecology, Volgograd);
- Prof. V. F. Kablov (Branch of Volgograd State Technical University, Volzhsk, Volgograd district);
- Prof. T. N. Lomova (Research Institute of Solutions, Ivanovo)
- Prof. G. A. Korablev (Scientific-Education Research Center of Chemical Physics and Mesoscopy, Udmurdian Research Center, Ural Branch of Russian Academy of Sciences, Izhevsk).

In all, he has scientific cooperation (publication of original papers, reviews, books and volumes) with 20 research centers abroad, 8 centers in CIS countries and 20 inside of Russia.

Zaikov left his position as a head of the laboratory on September, 2007 but he became head of the Polymer Division (PD) in IBP. PD included three laboratories (about 50 scientists).

Figure 1 shows the changing of amount of staff members in Dr. Zaikov's laboratory over time, and Fig. 2 shows the number of books, he published (mostly in English).

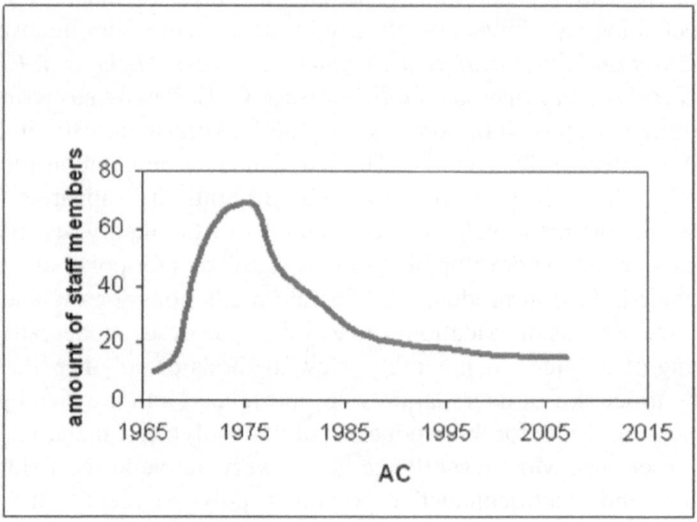

FIGURE 1 Number of staff members in laboratory of chemical resistance of polymers at different times.

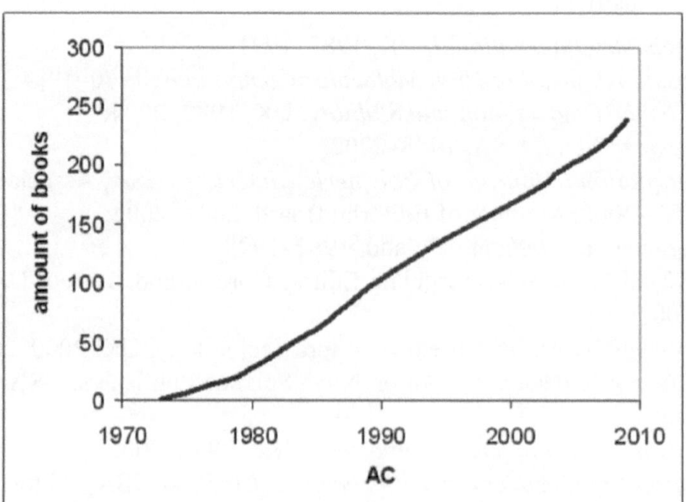

FIGURE 2 Integral number of books published by G. E. Zaikov.

G. E. Zaikov is an outstanding scientist with expertise in wide areas of chemistry: chemical and biological kinetics, chemistry and physics of polymers, history of chemistry, biochemistry. In addition to his position at the N. M. Emanuel Institute, he is a lecturer at the Moscow State Academy of Fine Chemical Technology, and he is researcher at Volzhsk Branch of Volgograd State Technological University. He taught his students from his own books: *Degradation and Stabilization of Polymers*, *Physical Methods in Chemistry*, and *Acid Rains and Environmental Problems*. G. E. Zaikov has written about 4000 original articles, 400 monographs (100 in Russian and 300 in English), and 350 chapters in 80 volumes. His Russian Science Citation Index number of scientific activity (Chirsch) is equal 30 units. It is apparent from this work that he has made valuable contributions to the theory and practice of polymers—aging and development of new stabilizers for polymers, organization of their industrial production, life-time predictions for use and storage, and the mechanisms of oxidation, ozonolysis, hydrolysis, biodegradation, and decreasing of polymer flammability. New methods of polymer modification using the processes of degradation were introduced into practice by Zaikov. These methods allow for the production of new polymeric materials with improved properties. Most recently, he is also very active in the field of semiconductors and electroconductive polymers, polymer blends, and polymer composites including nanocomposites.

G. E. Zaikov is a member of many editorial boards of journals published in Russia, Poland, Bulgaria, the U.S.A., and England. Below is a list of his activity in this field:

- *Chemistry International*, UK, 1987–1991;
- *Russian Journal of High Molecular Compounds*, 1970–1984;
- *Polymer Degradation and Stability*, UK, 1982–2004;
- *Polymer News*, USA, 1988–2002;
- *International Journal of Polymeric Materials*, USA, Associate Editor, 1989–2000; Member of Editorial Board: 2001–2002;
- *Polymers in Medicine*, Poland, 1982–1998;
- *Polymer Yearbook*, Associate Editor, Gordon and Breach, UK, 1985–2000;
- *Polymer Yearbook*, Co-editor, Rapra Technology, UK, 2000–2003;
- *Polymer Yearbook*, Co-editor, Nova Science Publishers, USA, 2005–to data;
- *Polymer and Polymer Composites*, UK, 1994–2000;
- *Journal of Chemical and Biochemical Kinetics*, USA, Editor-in-Chief, 1992–2000;
- *Russian Journal of Textile Chemistry*, 1992–to date;

- *Oxidation Communications*, Sofia, Bulgaria, 1994–to date;
- Associate Editor of series "Polymer Science and Engineering," Gordon and Breach Publ., USA, 1990–2000;
- Editor of series "Polymer Books for the 21st Century," Nova Science Publ., New York, USA, 1990–2006;
- Editor of series "Chemistry and Biochemistry," Nova Science Publishers, New York, USA, 2002–to date;
- Editor-in-Chief of series "New Concepts in Polymer Science", VSP International Sci. Publ., Leiden and Brill Academic Publishers, Amsterdam, the Netherlands, 1990 – 2004;
- Editor of series "Chemical and Biochemical Physics on the Edge of XXI Century," Nova Science Publ., USA 2000–2006;
- *Russian Polymer News Journal*, Associate Editor, New Jersey, USA, 1996–2003;
- *Journal of Balkan Tribological Association*, Sofia, Bulgaria, 2001–to date;
- *Polymer Plastic Technology and Engineering*, USA, 1997–2001;
- Member of Research Board and Advisers, American Biographical Institute, Inc., NC, USA, 2001–to date;
- Member of Editorial Board of *Polymer International*, 2004–to data;
- Member of Editorial Board of *Journal of Chemical Physics and Mesoscopy*, Russian Academy of Sciences, Izhevsk, Russia, 2004–to date;
- Member of Editorial Board of *Journal of Natural Fibers*, Poznan, Poland, 2005–to data;
- Member of Editorial Board of *D.I. Mendeleev Journal of Russian Chemical Society*, Moscow, 2006–to data;
- Member of Editorial Board of *Encyclopedia of Engineer-Chemist*, Moscow, 2006–to data;
- Member of Editorial Board of Journal of Coatings, Moscow, 1990–2000.

Dr. Zaikov is member of Academy of Creation (San Diego, USA – Moscow, Russia), International Academy of Sciences (Munich, Germany), American Chemical Society, Plastic Engineering Society (USA), and Royal Chemical Society (UK).

A Kazakh national proverb said: "After 60 years old, the brain is going back (to childhood)". We cannot say that the Kazakh proverb is not true. It is the wisdom of Kazakh nationalit, and it is right. We should agree with this. The question is: How fast does brain revert to childhood after 60 years old? It is very desirable that this speed of this movement should be not fast. It is a fact that Prof. Gennady E. Zaikov is as active now as he was 20 years ago when

he was 60 years old. Our conclusion is: The speed of movement of Zaikov's brain probably reverted but at very very small pace.

In the former Soviet Union (after academician N. M. Emanuel's death), Dr. Zaikov headed the team dealing with the problem of polymer aging in the U.S.S.R. and the Eastern European countries in cooperation with the Soviet Academy of Sciences. His present position is Head of Division, member of directorium, and deputy of department of the N. M. Emanuel Institute of Biochemical Physics, Russian Academy of Sciences; Professor of Polymer Chemistry in the Moscow State Academy of Fine Chemical Technology; Professor of Polymer Chemistry in Volzhsk Branch of Volgograd State Technical University. His fields of interest include chemical physics, chemical kinetics, flammability, degradation and stabilization of polymers, diffusion, polymer materials, kinetics in biology, history of chemistry, jokes.

A few words about the personal life of G. E. Zaikov. In addition to his parents (discussed previously), he has a sister, Zinaida E. Zaikova, who was also a teacher of mathematics in high school (she past away some years ago). Two of his sisters (Klara and Inna) died during the Stalin collectivization period at the end of the 1920s from starvation. Zaikov's wife, Marina Izrailevna Artsis, is a member of the N. M. Emanuel Institute of Biochemical Physics and has a PhD in chemistry. His son, Vadim G. Zaikov is a Sr. Research Chemist of Avery Dennison Co (Ohio, USA). He has a PhD in chemistry twice. The first one he received in USSR, the second one he received from the College of William and Mary (Williamsburg, VA) in the laboratory of Prof. William H. Starnes. His granddaughter, Alexandra (23 years old), is a student in Chicago (USA), and his grandson, Denis, (14 years old) is a schoolboy in Perry, Ohio (USA).

On his 80th birthday anniversary, G. E. Zaikov is in the prime of his life. Although support for scientists and research is now at a low point for many in Russia, he is hopeful that for the sake of his country and its future that this will improve (probably far into the future).

The practice of good science still exists in Russia, and G. E. Zaikov has been and is a significant contributor. We wish him a most happy birthday.

FIGURE 3 Professor, G. E. Zaikov (November, 9 2007) with his new book *Chemical and Biochemical Physics* (Nova Science Publishers, New York). Above him (right side) is the picture of his teacher Professor N. M. Emanuel (1915–1984) and his grandson's, Denis, drawing (left side).

FIGURE 4 Staff of the laboratory of chemical resistance of polymers: seated – Alexei A. Borodin (graduated student); first row (from the left to the right) – Lidia A. Zimina, Dr. Olga V. Alexeeva, Marina L. Konstantinova, Prof. Gennady E. Zaikov (Head of Division), Larisa L. Madyuskina; second row (from the left to the right) – Dr. Sergei M. Lomakin (Head of Laboratory), Prof. Stanislav D. Razumovskii, Prof. Vladimir M. Gol'dberg, Prof. Alexander A. Volod'kin, Dr. Nikolai N. Madyuskin, Dr. Marina I. Artsis, Vyacheslav V. Podmaster'ev. November, 9 2007. N.M. Emanuel Institute of Biochemical Physics Russian Academy of Sciences.

FIGURE 5 Prof. Gennady E. Zaikov on the Grand Canal in Venicie (2010) after international conference "Time of Polymers" in Ischia Island, Naples Bay, Naples.

FIGURE 6 Four scientists (1976, Moscow). From the left to the right: Winner Nobel Prize and Director of the Institute of Chemical Physics (ICP) Academy of Sciences of USSR, academician Nikolai Nikolaevich Semenov; Head of Chemical Division, Presidium of Academy of Sciences of USSR, Head of Department of Kinetics of Chemical and Biochemical Processes ICP, academician Nikolai Markovich Emanuel; Deputy of Head of Department of Kinetics of Chemical and Biochemical Processes ICP, head of laboratory of Chemical Resistance of Polymers Dr. of Science, Prof. Gennady Efremovich Zaikov and head of laboratory, Dr. of Science, Prof. Gunter Wagner, Institute of Organic Chemistry, Academy of Sciences of German Democratic Republic, Adlersdorf, Berlin.

CHAPTER 1

BACTERIAL POLY(3-HYDROXYBUTYRATE) AS A BIODEGRADABLE POLYMER FOR BIOMEDICINE

A. L. IORDANSKII, G. A. BONARTSEVA, T. A. MAKHINA, E. D. SKLYANCHUK, and G. E. ZAIKOV

CONTENTS

ABSTRACT

Bacterial poly(3-hydroxybutyrate) (PHB) as member of natural polymer family of polyalcanoates has been widely used in the innovative biomedical areas owing to relevant combination of biocompatibility, transport characteristics and biodegradation without toxic products formation. The review presents a comprehensive description of hydrolysis and enzymatic degradation during long-time period preferably under laboratory condition (in-vitro) and as medical implants (in-vivo). Besides, the focus of review is devoted to the literature comparison of different works in PHB degradation and its biomedical application including work of authors.

1.1 INTRODUCTION

Currently, the intensive development of biodegradable and biocompatible materials for medical implication provokes comprehensive interdisciplinary studies on biopolymer structures and functions. The well-known and applicable biodegradable polymers are polylactides (PLA), polyglycolides (PGA), and their copolymers, poly ε-caprolactone, poly(orthoesters), poly β-maleic acid, poly(propylene fumarate), polyalkylcyanoacrylates, polyorthoanhydrides, polyphosphazenes, poly(propylene fumarate), some natural polysaccharides (starch, chitosan, alginates, agarose, dextrane, chondroitin sulfate, hyaluronic acid), and proteins (collagen, silk fibroin, fibrin, gelatin, albumin). Since some of these polymers should be synthesized through chemical stages (e.g., via lactic and glycolic acids) it is not quite correct to define them as the biopolymers.

Besides biomedicine applications, the biodegradable biopolymers attract much attention as perspective materials in wide areas of industry, nanotechnologies, farming and packaging owing to the relevant combination of biomedical, transport, and physical-chemical properties. It is worth to emphasis that only medical area of these biopolymers includes implants and prosthesis, tissue engineering scaffolds, novel drug dosage forms in pharmaceutics, novel materials for dentistry and others.

Each potentially applicable biopolymer arranges a wide multidisciplinary network, which usually includes tasks of searching for efficacy ways of biosynthesis reactions; economical problems associated with large-scale production; academic studies of mechanical, physicochemical, biochemical properties of the polymer and material of interest; technology of preparation and using this biopolymer; preclinical and clinical trials of these materials and

products; a market analysis and perspectives of application of the developed products and many other problems.

Poly((R)-3-hydroxybutyrate) (PHB) is an illustrative example for the one of centers for formation of the above-mentioned scientific-technological network and a basis for the development of various biopolymer systems [1–2]. In recent decades an intense development of biomedical application of bacterial PHB in producing of biodegradable polymer implants and controlled drug release systems [3–6] needs for comprehensive understanding of the PHB biodegradation process. Examination of PHB degradation process is also necessary for development of novel friendly environment polymer packaging [7–9]. It is generally accepted that biodegradation of PHB both in living systems and in environment occurs via enzymatic and nonenzymatic processes that take place simultaneously under natural conditions. It is, therefore, important to understand both processes [6, 10]. Opposite to other biodegradable polymers (e.g., PGA and PLGA), PHB is considered to be moderately resistant to degradation in-vitro as well as to biodegradation in biological media. The rates of degradation are influenced by the characteristics of the polymer, such as chemical composition, crystallinity, morphology and molecular weight [11, 12]. In spite of that PHB application in-vitro and in-vivo has been intensively investigated, the most of the available data are often incomplete and sometimes even contradictory. The presence of conflicting data can be partially explained by the fact that biotechnologically produced PHB with standardized properties is relatively rare and is not readily available due to a wide variety of its biosynthesis sources and different manufacturing processes.

Above inconsistencies can be explained also by excess applied trend in PHB degradation research. At most of the papers observed in this review, PHB degradation process has been investigated in the narrow framework of development of specific medical devices. Depending on applied biomedical purposes biodegradation of PHB was investigated under different geometry: films and plates with various thickness [13–16], cylinders [17–19], monofilament threads [20–22] and micro and nanospheres [23, 24]. At these experiments PHB was used from various sources, with different molecular weight and crystallinity. Besides, different technologies of PHB devices manufacture affect such important characteristics as polymer porosity and surface structure [14, 15]. Reports regarding the complex theoretical research of mechanisms of hydrolysis, enzymatic degradation and biodegradation in-vivo of PHB processes are relatively rare [13–15, 16, 25–27] that attaches great value and importance to these investigations. Nevertheless, the effect of thickness, size and geometry of PHB device, molecular weight and crystallinity of PHB on the mechanism of PHB hydrolysis and biodegradation was not yet well clarified.

1.2 HYDROLYTIC AND ENZYMATIC DEGRADATION OF PHB

1.2.1 NONENZYMATIC HYDROLYSIS OF PHB IN-VITRO

Examination of hydrolytic degradation of natural poly((R)-3-hydroxybutyrate) in-vitro is a very important step for understanding of PHB biodegradation. There are several very profound and careful examinations of PHB hydrolysis that was carried out for 10–15 years [25–28]. Hydrolytic degradation of PHB was usually examined under standard experimental conditions simulating internal body fluid: in buffered solutions with pH=7.4 at 37 °C but at seldom cases the higher temperature (55 °C, 70 °C and more) and other values of pH (from 2 to 11) were selected.

The classical experiment for examination of PHB hydrolysis in comparison with hydrolysis of other widespread biopolymer, polylactic acid (PLA), was carried out by Koyama and Doi [25]. They compared films (10 × 10 mm size, 50 μm thickness, 5 mg initial mass) from PHB (M_n = 300,000, M_w = 650,000) with polylactic acid (M_n = 9000, M_w = 21,000) prepared by solvent casting and aged for 3 weeks to reach equilibrium crystallinity. They showed that hydrolytic degradation of natural PHB is the slow process. The mass of PHB film remained unchanged at 37 °C in 10 mM phosphate buffer (pH=7.4) over a period of 150 days, while the mass of the PLA film rapidly decreased with time and reached 17% of the initial mass after 140 days. The rate of decrease in the M_n of the PHB was also much slower than the rate of decrease in the M_n of PLA. The M_n of the PHB decreased approximately to 65% of the initial value after 150 days, while the M_n of the PLA decreased to 20% (M_n = 2000) at the same time. As PLA used at this research was with low molecular weight it is worth to compare these data with the data of hydrolysis investigation with the same initial M_n as for PHB. In other work the mass loss of two polymer films (PLA and PHB) with the same thickness (40 μm) and molecular weight (M_w = 450,000) was studied in-vitro. It was shown that the mass of PLA film decreased to 87%, whereas the mass of PHB film remained unchanged at 37 °C in 25 mM phosphate buffer (pH=7.4) over a period of 84 days, but after 360 days the mass of PHB film was 64.9% of initial one [29–31].

The cleavage of polyester chains is known to be catalyzed by the carboxyl end groups, and its rate is proportional to the concentrations of water and ester bonds that on the initial stage of hydrolysis are constant owing to the presence of a large excess of water molecules and ester bonds of polymer chains. Thus, the kinetics of nonenzymatic hydrolysis can be expressed by a simplified equation [32, 33]:

$$\ln M_n = \ln M_n^0 - k_h t \quad (1)$$

where M_n and M_n^0 are the number-average molecular weights of a polymer component at time t and zero, respectively and k_h is effective hydrolysis constant.

The average number of bond cleavage per original polymer molecule, N, is given by:

$$N = (M_n^0/M_n) - 1 = k_d M_m P_n^0 t, \quad (2)$$

where k_d is the effective rate constant of hydrolytic depolymerization, and P_n^0 is the initial number-average degree of polymerization at time zero, M_m is constant molecular mass of monomer. Thus, if the chain scission is completely random, the value of N is linearly dependent on time.

The molecular weight decrease with time is the distinguishing feature of mechanism under nonenzymatic hydrolysis condition in contrast to enzymatic hydrolysis condition of PHB when M_n values remained almost unchanged. It was supposed also that water-soluble oligomers of PHB with molecular mass about 3 kDa may accelerate the hydrolysis rate of PHB homo polymer [25]. In contrast, Freier et al. [14] showed that PHB hydrolysis was not accelerated by the addition of predegraded PHB: the rate of mass and M_w loss of blends (70/30) from high-molecular PHB (M_w = 641,000) and low-molecular PHB (M_w = 3000) was the same with degradation rate of pure high-molecular PHB. Meanwhile, the addition of amorphous atactic PHB (at PHB) (M_w = 10,000) to blend with high-molecular PHB caused significant acceleration of PHB hydrolysis: the relative mass loss of PHB/at PHB blends was 7% in comparison with 0% mass loss of pure PHB; the decrease of M_w was 88% in comparison with 48% M_w decrease of pure PHB [14, 34]. We have showed that the rate of hydrolysis of PHB films depends on M_w of PHB. The PHB films of high molecular weight (M_w = 450,000 and 1,000,000) degraded slowly as it was described above whereas films from PHB of low molecular weight (M_w = 150,000 and 300,000 kDa) lost weight relatively gradually and more rapidly [29–31].

To enhance the hydrolysis of PHB a higher temperature was selected for degradation experiments: 55 °C, 70 °C and more [25]. It was showed by the same research team that the weights of films (12 mm diameter, 65 μm thick) from PHB (M_n = 768 and 22 kDa, M_w = 1460 and 75 kDa) were unchanged at 55 °C in 10 mM phosphate buffer (pH=7.4) over a period of 58 days. The M_n value decreased from 768 to 245 kDa for 48 days. The film thickness increased from 65 to 75 μm for 48 days, suggesting that water permeated

the polymer matrix during the hydrolytic degradation. Examination of the surface and cross-section of PHB films before and after hydrolysis showed that surface after 48 days of hydrolysis was apparently unchanged, while the cross-section of the film exhibited a more porous structure (pore size < 0.5 μm). It was shown also that the rate of hydrolytic degradation is not dependent upon the crystallinity of PHB film. The observed data indicates that the nonenzymatic hydrolysis of PHB in the aqueous media proceeds via a random bulk hydrolysis of ester bonds in the polymer chain films and occurs throughout the whole film, since water permeates the polymer matrix [25, 26]. Moreover, as over the whole degradation time the first-order kinetics was observed and the molecular weight distribution was unimodal, a random chain scission mechanism is very probable both on the crystalline surfaces and in the amorphous regions of the biopolymer [14, 35, 36]. For synthetic amorphous at PHB it was shown that its hydrolysis is the two-step process. First, the random chain scission proceeds that accompanying with a molecular weight decrease. Then, at a molecular weight of about 10,000, mass loss begins [28].

The analysis of literature data shows a great spread in values of rate of PHB hydrolytic degradation in-vitro. It can be explained by different thickness of PHB films or geometry of PHB devices used for experiment as well as by different sources, purity degree and molecular weight of PHB (Table 18.1). At 37 °C and pH=7.4 the weight loss of PHB (unknown M_w) films (500 μm thick) was 3% after 40 days incubation [36–38], 0% after 52 weeks (364 days) and after 2 years (730 days) incubation (640 kDa PHB, 100 μm films) [14, 15], 0% after 150 days incubation (650 kDa PHB, 50 μm film) [25], 7.5% after 50 days incubation (279 kDa PHB, unknown thickness of films) [37], 0% after 3 months (84 days) incubation (450 kDa PHB, 40 μm films), 12% after 3 months (84 days) incubation (150 kDa PHB, 40 μm films) [29–30], 0% after 180 days incubation of monofilament threads (30 μm in diameter) from PHB (470 kDa) [22, 23]. The molecular weight of PHB dropped to 36% of the initial values after 2 years (730 days) of storage in buffer solution [15], to 87% of the initial values after 98 days [38], to 58% of the initial values after 84 days [29, 30] (Table 18.1).

TABLE 18.1 Nonenzymatic Hydrolysis of PHB In-Vitro*

Type of implant/ device	Initial M_w, kDa	Size/ Thickness, µm	Conditions	Relative mass loss, %	Relative loss of M_w, %	Time, Days	Links
Film	650	50	37 °C, pH=7.4	0	35	150	25
Film	640	100	37 °C, pH=7.4	0	64	730	15
Film	640	100	37 °C, pH=7.4	0	45	364	14
Film	450	40	37 °C, pH=7.4	0	42	84	29–30
Film	150	40	37 °C, pH=7.4	12	63	84	29–30
Film	450	40	37 °C, pH=7.4	35,1	-	360	31
Film	279	-	37 °C, pH=7.4	7.5	-	50	36
Plate	-	500	37 °C, pH=7.4	3	-	40	30
Plate	380	1000	37 °C, pH=7.4	0	-	28	43
Plate	380	2000	37 °C, pH=7.4	0	8	98	43
Thread	470	30	37 °C, pH=7.0	0	-	180	23
Thread	-	-	37 °C, pH=7.2	0	-	182	22
Micro-spheres	50	250–850	37 °C, pH=7.4	0	0	150	42
Thread	470	30	37 °C, pH=5.2	0	-	180	22
Film	279	-	37 °C, pH=10	100	-	28	36
Film	279	-	37 °C, pH=13	100	-	19	36
Film	650	50	55 °C, pH=7.4	0	68	150	25
Plate	380	2000	55 °C, pH=7.4	0	61	98	43
Film	640	100	70 °C, pH=7.4	-	55	28	14
Film	150	40	70 °C, pH=7.4	39	96	84	29–30
Film	450	40	70 °C, pH=7.4	12	92	84	29–30
Micro-spheres	50	250–850	85 °C, pH=7.4	50	68	150	42
Micro-spheres	600	250–850	85 °C, pH=7.4	25	-	150	42

* Collected by Bonartsev.

In acidic or alkaline aqueous media PHB degrades more rapidly: 0% degradation after 140 days incubation in 0.01 M NaOH (pH=11) (200 kDa, 100 μm film thickness) with visible surface changing [39], 0% degradation after 180 days incubation of PHB threads in phosphate buffer (pH=5.2 and 5.9) [23], complete PHB films biodegradation after 19 days (pH=13) and 28 days (pH=10) [37]. It was demonstrated that after 20 weeks of exposure to NaOH solution, the surfaces of PHB samples became rougher due to, along with an increased density in their surface layers. From these results, one may surmise that the nonenzymatic degradation of PHAs progresses on their surfaces before noticeable weight loss occurring as illustrated in Ref. [39] and by the authors in Fig.1.1.

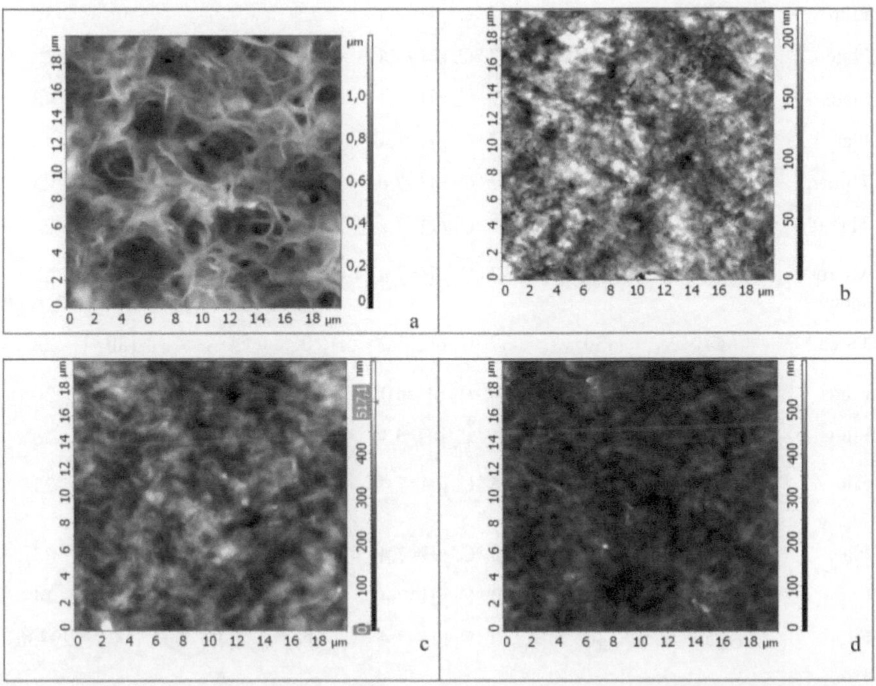

FIGURE 1.1 AFM topographic images of PHB films (170 kDa) with a scan size of 18×18 μm: (a) the rough surface of fresh-prepared sample (exposed to air); (b) the smooth surface of fresh-prepared sample (exposed to glass); (c) the sample exposed to phosphate buffer at 310K for 83 days; (d) the sample exposed to phosphate buffer at 343 K for 83 days. General magnificence is 300.

It was shown also that treatment of PHB film with 1 M NaOH caused a reduce in pore size on film surface from 1–5 μm to around 1 μm that indicates a partially surface degradation of PHB in alkaline media [40, 41]. At higher temperature no weight loss of PHB films and threads was observed after 98 and 182 days incubation in phosphate buffer (pH=7.2) at 55 °C and 70 °C, respectively [22], 12% and 39% of PHB (450 and 150 kDa, respectively) films after 84 days incubation at 70 °C [35, 40], 50% and 25% after 150 days incubation of microspheres (250–850 μm diameter) from PHB (50 kDa and 600 kDa, respectively) [42].

During degradation of PHB monofilament threads, films and plates the change of mechanical properties was observed under different conditions in-vitro [22, 43]. It was shown that a number of mechanical indices of threads became worse: load at break lost 36%, strain at break lost 33%, Young's modulus didn't change, tensile strength lost 42% after 182 days incubation in phosphate buffer (pH=7.2) at 70 °C. But at 37 °C the changes were more complicated: at first load at break increased from 440 g to 510 g (16%) at 90th day and then decreased to the initial value on 182nd day, strain at break increased rapidly from 60 to 70% (in 17%) at 20th day and then gradually increased to 75% (in 25%) at 182nd day, Young's modulus didn't change [22]. For PHB films it was demonstrated a gradual 32% decrease in Young's modulus and 77% fall in tensile strength during 120 days incubation in phosphate buffer (pH=7.4) at 37 °C [43]. For PHB plates more complicated changes were observed: at first tensile strength dropped in 13% for 1st day and then increased to the initial value at 28th day, Young's modulus dropped in 32% for 1st day and then remain unchanged up to 28th day, stiffness decreased sharply also in 40% for 1st day and then remain unchanged up to 28th day [44].

1.2.2 ENZYMATIC DEGRADATION OF PHB IN-VITRO

The examination of enzymatic degradation of PHB in-vitro is the following important step for understanding of PHB performance in animal tissues and in environment. The most papers observed degradation of PHB by depolymerases of its own bacterial producers. The degradation of PHB in-vitro by depolymerase was thoroughly examined and mechanism of enzymatic PHB degradation was perfectly clarified by Doi Y [25–26]. At these early works it was shown that 68–85% and 58% mass loss of PHB (Mw = 650–768 and 22 kDa, respectively) films (50–65 μm thick) occurred for 20 h under incubation at 37°C in phosphate solution (pH=7.4) with depolymerase (1.5–3 μg/mL) isolated from *A. fecalis*. The rate (k_e) of enzymatic degradation of films from PHB (M_n=768 and 22 kDa) was 0.17 and 0.15 mg/h, respectively.

The thickness of polymer films dropped from 65 to 22 μm (32% of initial thickness) during incubation. The scanning electron microscopy examination showed that the surface of the PHB film after enzymatic degradation was apparently blemished by the action of PHB depolymerase, while no change was observed inside the film. Moreover, the molecular weight of PHB remained almost unchanged after enzymatic hydrolysis: the M_n of PHB decreased from 768 to 669 kDa or unchanged (22 kDa) [25–26].

The extensive literature data on enzymatic degradation of PHB by specific PHB depolymerases was collected in detail in review of Sudesh, Abe and Doi [45]. We would like to summarize some the most important data. But at first it is necessary to note that PHB depolymerase is very specific enzyme and the hydrolysis of polymer by depolymerase is the unique process. But in animal tissues and even in environment the enzymatic degradation of PHB is occurred mainly by nonspecific esterases [24, 46]. Thus, in the frameworks of this review, it is necessary to observe the fundamental mechanisms of PHB enzymatic degradation.

The rate of enzymatic erosion of PHB by depolymerase is strongly dependent on the concentration of the enzyme. The enzymatic degradation of solid PHB polymer is heterogeneous reaction involving two steps, namely, adsorption and hydrolysis. The first step is adsorptions of the enzyme onto the surface of the PHB material by the binding domain of the PHB depolymerase, and the second step is hydrolysis of polyester chains by the active site of the enzyme. The rate of enzymatic erosion for chemosynthetic PHB samples containing both monomeric units of (R) – and (S)-3-hydrohybutyrate is strongly dependent on both the stereo composition and on the tacticity of the sample as well as on substrate specificity of PHB depolymerase. The water-soluble products of random hydrolysis of PHB by enzyme showed a mixture of monomers and oligomers of (R)-3-hydrohybutirate. The rate of enzymatic hydrolysis for melt-crystallized PHB films by PHB depolymerase decreased with an increase in the crystallinity of the PHB film, while the rate of enzymatic degradation for PHB chains in an amorphous state was approximately 20 times higher than the rate for PHB chains in a crystalline state. It was suggested that the PHB depolymerase predominantly hydrolyzes polymer chains in the amorphous phase and then, subsequently, erodes the crystalline phase. The surface of the PHB film after enzymatic degradation was apparently blemished by the action of PHB depolymerase, while no change was observed inside the film. Thus, depolymerase hydrolyzes of the polyester chains in the surface layer of the film and polymer erosion proceeds in surface layers, while dissolution, the enzymatic degradation of PHB are affected by many factors as monomer composition, molecular weight and degree of crystallinity [45].

The PHB polymer matrix ultra structure [47] plays also very important role in enzymatic polymer degradation [48].

At the next step it is necessary to observe enzymatic degradation of PHB under the conditions that modeled the animal tissues and body fluids containing nonspecific esterases. in-vitro degradation of PHB films in the presence of various lipases as nonspecific esterases was carried out in buffered solutions containing lipases [18, 49, 50], in digestive juices (e.g., pancreatin) [14], simulated body fluid [51] biological media (serum, blood etc.) [23] and crude tissue extracts containing a mixture of enzymes [24] to examine the mechanism of nonspecific enzymatic degradation process. It was noted that a Ser…His…Asp triad constitutes the active center of the catalytic domain of both PHB depolymerase [52] and lipases [53]. The serine is part of the pentapeptide Gly X1-Ser-X2-Gly, which has been located in all known PHB depolymerases as well as in lipases, esterase and serine proteases [52].

On the one hand, it was shown that PHB was not degraded for 100 days with a quantity of lipases isolated from different bacteria and fungi [49, 50]. On the other hand, the progressive PHB degradation by lipases was shown [18, 40, 41]. PHB enzymatic biodegradation was studied also in biological media: it was shown that with pancreatin addition no additional mass loss of PHB was observed in comparison with simple hydrolysis [14], the PHB degradation process in serum and blood was demonstrated to be similar to hydrolysis process in buffered solution [31], whereas progressive mass loss of PHB sutures was observed in serum and blood: 16% and 25%, respectively, after 180 days incubation [23], crude extracts from liver, muscle, kidney, heart and brain showed the activity to degrade the PHB: from 2% to 18% mass loss of PHB microspheres after 17 h incubation at pH 7.5 and 9.5 [24]. The weight loss of PHB (M_w = 285,000) films after 45 days incubation simulated body fluid was about 5% [51]. The degradation rate in solution with pancreatin addition, obtained from the decrease in M_w of pure PHB, was accelerated about threefold: 34% decrease in M_w after incubation for 84 days in pancreatin (10 mg/mL in Sorensen buffer) vs. 11% decrease in M_w after incubation in phosphate buffer [14].The same data was obtained for PHB biodegradation in buffered solutions with porcine lipase addition: 72% decrease in M_w of PHB (M_w = 450,000) after incubation for 84 days with lipase (20 U/mg, 10 mg/mL in Tris-buffer) vs. 39% decrease in M_w after incubation in phosphate buffer [18]. This observation is in contrast to enzymatic degradation by PHB depolymerases, which was reported to proceed on the surface of the polymer film with an almost unchanged molecular weight [24, 25]. It has been proposed that for depolymerases the relative size of the enzyme compared with the void space in solvent cast films is the limiting factor for diffusion into the polymer

matrix [54] whereas lipases can penetrate into the polymer matrix through pores in PHB film [40, 41]. It was shown that lipase (0.1 g/L in buffer) treatment for 24 h caused significant morphological change in PHB film surface: transferring from native PHB film with many pores ranging from 1 to 5 μm in size into a pore free surface without producing a quantity of hydroxyl groups on the film surface. It was supposed that the pores had a fairly large surface exposed to lipase, thus it was degraded more easily [40, 41]. It indicates also that lipase can partially penetrate into pores of PHB film but the enzymatic degradation proceeds mainly on the surface of the coarse polymer film, which is achievable for lipase. Two additional effects reported for depolymerases could be of importance. It was concluded that segmental mobility in amorphous phase and polymer hydrophobicity play an important role in enzymatic PHB degradation by nonspecific esterases [14]. Significant impairment of the tensile strength and other mechanical properties were observed during enzymatic biodegradation of PHB threads in serum and blood. It was shown that load at break lost 29%, Young's modulus lost 20%, and tensile strength didn't change after 180 days of threads incubation, the mechanical properties changed gradually [23].

1.2.3 BIODEGRADATION OF PHB BY SOIL MICROORGANISMS

Polymers exposed to the environment are degraded through their hydrolysis, mechanical, thermal, oxidative, and photochemical destruction, and biodegradation [7, 38, 55, 56]. One of the valuable properties of PHB is its biodegradability, which can be evaluated using various field and laboratory tests. Requirements for the biodegradability of PHB may vary in accordance with its applications. The most attractive property of PHB with respect to ecology is that it can be completely degraded by microorganisms finally to CO_2 and H_2O. This property of PHB allows to manufacture biodegradable polymer objects for various applications (Fig. 1.2) [4].

The degradation of PHB and its composites in natural ecosystems, such as soil, compost, and bodies of water, was described in a number of publications [4, 38, 55, 56]. Maergaert et al. [55] isolated from soil more than 300 microbial strains capable of degrading PHB in-vitro. The bacteria detected on the degraded PHB films were dominated by the genera *Pseudomonas*, *Bacillus, Azospirillum, Mycobacterium,* and *Streptomyces,* etc. The samples of PHB have been tested for fungicidity and resistance to fungi by estimating the growth rate of test fungi from the genera *Aspergillus, Aureobasidium, Chaetomium, Paecilomyces, Penicillum, Trichoderma* under optimal growth condi-

tions. PHB film did not exhibit neither fungicide properties, nor the resistance to fungal damage, and served as a good substrate for fungal growth [57].

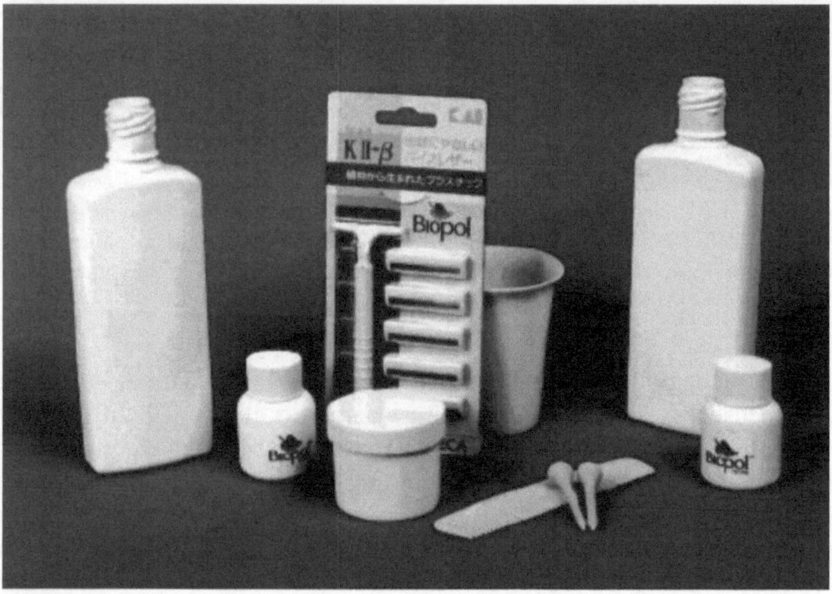

FIGURE 1.2 Moulded PHB objects for various applications (in soil burial or composting experiments, such objects biodegrade in about three months).

It was studied biodegradability of PHB films under aerobic, micro aerobic and anaerobic condition in the presence and absence of nitrate by microbial populations of soil, sludge from anaerobic and nitrifying/denitrifying reactors, and sediment of a sludge deposit site, as well as to obtain active denitrifying enrichment culture degrading PHB (Fig. 1.3) [58]. Changes in molecular mass, crystallinity, and mechanical properties of PHB have been studied. A correlation between the PHB degradation degree and the molecular weight of degraded PHB was demonstrated. The most degraded PHB exhibited the highest values of the crystallinity index. As it has been shown by Spyros et al. [59], PHAs contain amorphous and crystalline regions, of which the former are much more susceptible to microbial attack. If so, the microbial degradation of PHB must be associated with a decrease in its molecular weight and an increase in its crystallinity, which was really observed in the experiments. Moreover, microbial degradation of the amorphous regions of PHB

films made them more rigid. However, further degradation of the amorphous regions made the structure of the polymer much looser [58].

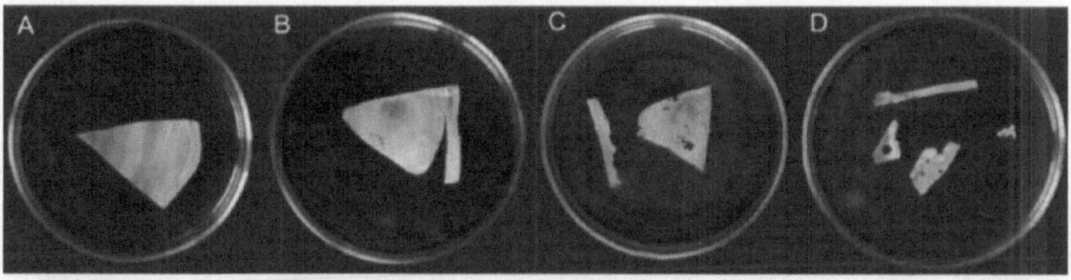

FIGURE 1.3 Undegraded PHB film (a) and PHB films with different degrees of degradation after 2 months incubation in soil suspension: anaerobic conditions without nitrate (b), micro aerobic conditions without nitrate (c), and micro aerobic conditions with nitrate (d).

PHB biodegradation in the enriched culture obtained from soil on the medium used to cultivate denitrifying bacteria (Gil'tai medium) has been also studied. The dominant bacterial species, *Pseudomonas fluorescens* and *Pseudomonas stutzeri*, have been identified in this enrichment culture. Under denitrifying conditions, PHB films were completely degraded for seven days. Both the film weight and M_w of PHB decreased with time. In contrast to the data of Doi et al. [26] who found that M_w of PHB remained unchanged upon enzymatic biodegradation in an aquatic solution of PHB- depolymerase from *Alcaligenes fecalis,* in our experiments, the average viscosity molecular weight of the higher and lower-molecular polymers decreased gradually from 1540 to 580 kDa and from 890 to 612 kDa, respectively. As it was shown at single PHB crystals [47] the "exo"-type cleavage of the polymer chain, that is, a successive removal of the terminal groups, is known to occur at a higher rate than the "endo"-type cleavage, that is, a random breakage of the polymer chain at the enzyme-binding sites. Thus, the former type of polymer degradation is primarily responsible for changes in its average molecular weight. However, the "endo"-type attack plays the important role at the initiation of biodegradation, because at the beginning, a few polymer chains are oriented so that their ends are accessible to the effect of the enzyme [60]. Biodegradation of the lower-molecular polymer, which contains a higher number of terminal groups, is more active, probably, because the "exo" type degradation is more active in lower than in higher molecular polymer [58, 61].

1.2.4 BIODEGRADATION OF PHB IN-VIVO IN ANIMAL TISSUES

The first scientific works on biodegradation of PHB in-vivo in animal tissues were carried out 15–20 years ago by Miller et al. and Saito et al. [22, 24]. They are high-qualitative researches that disclosed many important characteristics of this process. As it was noted above the both enzymatic and nonenzymatic processes of biodegradation of PHB in-vivo can occur simultaneously under normal conditions. But it doesn't mean that polymer biodegradation in-vivo is a simple combination of nonenzymatic hydrolysis and enzymatic degradation. Moreover, in-vivo biodegradation (decrease of molecular weight and mass loss) of PHB is a controversial subject in the literature. As it was noted above for in-vitro PHB hydrolysis, the main reason for the controversy is the use of samples made by various processing technologies and the incomparability of different implantation and animal models. The most of researches on PHB biodegradation was carried out with use of prototypes of various medical devices on the base of PHB: solid films and plates [13, 16, 18, 31, 62], porous patches [14, 15], porous scaffolds [63], electro spun microfiber mats [64], nonwoven patches consisted of fibers [65–69], screws [31], cylinders as nerve guidance channels and conduits [16, 20, 21], monofilament sutures [22, 23], cardiovascular stents [70], microspheres [24, 71]. in-vivo biodegradation was studied on various laboratory animals: rats [14, 18, 20–24, 64], mice [16, 75], rabbits [13, 62, 70, 72], minipigs [15], cats [20], calves [65], sheep [66–68], and even at clinical trials on patients [69]. It is obviously that these animals differ in level of metabolism very much: for example, only weight of these animals differs from 10–20 g (mice) to 50 kg (calves). The implantation of devices from PHB was carried out through different ways: subcutaneously [13, 16, 18, 22, 23, 72], intraperitoneally on a bowel [14], sub periostally on the osseous skull [15–62], nerve wrap-around [19–21], intramuscularly [71, 72], into the pericardium [66–69], into the atrium [65] and intravenously [24]. The terms of implantation were also different: 2.5 h, 24 h, 13 days, 2 months [24]; 7, 14, 30 days [21], 2, 7, 14, 21, 28, 55, 90, 182 days [22]; 1, 3, 6 months [13, 16, 19]; 3, 6, 12 months [65]; 6, 12 months [66]; 6, 24 months [69]; 3, 6, 9, 12, 18, 24 months [68].

The most entire study of PHB in-vivo biodegradation was fulfilled by Gogolewski et al. and Qu et al. [13, 16]. It was shown that PHB lost about 1.6% (injection-molded film, 1.2 mm thick, M_w of PHB = 130 kDa) [16] and 6% (solvent-casting film, 40 μm thick, M_w = 534 kDa) [13] of initial weight after 6 months of implantation. But the observed small weight loss was partially due to the leaching out of low molecular weight fractions and impurities

present initially in the implants. The M_w of PHB decreased from 130,000 to 74,000 (57% of initial M_w) [16] and from 534,000 to 216,000 (40% of initial Mw) [13] after 6 months of implantation. The poly dispersity of PHB polymers narrowed during implantation. PHB showed a constant increase in crystallinity (from 60.7 to 64.8%) up to 6 months [16] or an increase (from 65.0 to 67.9%) after 1 month and a fall again (to 64.5%) after 6 month of implantation [13], which suggests the degradation process had not affected the crystalline regions. This data is in accordance with data of PHB hydrolysis [25] and enzymatic PHB degradation by lipases in-vitro [14] where M_w decrease was observed. The initial biodegradation of amorphous regions of PHB in-vivo is similar to PHB degradation by depolymerase [45].

Thus, the observed biodegradation of PHB showed coexistence of two different degradation mechanisms in hydrolysis in the polymer: enzymatically or nonenzymatically catalyzed degradation. Although nonenzymatical catalysis occurred randomly in homopolymer, indicated by M_w loss rate in PHB, at some point in a time, a critical molecular weight is reached whereupon enzyme-catalyzed hydrolysis accelerated degradation at the surface because easier enzyme/polymer interaction becomes possible. However, considering the low weight loss of PHB, the critical molecular weight appropriate for enzymes predominantly does not reach, yet resulting low molecular weight and crystallinity in PHB could provide some sites for the hydrolysis of enzymes to accelerate the degradation of PHB [13, 16]. Additional data revealing the mechanism of PHB biodegradation in animal tissues was obtained by Kramp et al. [62] in long-term implantation experiments. A very slow, clinically not recordable degradation of films and plates was observed during 20 month (much more than in experiments mentioned above). A drop in the PHB weight loss evidently took place between the 20th and 25th month. Only initial signs of degradation were to be found on the surface of the implant until 20 months after implantation but no more test body could be detected after 25 months [62]. The complete biodegradation in-vivo in the wide range from 3 to 30 months of PHB was shown by other researches [65, 67–69, 73], whereas almost no weight loss and surface changes of PHB during 6 months of biodegradation in-vivo was shown [16, 22]. Residual fragments of PHB implants were found after 30 months of the patches implantation [66, 68]. A reduction of PHB patch size in 27% was shown in patients after 24 months after surgical procedure on pericardial closure with the patch [69]. Significantly more rapid biodegradation in-vivo was shown by other researches [13, 20, 23, 43, 65]. It was shown that 30% mass loss of PHB sutures occurred gradually during 180 days of in-vivo biodegradation with minor changes in the microstructure on the surface and in volume of sutures [23]. It was shown that PHB

nonwoven patches (made to close a trial septal defect in calves) was slowly degraded by polynucleated macrophages, and 12 months postoperatively no PHB device was identifiable but only small particles of polymer were still seen. The absorption time of PHB patches was long enough to permit regeneration of a normal tissue [65]. The PHB sheets progressive biodegradation was demonstrated qualitatively at 2, 6 and 12 months after implantation as weakening of the implant surface, tearing/cracking of the implant, fragmentation and a decrease in the volume of polymer material [21, 43, 72]. The complete biodegradation of PHB (M_w = 150–1000 kDa) thin films (10–50 μm) for 3–6 months was shown and degradation process was described. The process of PHB biodegradation consists of several phases. At initial phase PHB films was covered by fibrous capsule. At second phase capsulated PHB films very slowly lost weight with simultaneous increase of crystallinity and decrease of M_w and mechanical properties of PHB. At third phase PHB films were rapidly disintegrated and then completely degraded. At 4th phase empty fibrous capsule resolved (Fig. 1.4) [18, 31]. Interesting data were obtained for biodegradation in-vivo of PHB microspheres (0.5–0.8 μm in diameter). It was demonstrated indirectly that PHB loss about 8% of weight of microspheres accumulated in liver after 2 month of intravenous injection. It was demonstrated also a presence of several types PHB degrading enzymes in the animal tissues extracts [24].

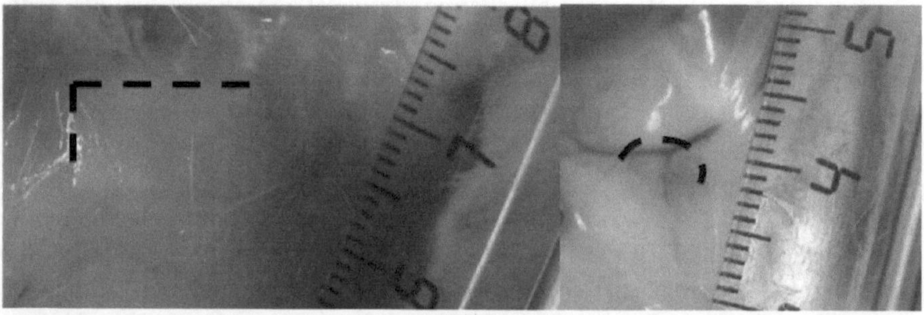

FIGURE 1.4 Biodegradation of PHB films in-vivo. Connective-tissue capsule with PHB thin films (outlined with broken line) 2 weeks (98% residual weight of the film) (left photograph) and 3 months (0% residual weight of the film) (left photograph) after subcutaneous implantation.

Some researchers studied a biodegradation of PHB threads with a tendency of analysis of its mechanical properties in-vivo [22, 23]. It was shown that at first load at break index decreased rapidly from 440 g to 390 g (12%) at 15th day and then gradually increased to the initial value at 90th and remain

almost unchanged up to 182nd day [22] or gradually decreased in 27% during 180 days [23], strain at break decreased rapidly from 60 to 50% (in 17% of initial value) at 10th day and then gradually increased to 70% (in 17% of initial value) at 182nd day [22] or didn't change significantly during 180 days [23].

It was demonstrated that the primary reason of PHB biodegradation in-vivo was a lysosomal and phagocytic activity of polynucleated macrophages and giant cells of foreign body reaction. The activity of tissue macrophages and nonspecific enzymes of body liquids made a main contribution to significantly more rapid rate of PHB biodegradation in-vivo in comparison with rate of PHB hydrolysis in-vitro. The PHB material was encapsulated by degrading macrophages. Presence of PHB stimulated uniform macrophage infiltration, which is important for not only the degradation process but also the restoration of functional tissue. The long absorption time produced a foreign-body reaction, which was restricted to macrophages forming a peripolymer layer [23, 65, 68, 72]. Very important data that clarifies the tissue response that contributes to biodegradation of PHB was obtained by Lobler [46]. It was demonstrated an significant increase of expression of two specific lipases after 7 and 14 days of PHB contact with animal tissues. Moreover, liver specific genes were induced with similar results. It is striking that pancreatic enzymes are induced in the gastric wall after contact with biomaterials [46]. Saito et al. [24] suggested the presence of at least two types of degradative enzymes in rat tissues: liver serine esterases with the maximum of activity in alkaline media (pH=9.5) and kidney esterases with the maximum of activity in neutral media. The mechanism of PHB biodegradation by macrophages was demonstrated at cultured macrophages incubated with particles of low-molecular weight PHB [74]. It was shown that macrophages and, to a lesser level, fibroblasts have the ability to take up (phagocytize) PHB particles (1–10 μm). At high concentrations of PHB particles (>10 μg/mL) the phagocytosis is accompanied by toxic effects and alteration of the functional status of the macrophages but not the fibroblasts. This process is accompanied by cell damage and cell death. The elevated production of nitric oxide (NO) and tumor necrosis factor-α (TNF-α) by activated macrophages was observed. It was suggested that the cell damage and cell death may be due to phagocytosis of large amounts of PHB particles; after phagocytosis, polymer particles may fill up the cells, and cause cell damage and cell death. It was demonstrated also that phagocytized PHB particles disappeared in time due to an active PHB biodegradation process (Fig. 1.5) [74].

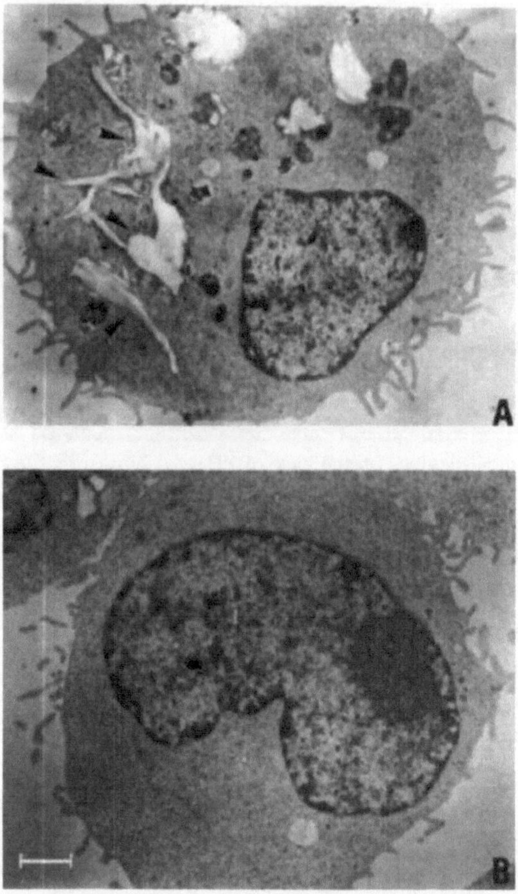

FIGURE 1.5 Phagocytosis of microparticles of PHB in macrophages. TEM analysis of cultured macrophages in the presence (a) or absence (b) of 2 μg PHB microparticles/mL for 24 h. Bar in B represents 1μm, for A and B.

1.3 PHB APPLICATIONS

1.3.1 MEDICAL IMPLANTS AND DEVICES ON THE BASE OF PHB AND ITS BIOCOMPATIBILITY

The perspective area of PHB application is development of implanted medical devices for dental, cranio-maxillofacial, orthopedic, cardiovascular, hernioplastic and skin surgery [3, 6]. A number of potential medical devices on the base of PHB: bioresorbable surgical sutures [6, 22, 23, 31, 75, 76], bio-

degradable screws and plates for cartilage and bone fixation [6, 24, 31, 62], biodegradable membranes for periodontal treatment [6, 31, 77, 78], surgical meshes with PHB coating for hernioplastic surgery [6, 31], wound coverings [79], patches for repair of a bowel, pericardial and osseous defects [14, 15, 65–69], nerve guidance channels and conduits [20, 21], cardiovascular stents [80] etc. was developed (Fig. 1.6).

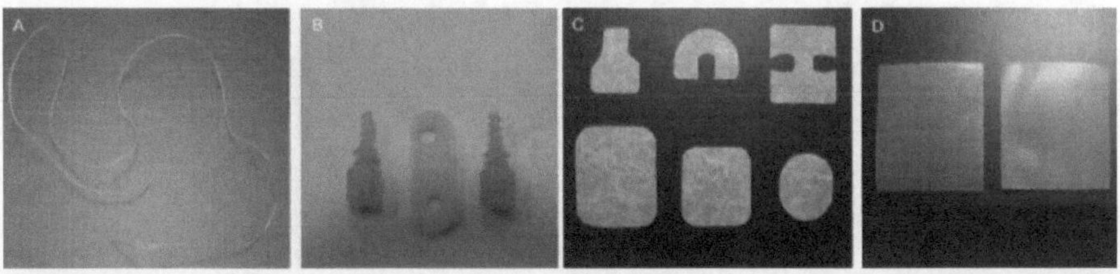

FIGURE 1.6 Medical devices on the base of PHB (a) bioresorbable surgical suture; (b) biodegradable screws and plate for cartilage and bone fixation; (c) biodegradable membranes for periodontal treatment; (d) surgical meshes with PHB coating for hernioplastic surgery, pure (left) and loaded with antiplatelet drug, dipyridamole (right).

The tissue reaction in-vivo to implanted PHB films and medical devices was studied. In most cases a good biocompatibility of PHB was demonstrated. In general, no acute inflammation, abscess formation, or tissue necrosis was observed in tissue surrounding of the implanted PHB materials. In addition, no tissue reactivity or cellular mobilization occurred in areas remote from the implantation site [13, 16, 31, 71]. On the one hand, it was shown that PHB elicited similar mild tissue response as PLA did [16], but on the other hand the use of implants consisting of polylactic acid, polyglicolic acid and their copolymers is not without a number of squeal related with the chronic inflammatory reactions in tissue [81–85].

Subcutaneous implantation of PHB films for 1 month has shown that the samples were surrounded by a well-developed, homogeneous fibrous capsule of 80–100 μm in thickness. The vascularized capsule consists primarily of connective tissue cells (mainly, round, immature fibroblasts) aligned parallel to the implant surface. A mild inflammatory reaction was manifested by the presence of mononuclear macrophages, foreign body cells, and lymphocytes. Three months after implantation, the fibrous capsule has thickened to 180–200 μm due to the increase in the amount of connective tissue cells and a few collagen fiber deposits. A substantial decrease in inflammatory cells was observed after 3 months, tissues at the interface of the polymer were densely organized to form bundles. After 6 months of implantation, the number of in-

flammatory cells had decreased and the fibrous capsule, now thinned to about 80–100 μm, consisted mainly of collagen fibers, and a significantly reduced amount of connective tissue cells. A little inflammatory cells effusion was observed in the tissue adherent to the implants after 3 and 6 months of implantation [13, 16]. The biocompatibility of PHB has been demonstrated in-vivo under subcutaneous implantation of PHB films. Tissue reaction to films from PHB of different molecular weight (300, 450, 1000 kDa) implanted subcutaneously was relatively mild and didn't change from tissue reaction to control glass plate [18, 31].

At implantation of PHB with contact to bone the overall tissue response was favorable with a high rate of early healing and new bone formation with some indication of an osteogenic characteristic for PHB compared with other thermoplastics, such as polyethylene. Initially there was a mixture of soft tissue, containing active fibroblasts, and rather loosely woven osteonal bone seen within 100 μm of the interface. There was no evidence of a giant cell response within the soft tissue in the early stages of implantation. With time this tissue became more orientated in the direction parallel to the implant interface. The dependence of the bone growth on the polymer interface is demonstrated by the new bone growing away from the interface rather than towards it after implantation of 3 months. By 6 months postimplantation the implant is closely encased in new bone of normal appearance with no interposed fibrous tissue. Thus, PHB-based materials produce superior bone healing [43].

Regeneration of a neo intima and a neomedia, comparable to native arterial tissue, was observed at 3–24 months after implantation of PHB nonwoven patches as transannular patches into the right ventricular outflow tract and pulmonary artery. In the control group, a neointimal layer was present but no neomedia comparable to native arterial tissue. Three layers were identified in the regenerated tissue: neointima with a endothelium-like lining, neomedia with smooth muscle cells, collagenous and elastic tissue, and a layer with polynucleated macrophages surrounding islets of PHB, capillaries and collagen tissue. Lymphocytes were rare. It was concluded that PHB nonwoven patches can be used as a scaffold for tissue regeneration in low-pressure systems. The regenerated vessel had structural and biochemical qualities in common with the native pulmonary artery [68]. Biodegradable PHB patches implanted in a trial septal defects promoted formation of regenerated tissue that macroscopically and microscopically resembled native a trial septal wall. The regenerated tissue was found to be composed of three layers: monolayer with endothelium-like cells, a layer with fibroblasts and some smooth-muscle cells, collagenous tissue and capillaries, and a third layer with phagocytizing cells isolating and degrading PHB. The neointima contained a complete

endothelium-like layer resembling the native endothelial cells. The patch material was encapsulated by degrading macrophages. There was a strict border between the collagenous and the phagocytizing layer. Presence of PHB seems to stimulate uniform macrophage infiltration, which was found to be important for the degradation process and the restoration of functional tissue. Lymphocytic infiltration as foreign-body reaction, which is common after replacement of vessel wall with commercial woven Dacron patch, was wholly absent when PHB. It was suggested that the absorption time of PHB patches was long enough to permit regeneration of a tissue with sufficient strength to prevent development of shunts in the atrial septal position [65]. The prevention of postoperative pericardial adhesions by closure of the pericardium with absorbable PHB patch was demonstrated. The regeneration of mesothelial layer after implantation of PHB pericardial patch was observed. The complete regeneration of mesothelium, with morphology and biochemical activity similar to findings in native mesothelium, may explain the reduction of postoperative pericardial adhesions after operations with insertion of absorbable PHB patches [67]. The regeneration of normal filament structure of restored tissues was observed by immune histochemical methods after PHB devices implantation [66]. The immune histochemical demonstration of cytokeratine, an intermediate filament, which is constituent of epithelial and mesodermal cells, agreed with observations on intact mesothelium. Heparin sulfate proteoglycan, a marker of basement membrane, was also identified [66]. However, in spite of good tissue reaction to implantation of cardiovascular PHB patches, PHB endovascular stents in the rabbit iliac arterial caused intensive inflammatory vascular reactions [80].

PHB patches for the gastrointestinal tract were tested using animal model. Patches made from PHB sutured and PHB membranes were implanted to close experimental defects of stomach and bowel wall. The complete regeneration of tissues of stomach and bowel wall was observed at 6 months after patch implantation without strong inflammatory response and fibrosis [14, 86].

Recently an application of biodegradable nerve guidance channels (conduits) for nerve repair procedures and nerve regeneration after spinal cord injury was demonstrated. Polymer tubular structures from PHB can be modulated for this purpose. Successful nerve regeneration through a guidance channel was observed as early as after 1 month. Virtually all implanted conduits contained regenerated tissue cables centrally located within the channel lumen and composed of numerous myelinated axons and Schwann cells. The inflammatory reaction had not interfered with the nerve regeneration process. Progressive angiogenesis was present at the nerve ends and through the walls

of the conduit. The results demonstrate good-quality nerve regeneration in PHB guidance channels [21, 87].

Biocompatibility of PHB was evaluated by implanting microspheres from PHB (M_w = 450 kDa) into the femoral muscle of rats. The spheres were surrounded by one or two layers of spindle cells, and infiltration of inflammatory cells and mononuclear cells into these layers was recognized at 1 week after implantation. After 4 weeks, the number of inflammatory cells had decreased and the layers of spindle cells had thickened. No inflammatory cells were seen at 8 weeks, and the spheres were encapsulated by spindle cells. The toxicity of PHB microspheres was evaluated by weight change and survival times in L1210 tumor-bearing mice. No differences were observed in the weight change or survival time compared with those of control. These results suggest that inflammation accompanying microsphere implantation is temporary as well as toxicity to normal tissues is minimal [71].

The levels of tissue factors, inflammatory cytokines, and metabolites of arachidonic acid were evaluated. Growth factors derived from endothelium and from macrophages were found. These factors most probably stimulate both growth and regeneration occurring when different biodegradable materials were used as grafts [46, 65, 67, 86]. The positive reaction for thrombomodulin, a multifunctional protein with anticoagulant properties, was found in both mesothelial and endothelial cells after pericardial PHB patch implantation. Prostacycline production level, which was found to have cyto protective effect on the pericardium and prevent adhesion formation, in the regenerated tissue was similar to that in native pericardium [65, 67]. The PHB patch seems to be highly biocompatible, since no signs of inflammation were observed macroscopically and also the level of inflammation associated cytokine mRNA did not change dramatically, although a transient increase of interleukin-1β and interleukin-6 mRNA through days 1–7 after PHB patch implantation was detected. In contrast, tumor necrosis factor-α mRNA was hardly detectable throughout the implantation period, which agrees well with an observed moderate fibrotic response [46, 86].

1.3.2 PHB AS TISSUE ENGINEERING MATERIAL AND PHB IN-VITRO BIOCOMPATIBILITY

Biopolymer PHB is promising material in tissue engineering due to high biocompatibility in-vitro. Cell cultures of various origins including murine and human fibroblasts [15, 40, 78, 88–92], human mesenchymal stem cells [93], rabbit bone marrow cells (osteoblasts) [36, 89, 94], human osteogenic sarcoma cells [95], human epithelial cells [90, 95], human endothelial cells [96,

97], rabbit articular cartilage chondrocytes [98, 99] and rabbit smooth muscle cells [100], human neurons (schwann cells) [101] in direct contact with PHB when cultured on polymer films and scaffolds exhibited satisfactory levels of cell adhesion, viability and proliferation. Moreover, it was shown that fibroblasts, endothelium cells, and isolated hepatocytes cultured on PHB films exhibited high levels of cell adhesion and growth (Fig. 1.7) [102]. A series of 2D and 3D PHB scaffolds was developed by various methods: polymer surface modification [97], blending [36, 78, 88, 90, 93, 99, 103], electrospinning [104–106], salt leaching [36, 94, 107, 108], microspheres fusion [109], forming on porous mold [110], laser cutting [111].

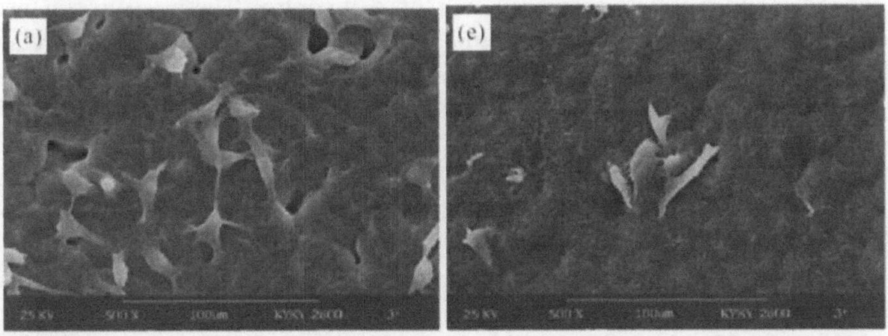

FIGURE 1.7 Scanning electron microscopy image of 2 days growth of fibroblast cells on films made of (a) PHB; (e) PLA; (500×). Cell density of fibroblasts grown on PHB film is significantly higher vs. cell density of fibroblasts grown on PLA film.

It was shown also that cultured cells produced collagen II and glycosaminoglycan, the specific structural biopolymers formed the extracellular matrix [95, 98, 99]. A good viability and proliferation level of macrophages and fibroblasts cell lines was obtained under culturing in presence of particles from short-chain low-molecular PHB [74]. However, it was shown that cell growth on the PHB films was relatively poor: the viable cell number ranged from 1×10^3 to 2×10^5 [40, 89, 99]. An impaired interaction between PHB matrix and cytoskeleton of cultured cells was also demonstrated [95]. It was reported that a number of polymer properties including chemical composition, surface morphology, surface chemistry, surface energy and hydrophobicity play important roles in regulating cell viability and growth [112]. The investigation showed that this biomaterial can be used to make scaffolds for in-vitro proliferating cells [40, 89, 98].

The most widespread methods to manufacture the PHB scaffolds for tissue engineering by means of improvement of cell adhesion and growth on

polymer surface are change of PHB surface properties and microstructure by salt-leaching methods and enzymatic/chemical/physical treatment of polymer surface [40, 89, 98, 113]. Adhesion to polymer substrates is one of the key issues in tissue engineering, because adhesive interactions control cell physiology. One of the most effective techniques to improve adhesion and growth of cells on PHB films is treatment of polymer surface with enzymes, alkali or low-pressure plasma [40,113]. Lipase treatment increases the viable cell number on the PHB film from 100 to 200 times compared to the untreated PHB film. NaOH treatment on PHB film also indicated an increase of 25 times on the viable cell number compared with the untreated PHB film [40]. It was shown that treatment of PHB film surface with low-pressure ammonia plasma improved growth of human fibroblasts and epithelial cells of respiratory mucosa due to increased hydrophilicity (but with no change of microstructure) of polymer surface [90]. It was suggested that the improved hydrophilicity of the films after PHB treatment with lipases, alkali and plasma allowed cells in its suspension to easily attach on the polymer films compared to that on the untreated ones. The influence of hydrophilicity of biomaterial surface on cell adhesion was demonstrated earlier [114].

But a microstructure of PHB film surface can be also responsible for cell adhesion and cell growth [115–117]. Therefore, noticed above modification of polymer film surface after enzymatic and chemical treatment (in particular, reduced pore size and a surface smoothing) is expected to play an important role for enhanced cell growth on the polymer films [40]. Different cells prefer different surface. For example, osteoblasts preferred rougher surfaces with appropriate size of pores [115, 116] while fibroblast prefer smoother surface, yet epithelial cells only attached to the smoothest surface [117]. This appropriate roughness affects cell attachment as it provides the right space for osteoblast growth, or supplies solid anchors for Fila podia. A scaffold with appropriate size of pores provided better surface properties for anchoring type II collagen filaments and for their penetration into internal layers of the scaffolds implanted with chondrocytes. This could be illuminated by the interaction of extracellular matrix proteins with the material surface. Moreover, the semicrystalline surface PHB ultra structure can be connected with protein adsorption and cell adhesion [91, 92, 118]. The appropriate surface properties may also promote cell attachment and proliferation by providing more spaces for better gas/nutrients exchange or more serum protein adsorption [36, 94, 98]. Additionally, Sevastianov et al. [119] found that PHB films in contact with blood did not activate the hemostasis system at the level of cell response, but they did activate the coagulation system and the complement reaction.

The high biocompatibility of PHB may be due to several reasons. First of all, PHB is a natural biopolymer involved in important physiological functions both prokaryotes and eukaryotes. PHB from bacterial origin has property of stereo specificity that is inherent to biomolecules of all living things and consists only from residues of D(-)-3-hydrohybutyric acid [120–122]. Low molecular weight PHB (up to 150 resides of 3-hydrohybutyric acid), complexed to other macromolecules (cPHB), was found to be a ubiquitous constituent of both prokaryotic and eukaryotic organisms of nearly all phyla [123–127]. Complexed cPHB was found in a wide variety of tissues and organs of mammals (including human): blood, kidney, vessels, nerves, vessels, eye, brain, as well as in organelles, membrane proteins, lipoproteins, and plaques. cPHB concentration ranged from 3–4 μg/g wet tissue weight in nerves and brain to 12 μg/g in blood plasma [128, 129]. In humans, total plasma cPHB ranged from 0.60 to 18.2 mg/L, with a mean of 3.5 mg/L. [129]. It was shown that cPHB is a functional part of ion channels of erythrocyte plasma membrane and hepatocyte mitochondria membrane [130, 131]. The singular ability of cPHB to dissolve salts and facilitate their transfer across hydrophobic barriers defines a potential physiological niche for cPHB in cell metabolism [125]. However, a mechanism of PHB synthesis in eukaryotic organisms is not well clarified that requires additional studies. Nevertheless, it could be suggested that cPHB is one of products of symbiotic interaction between animals and gut microorganisms. It was shown, for example, that E-coli is able to synthesize low molecular weight PHB and cPHB plays various physiological roles in bacteria cell [127, 132].

Intermediate product of PHB biodegradation, D(-)-3-hydroxybutyric acid is also a normal constituent of blood at concentrations between 0.3 and 1.3 mM and contains in all animal tissues [133, 134]. As it was noted above PHB has a rather low degradation rate in the body in comparison to, e.g., poly(lactic-co-glycolic) acids, that prevent from increase of 3-hydroxybutyric acid concentration in surrounding tissues [13, 16], whereas polylactic acid release, following local pH decrease in implantation area and acidic chronic irritation of surrounding tissues is a serious problem in application of medical devices on the base of poly(lactic-co-glycolic) acids [135, 136]. Moreover, chronic inflammatory response to polylactic and polyglycolic acids that was observed in a number of cases may be induced by immune response to water-soluble oligomers that released during degradation of synthetic polymers [136–138].

1.3.3 NOVEL DRUG DOSAGE FORMS ON THE BASE OF PHB

An improvement of medical materials on the base of biopolymers by encapsulating different drugs opens up the wide prospects in applications of new devices with pharmacological activity. The design of injection systems for sustained drug delivery in the forms of microspheres or microcapsules prepared on the base of biodegradable polymers is extremely challenging in the modern pharmacology. The fixation of pharmacologically active component with the biopolymer and following slow drug release from the microparticles provides an optimal level of drug concentration in local target organ during long-term period (up to several months). At curative dose the prolonged delivery of drugs from the systems into organism permits to eliminate the shortcomings in perioral, injectable, aerosol, and the other traditional methods of drug administration. Among those shortcomings hypertoxicity, instability, pulsative character of rate delivery, ineffective expenditure of drugs should be pointed out. Alternatively, applications of therapeutical polymer systems provide orderly and purposefully the deliverance for an optimal dose of agent that is very important at therapy of acute or chronic diseases [1, 3, 6, 139]. An ideal biodegradable microsphere formulation would consist of a free-flowing powder of uniform-sized microspheres less than 125 μm in diameter and with a high drug loading. In addition, the drug must be released in its active form with an optimized profile. The manufacturing method should produce the microspheres, which are reproducible, scalable, and benign to some often-delicate drugs, with a high encapsulation efficiency [140, 141].

PHB as biodegradable and biocompatible is a promising material for producing of polymer systems for controlled drug release. A number of drugs with various pharmacological activities were used for development of polymer controlled release systems on the base of PHB and its copolymers: model drugs (2,7-dichlorofluorescein [142], dextran-FITC [143], methyl red [122, 144, 145], 7-hydroxethyltheophylline [146, 147]), antibiotics and antibacterial drugs (rifampicin [148, 149], tetracycline [150], cefoperazone and gentamicin [151], sulperazone and duocid [152–155], sulbactam and cefoperazone [156], fusidic acid [157], nitrofural [158]), anticancer drugs (5-fluorouracil [159], 2,'3'-diacyl-5-fluoro-2'-deoxyuridine [71], paclitaxel [160, 161], rubomycin [162], chlorambucil and etoposide [163]), antiinflammatory drug (indomethacin [164], flurbiprofen [165], ibuprofen [166]), analgesics (tramadol [167], vasodilator and antithrombotic drugs (dipyridamole [168, 169], nitric oxide donor [170, 171], nimodipine [174], felodipine [175]), proteins (hepatocyte growth factor [176], mycobacterial protein for vaccine development

[177], bone morphogenetic protein 7 [178]). Various methods for manufacture of drug-loaded PHB matrices and microspheres were used: films solvent casting [144–147], emulsification and solvent evaporation [148–164], spray drying [172], layer-by-layer self-assembly [165], supercritical antisolvent precipitation [173]. The biocompatibility and pharmacological activity of some of these systems was studied [71, 148, 154–156, 160, 161, 167, 171]. But only a few drugs were used for production of drug controlled release systems on the base of PHB homopolymer: 7-hydroxethyltheophylline, methyl red, 2,'3'-diacyl-5-fluoro-2'-deoxyuridine, rifampicin, tramadol, indomethacin, dipyridamole and paclitaxel [71, 154–156, 160–171]. The latest trend in PHB drug delivery systems study is PHB nanoparticles development loaded with different drugs [179–180].

The first drug-sustained delivery system on the base of PHB was developed by Korsatko W. et al., who observed a rapid release of encapsulated drug, 7-hydroxethyltheophylline, from tablets of PHB (M_w = 2000 kDa), as well as weight losses of PHB tablets containing the drug after subcutaneous implantation. It was suggested that PHB with molecular weight greater than 100 kDa was undesirable for long-term medication dosage [146].

Pouton C.W. and Akhtar S. describing the release of low molecular drugs from PHB matrices reported that the latter have the tendencies to enhance water penetration and pore formation [181]. The entrapment and release of the model drug, methyl red (MR), from melt-crystallized PHB was found to be a function of polymer crystallization kinetics and morphology whereas overall degree of crystallinity was shown to cause no effect on drug release kinetics. MR released from PHB films for 7 days with initial phase of rapid release ("burst effect") and second phase with relatively uniform release. Release profiles of PHB films crystallized at 110°C exhibited a greater burst effect when compared to those crystallized at 60°C. This was explained by better trapping of drug within polymeric spherulites with the more rapid rates of PHB crystallization at 110°C [144, 145].

Kawaguchi et al. [71] showed that chemical properties of drug and polymer molecular weight had a great impact on drug delivery kinetics from PHB matrix. Microspheres (100–300 μm in diameter) from PHB of different molecular weight (65,000, 135,000, and 450,000) were loaded with prodrugs of 5-fluoro-2'deoxyuridine (FdUR) synthesized by esterification with aliphatic acids (propionate, butyrate, and pentanoate). Prodrugs have different physicochemical properties, in particular, solubility in water (from 70 mg/mL for FdUR to 0.1 mg/mL for butyryl-FdUR). The release rates from the spheres depended on both the lipophilicity of the prodrug and the molecular weight of the polymer. Regardless of the polymer, the relative release rates were propio-

nyle-FdUR > butyryl-FdUR > pentanoyl-FdUR. The release of butyryl-FdUR and pentanoyl-FdUR from the spheres consisting of low-molecular-weight polymer (M_w = 65,000) was faster than that from the spheres of higher molecular weight (M_w = 135,000 or 450,000). The effect of drug content on the release rate was also studied. The higher the drug content in the PHB microspheres, the faster was the drug release. The release of FdUR continued for more than 5 days [71].

Kassab [149] developed a well-managed technique for manufacture of PHB microspheres loaded with drugs. Microspheres were obtained within a size of 5–100 μm using a solvent evaporation method by changing the initial polymer/solvent ratio, emulsifier concentration, stirring rate, and initial drug concentration. The drug overloading of up to 0.41 g rifampicin/g PHB were achieved. Drug release was rapid: the maximal duration of rifampicin delivery was 5 days. Both the size and drug content of PHB microspheres were found to be effective in controlling the drug release from polymer microspheres [149].

The sustained release of analgesic drug, tramadol, from PHB microspheres was demonstrated by Salman et al. [167]. It was shown that 58% of the tramadol (the initial drug content in PHB matrix = 18%) was released from the microspheres (7.5 μm in diameter) in the first 24 h. Drug release decreased with time. From 2 to 7 days the drug release was with zero-order rate. The entire amount of tramadol was released after 7 days [167].

The kinetics of different drug release from PHB micro and nanoparticles loaded with dipyridamole, indomethacin and paclitaxel was studied [160, 161, 164, 168, 169]. It was found that the release occurs via two mechanisms, diffusion and degradation, operating simultaneously. Vasodilator and antithrombotic drug, dipyridamole, and antiinflammatory drug, indomethacin, diffusion processes determine the rate of the release at the early stages of the contact of the system with the environment (the first 6–8 days). The coefficient of the release diffusion of a drug depends on its nature, the thickness of the PHB films containing the drug, the weight ratio of dipyridamole and indomethacin in polymer, and the molecular weight of PHB. Thus, it is possible to regulate the rate of drug release by changing of molecular weight of PHB (e.g., refer Ref. [164]). The biodegradable microspheres on the base of PHB designed for controlled release of dipyridamole and paclitaxel were kinetically studied. The profiles of release from the microspheres with different diameters present the progression of nonlinear and linear stages. Diffusion kinetic equation describing both linear (PHB hydrolysis) and nonlinear (diffusion) stages of the dipyridamole and paclitaxel release profiles from the spherical subjects has been written down as the sum of two terms: desorption from the homogeneous

sphere in accordance with diffusion mechanism and the zero-order release. In contrast to the diffusivity dependence on microsphere size, the constant characteristics of linearity are scarcely affected by the diameter of PHB microparticles. The view of the kinetic profiles as well as the low rate of dipyridamole and paclitaxel release are in satisfactory agreement with kinetics of weight loss measured in-vitro for the PHB films and observed qualitatively for PHB microspheres. Taking into account kinetic results, it was supposed that the degradation of PHB microspheres is responsible for the linear stage of dipyridamole and paclitaxel release profiles (Fig 1.8) [24, 160, 161, 168, 169].

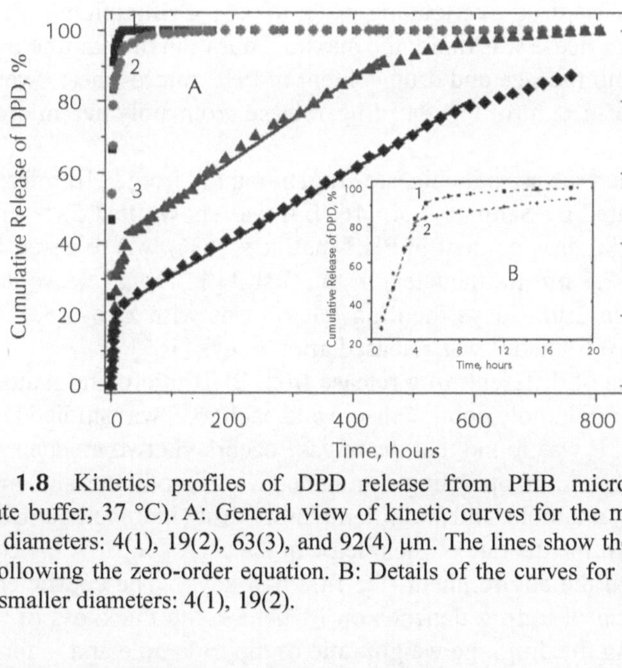

FIGURE 1.8 Kinetics profiles of DPD release from PHB microspheres in-vitro (phosphate buffer, 37 °C). A: General view of kinetic curves for the microspheres with different diameters: 4(1), 19(2), 63(3), and 92(4) μm. The lines show the second stage of release following the zero-order equation. B: Details of the curves for the microspheres with the smaller diameters: 4(1), 19(2).

The biocompatibility and pharmacological activity of advanced drug delivery systems on the base of PHB was studied [71, 148, 167–169]. It was shown that implanted PHB microspheres loaded with paclitaxel caused the mild tissue reaction. The inflammation accompanying implantation of PHB matrices is temporary and additionally toxicity relative to normal tissues is minimal [169]. No signs of toxicity were observed after administration of PHB microspheres loaded with analgesic, tramadol, [167]. A single intraper-

itoneal injection of PHB microspheres containing anticancer prodrugs, butyryl-FdUR and pentanoyl- FdUR, resulted in high antitumor effects against P388 leukemia in mice over a period of five days [71]. Embolization with PHB microspheres in-vivo at dogs as test animals has been studied by Kasab et al. Renal angiograms obtained before and after embolization and also the histopathological observations showed the feasibility of using these microspheres as an alternative chemoembolization agent [148]. Epidural analgesic effects of tramadol released from PHB microspheres were observed for 21 h, whereas an equal dose of free tramadol was effective for less than 5 h. It was suggested that controlled release of tramadol from PHB microspheres in-vivo is possible, and pain relief during epidural analgesia is prolonged by this drug formulation compared with free tramadol [167].

The observed data indicate the wide prospects in applications of drug-loaded medical devices and microspheres on the base of PHB as implantable and injectable therapeutic systems in medicine for treatment of various diseases: cancer, cardio-vascular diseases, tuberculosis, osteomyelitis, arthritis, etc. [6].

1.4 CONCLUSIONS

The natural PHB is unique biodegradable thermoplastics of considerable commercial importance. With this review, we have attempted to systematically evaluate the impact of physicochemical factors on the hydrolysis and the biodegradation of natural PHB both in-vitro and in-vivo. Clearly, the degradation behavior observed is very dependent upon both physicochemical conditions. Geometry and structural and microbial properties. If these conditions of (bio)degradation are known, the systems on the base PHB can be designed in such biomedicine areas as medical devices (Section 1.3.1), tissue scaffolds in bioengineering (Section 1.3.2) and development of novel biodegradable therapeutic systems for drug delivery.

ACKNOWLEDGMENTS

The work was supported by the Russian Foundation for Basic Research (grant no. 14–03–00405-a) and the Russian Academy of Sciences under the program "Construction of New Generation Macromolecular Structures" (03/OC-14).

KEYWORDS

- **Biodegradability**
- **Biomedical applications**
- **Nonenzymatic hydrolysis**
- **Poly(3-hydroxybutyrate)**
- **Surface degradation**
- **Weight loss**

REFERENCES

1. Rice, J. J., Martino, M. M., De Laporte, L., Tortelli, F., Briquez, P. S., & Hubbell, J. A. (2013). *Adv. Healthcare Mater, 2,* 57–71. DOI: 10.1002adhm.201200197. Engineering the Regenerative Microenvironment with Biomaterials.
2. Khademhosseini, A., & Peppas, N. A. (2013). *Adv. Healthcare Mater*, *2*, 10–12 DOI: 10.1002/adhm.201200444. Micro and Nano Engineering of Biomaterials for Healthcare Applications.
3. Chen, G. Q., & Wu, Q. (2005). *Biomaterials, 26(33)*, 6565–6578. The Application of Polyhydroxy Alkanoates as Tissue Engineering Materials.
4. Lenz, R. W., & Marchessault, R. H. (2005). *Bio macromolecules, 6(1)*, 1–8. Bacterial Polyesters: Biosynthesis, Biodegradable Plastics and Biotechnology.
5. Anderson, A. J., & Dawes, E. A. (1990). *Microbiological Reviews, 54(4)*, 450–472. Occurrence, Metabolism, Metabolic Role, and Industrial Uses of Bacterial Polyhydroxyalkanoates.
6. Bonartsev, A. P., Bonartseva, G. A., Shaitan, K. V., & Kirpichnikov, M. P. (2011). *Biochemistry* (Moscow) *Supplement Series B*: *Biomedical Chemistry, 5(1),* 10–21. Poly (3–hydroxybutyrate) and poly(3–hydroxybutyrate) based biopolymer systems.
7. Jendrossek, D., & Handrick, R. (2002). *Annu Rev Microbiol, 56*, 403–432. Microbial Degradation of Polyhydroxyalkanoates.
8. Fabra, M. J., Lopez–Rubio, A., & Lagaron, J. M. (2013). *Food Hydrocolloids, 32*, 106–114. DOI http:dx.doi.org/10.1016/j.foodhyd.2012.12.007. High Barrier Polyhydroxyalcanoate Food Packaging Film by Means of Nanostructured Electrospun Interlayer's of Zein.
9. Kim, D. Y., & Rhee, Y. H. (2003). *Appl. Microbiol Biotechnol 61*, 300–308. Biodegradation of Microbial and Synthetic Polyesters by Fungi.
10. Marois, Y., Zhang, Z., Vert, M., Deng, X., Lenz, R., & Guidoin, R. J. (1999). *Biomater. Sci.* Polym. Ed *10*, 483–499. Hydrolytic and Enzymatic Incubation of Polyhydroxyoctanoate (PHO): A Short–Term in Vitro Study of Degradable Bacterial Polyester.
11. Abe, H., & Doi, Y. (2002). *Biomacromolecules 3(1)*, 133–138. Side–Chain Effect of Second Monomer Units on Crystalline Morphology, Thermal Properties, and Enzymatic Degradability for Random Copolyesters of (R)–3–Hydroxybutyric Acid with (R)–3–Hydroxyalkanoic Acids.
12. Renstad, R., Karlsson, S., & Albertsson, A. C. (1999). *Polym. Degrad. Stab 63*, 201–211. The influence of processing induced differences in molecular structure on the biological and non–Biological Degradation of poly (3–hydroxybutyrate–co–3–hydroxyvalerate), P (3–HB–co–3–HV).

13. Qu, X. H., Wu, Q., Zhang, K. Y., & Chen, G. Q. (2006). *Biomaterials, 27(19)*, 3540–3548. In Vivo Studies of Poly (3–Hydroxybutyrate–Co–3–Hydroxyhexanoate) Based Polymers: Biodegradation and Tissue Reactions.
14. Freier, T., Kunze, C., Nischan, C., Kramer, S., Sternberg, K., Sass, M., Hopt, U. T., & Schmitz, K. P. (2002). *Biomaterials. 23(13)*, 2649–2657. In Vitro and in Vivo Degradation Studies for Development of a Biodegradable Patch Based on Poly (3–Hydroxybutyrate).
15. Kunze, C., Edgar Bernd, H., Androsch, R., Nischan, C., Freier, T., Kramer, S., Kramp, B., Schmitz, K. P. (Jan 2006). *Biomaterials, 27(2)*, 192–201. In Vitro and in Vivo Studies on Blends of Isotactic and Atactic Poly (3–Hydroxybutyrate) for Development of a Dura Substitute Material.
16. Gogolewski, S., Jovanovic, M., Perren, S. M., Dillon, J. G., & Hughes, M. K. (1993). *J Biomed. Mater. Res. 27(9)*, 1135–1148. Tissue Response and in Vivo Degradation of Selected Polyhydroxyacids: Polylactides (PLA), Poly (3–Hydroxybutyrate) (PHB), and Poly (3–Hydroxybutyrate–Co–3–Hydroxyvalerate) (PHB/VA).
17. Boskhomdzhiev, A. P., Bonartsev, A. P., Ivanov, E. A., Makhina, T. K., Myshkina, V. L., Bagrov, D. V., Filatova, E. V., Bonartseva, G. A., & Iordanskii, A. L. (2010). *International Polymer Science and Technology 37(11)*, 25–30. Hydrolytic Degradation of Biopolymer Systems Based on Poly (3–Hydroxybutyrate. Kinetic and Structural Aspects.
18. Boskhomdzhiev, A. P., Bonartsev, A. P., Makhina, T. K., Myshkina, V. L., Ivanov, E. A., Bagrov, D. V., Filatova, E. V., Iordanskiĭ, A. L., & Bonartseva, G. A. (2010). *Biochemistry* (Moscow) *Supplement Series B*: *Biomedical Chemistry, 4(2)*, 177–183. Biodegradation Kinetics of Poly (3–hydroxybutyrate) based biopolymer systems.
19. Borkenhagen, M., Stoll, R. C., Neuenschwander, P., Suter, U. W., & Aebischer, P. (1998). *Biomaterials 19 (23)*, 2155–2165. In Vivo Performance of a New Biodegradable Polyester Urethane System Used as a Nerve Guidance Channel.
20. Hazari, A., Johansson–Ruden, G., Junemo–Bostrom, K., Ljungberg, C., Terenghi, G., Green, C., & Wiberg, M. (1999). *Journal of Hand Surgery* (*British and European) 24B (3),* 291–295. A New Resorbable Wrap–Around Implant as an Alternative Nerve Repair Technique.
21. Hazari, A., Wiberg, M., Johansson–Rudén, G., Green, C., & Terenghi, G. A. (1999). *British Journal of Plastic Surgery, 52*, 653–657. Resorbable Nerve Conduit as an Alternative to Nerve Auto graft.
22. Miller, N. D., & Williams, D. F. (1987 Mar). *Biomaterials 8(2)*, 129–137. On the Biodegradation of Poly–Beta–Hydroxybutyrate (PHB) Homopolymer and Poly–Beta–Hydroxybutyrate–Hydroxyvalerate Copolymers.
23. Shishatskaya, E. I., Volova, T. G., Gordeev, S. A., & Puzyr, A. P. (2005). *J Biomater Sci Polym Ed. 16(5)*, 643–657. Degradation of P (3HB) and P (3HB–co–3HV) in biological media.
24. Bonartsev, A. P., Livshits, V. A., Makhina, T. A., Myshkina, V. L., Bonartseva, G. A., & Iordanskii1 A. L. (2007). *Express Polymer Letters. 1(12)*, 797–803. DOI: 10.3144/expresspolymlett.2007.110 Controlled Release Profiles of Dipyridamole from Biodegradable Microspheres on the Base of Poly (3–hydroxybutyrate).
25. Koyama, N., & Doi, Y. (1995). Morphology and biodegradability of a binary blend of poly ((R)–3–hydroxybutyric acid) and poly ((R, S)–lactic acid) *Can. J. Microbiol., 41(Suppl. 1),* 316–322
26. Doi, Y., Kanesawa, Y., Kunioka, M., & Saito, T. (1990a). Biodegradation of microbial copolyesters: poly(3–hydroxybutyrate–co–3–hydroxyvalerate) and poly(3–hydroxybutyrate–co–4– hydroxybutyrate). *Macromolecules, 23*, 26–31.

27. Holland, S. J., Jolly, A. M., Yasin, M., & Tighe, B. J. (1987). Polymers for biodegradable medical devices II Hydroxybutyrate Hydroxyvalerate Copolymers: Hydrolytic Degradation Studies. *Biomaterials, 8(4)*, 289–295.
28. Kurcok, P., Kowalczuk, M., Adamus, G., Jedlinrski, Z., & Lenz, R. W. (1995). Degradability of Poly (b–hydroxybutyrate)s. Correlation with Chemical Microstructure. *JMS–Pure Appl. Chem. A32*, 875–880.
29. Bonartsev, A. P., Boskhomodgiev, A. P., Iordanskii, A. L., Bonartseva, G. A., Rebrov, A. V., Makhina, T. K., Myshkina, V. L., Yakovlev, S. A., Filatova, E. A., Ivanov, E. A., Bagrov, D. V., & Zaikov, G. E. (2012). Hydrolytic Degradation of Poly (3–hydroxybutyrate), Polylactide and their Derivatives: Kinetics, Crystallinity, and Surface Morphology. *Molecular Crystals and Liquid Crystals, 556(1)*, 288–300.
30. Bonartsev, A. P., Boskhomodgiev, A. P., Voinova, V. V., Makhina, T. K., Myshkina, V. L., Yakovlev, S. A., Zharkova, I. I., Zernov, A. L., Filatova, E. A., Bagrov, D. V., Rebrov, A. V., Bonartseva, G. A., & Iordanskii, A. L. (2012). Hydrolytic Degradation of Poly (3–hydroxybutyrate) and its Derivates: Characterization and Kinetic Behavior. *Chemistry and Chemical Technology, T.6, N.4*, 385–392.
31. Bonartsev, A. P., Myshkina, V. L., Nikolaeva, D. A., Furina, E. K., Makhina, T. A., Livshits, V. A., Boskhomdzhiev, A. P., Ivanov, E. A., Iordanskii, A. L., & Bonartseva, G. A. (2007). Biosynthesis, Biodegradation, and Application of Poly (3–Hydroxybutyrate) and Its Copolymers Natural Polyesters Produced By Diazotrophic Bacteria *Communicating Current Research and Educational Topics and Trends in Applied Microbiology*, Ed: A. Méndez–Vilas, Formatex, Spain, *1*, 295–307.
32. Cha, Y., & Pitt, C. G. (1990). The biodegradability of polyester blends. *Biomaterials 11(2)*, 108–112.
33. Schliecker, G., Schmidt, C., Fuchs, S., Wombacher, R., & Kissel, T. (2003). Hydrolytic degradation of poly (lactide–co–glycolide) films: effect of oligomers on degradation rate and crystallinity. Int. *J. Pharm. 266(1–2)*, 39–49.
34. Scandola, M., Focarete, M. L., Adamus, G., Sikorska, W., Baranowska, I., Swierczek, S., Gnatowski, M., Kowalczuk, M., & Jedlinr ski, Z. (1997). Polymer blends of natural poly (3–hydroxybutyrate–co–hydroxyvalerate) and a synthetic a tactic poly (3–hydroxybutyrate). Characterization and Biodegradation Studies *Macromolecules, 30*, 2568–2574.
35. Doi, Y., Kanesawa, Y., Kawaguchi, Y., & Kunioka, M. (1989). Hydrolytic Degradation of Microbial Poly (hydroxyalkanoates). Makrom Chem. Rapid. *Commun 10*, 227–230.
36. Wang, Y. W., Yang, F., Wu, Q., Cheng, Y. C., Yu, P. H., Chen, J., & Chen, G. Q. (2005). Evaluation of Three Dimensional Scaffolds Made of Blends of Hydroxyapatite and Poly (3hydroxybutyrate–Co–3–hydroxyhexanoate) for Bone Reconstruction *Biomaterials 26(8)*, 899–904(a)
37. Muhamad, I. I., Joon, L. K., & Noor, M. A.M. (2006). Comparing the degradation of poly–β–(hydroxybutyrate), poly–β–(hydroxybutyrate–co–valerate)(PHBV) and PHBV/ Cellulose triacetate blend. *Malaysian Polymer Journal, 1*, 39–46.
38. Mergaert, J., Webb, A., Anderson, C., & Wouters, A. (1993). *Swings J. Microbial Degradation of Poly(3–hydroxybutyrate) and poly(3–hydroxybutyrate–co–3–hydroxyvalerate) in soils Applied and environmental microbiology, 59(10)*, 3233–3238.
39. Choi, G. G., Kim, H. W., & Rhee, Y. H. (2004). Enzymatic and non enzymatic degradation of poly (3–hydroxybutyrate–co–3–hydroxyvalerate) Co polyesters produced by Alcaligenes sp. MT–16 *The Journal of Microbiology, 42(4)*, 346–352.
40. Yang, X., Zhao, K., & Chen, G. Q. (2002). Effect of Surface Treatment on the Biocompatibility of Microbial Polyhydroxyalkanoates *Biomaterials 23(5)*, 1391–1397.

41. Zhao, K., Yang, X., Chen, G. Q., & Chen, J. C. (2002). Effect of Lipase Treatment on the Biocompatibility of Microbial Polyhydroxyalkanoates *J. Material Science: Materials in Medicine. 13*, 849–854.
42. Wang, H. T., Palmer, H., Linhardt, R. J., Flanagan, D. R., & Schmitt, E. (1990). Degradation of Poly(ester) microspheres. *Biomaterials 11(9)*, 679–685.
43. Doyle, C., Tanner, E. T., & Bonfield, W. (1991). In Vitro and in Vivo Evaluation of Polyhydroxybutyrate and of Polyhydroxybutyrate Reinforced with Hydroxyapatite *Biomaterials 12*, 841–847
44. Coskun, S., Korkusuz, F., & Hasirci, V. (2005). Hydroxyapatite reinforced Poly (3–hydroxybutyrate) and Poly (3–hydroxybutyrate–co–3–hydroxyvalerate) based degradable composite bone plate. *J. Biomater. Sci. Polymer Edn, 16(12),* 1485–1502.
45. Sudesh, K., Abe, H., & Doi, Y. (2000). Synthesis, Structure and Properties of Polyhydroxyalkanoates: Biological Polyesters. *Prog Polym Sci. 25*, 1503–1555
46. Lobler, M., Sass, M., Kunze, C., Schmitz, K. P., & Hopt, U. T. (2002). Biomaterial Patches Sutured onto the Rat Stomach Induce a Set of Genes Encoding Pancreatic Enzymes. *Biomaterials, 23*, 577–583.
47. Bagrov, D. V., Bonartsev, A. P., Zhuikov, V. A., Myshkina, V. L., Makhina, T. K., Zharkova, I. I., Yakovlev, S. G., Voinova, V. V., Boskhomdzhiev, A. P., Bonartseva, G. A., & Shaitan, K. V. (2012). Amorphous and Semicrystalline Phases in Ultrathin Films of Poly (3–hydroxybutirate) Tech Connect World NTSI Nanotech 2012 Proceedings, ISBN 978–1–4665–6274–5, *1,* 602–605
48. Kikkawa, Y., Suzuki, T., Kanesato, M., Doi, Y., & Abe, H. (2009). Effect of phase structure on enzymatic degradation in poly (L–lactide)/atactic Poly (3–hydroxybutyrate) blends with different miscibility *Bio macromolecules 10(4)*, 1013–1018.
49. Tokiwa, Y., Suzuki, T., & Takeda, K. (1986). Hydrolysis of Polyesters by Rhizopus Arrhizus Lipase *Agric. Biol. Chem. 50*, 1323–1325.
50. Hoshino, A. & Isono, Y. (2002). Degradation of aliphatic polyester films by commercially available lipases with special reference to rapid and complete degradation of poly (L–lactide) film by lipase PL derived from Alcaligenes sp. *Biodegradation 13* 141–147.
51. Misra, S. K., Ansari, T., Mohn, D., Valappil, S. P., Brunner, T. J., Stark, W. J., Roy, I., Knowles, J. C., Sibbons, P. D., Jones, E. V., Boccaccini, A. R., & Salih, V. (2010). Effect of Nano particulate Bioactive Glass Particles on Bioactivity and Cytocompatibility of Poly (3–Hydroxybutyrate) *Composites J. R. Soc. Interface 7(44)*, 453–465
52. Jendrossek, D., Schirmer, A., & Schlegel, H. G. (1996). Biodegradation of Polyhydroxyalkanoic Acids *Appl Microbiol Biotechnol, 46,* 451–463.
53. Winkler, F. K., D'Arcy, A., & Hunziker, W. (1990). Structure of Human Pancreatic Lipase *Nature, 343*, 771–774.
54. Jesudason, J. J., Marchessault, R. H., & Saito, T. (1993). Enzymatic degradation of poly ([R, S] β–hydroxybutyrate) *Journal of environmental polymer degradation 1(2),* 89–98
55. Mergaert, J., Anderson, C., Wouters, A., Swings, J., & Kersters, K. (1992). Biodegradation of polyhydroxyalkanoates FEMS Microbiol Rev. 9(2–4): 317–321.
56. Tokiwa, Y., & Calabia, B. P., Degradation of microbial polyesters. Biotechnol Lett. (2004) 26(15):1181–1189.
57. Mokeeva, V., Chekunova, L., Myshkina, V., Nikolaeva, D., Gerasin, V., & Bonartseva, G. (2002). Biodestruction of poly(3–hydroxybutyrate) by microscopic fungi: tests of polymer on resistance to fungi and fungicidal properties. *Mikologia and Fitopatologia 36(5),* 59–63.
58. Bonartseva, G. A., Myshkina, V. L., Nikolaeva, D. A., Kevbrina, M. V., Kallistova, A. Y., Gerasin, V. A., Iordanskii, A. L., & Nozhevnikova, A. N., Aerobic and anaerobic microbial

degradation of poly–beta–hydroxybutyrate produced by Azotobacter chroococcum. Appl Biochem Biotechnol. (2003) 109(1–3): 285–301.
59. Spyros, A., Kimmich, R., Briese, B. H., & Jendrossek, D. (1997). 1H NMR Imaging Study of Enzymatic Degradation in Poly(3–hydroxybutyrate) and Poly(3–hydroxybutyrate–co–3–hydroxyvalerate). Evidence for Preferential Degradation of the Amorphous Phase by PHB Depolymerase B from Pseudomonas lemoignei *Macromolecules, 30(26),* 8218–8225
60. Hocking, P. J., Marchessault, R. H., Timmins, M. R., Lenz, R. W., & Fuller, R. C., Enzymatic Degradation of Single Crystals of Bacterial and Synthetic Poly(–hydroxybutyrate) Macromolecules, (1996) 29(7), 2472–2478.
61. Bonartseva, G. A., Myshkina, V. L., Nikolaeva, D. A., Rebrov, A. V., Gerasin, V. A., & Makhina, T. K., [The biodegradation of poly–beta–hydroxybutyrate (PHB) by a model soil community: the effect of cultivation conditions on the degradation rate and the physicochemical characteristics of PHB] Mikrobiologiia (2002) 71(2):258–263. Russian
62. Kramp, B., Bernd, H. E., Schumacher, W. A., Blynow, M., Schmidt, W., Kunze, C., Behrend, D., & Schmitz, K. P., [Poly–beta–hydroxybutyric acid (PHB) films and plates in defect covering of the osseus skull in a rabbit model]. Laryngorhinootologie, (2002) 81(5):351–356. [Article in German]
63. Misra, S. K., Ansari, T. I., Valappil, S. P., Mohn, D., Philip, S. E., Stark, W. J., Roy, I., Knowles, J. C., Salih, V., & Boccaccini, A. R., Poly(3–hydroxybutyrate) multifunctional composite scaffolds for tissue engineering applications. Biomaterials, (2010) 31(10), 2806–2815.
64. Kuppan, P., Vasanthan, K. S., Sundaramurthi, D., Krishnan, U. M., & Sethuraman, S. (2011). Development of poly(3–hydroxybutyrate–co–3–hydroxyvalerate) fibers for skin tissue engineering: effects of topography, mechanical, and chemical stimuli. *Biomacromolecules, 12(9)*, 3156–3165
65. Malm, T., Bowald, S., Karacagil, S., Bylock, A., & Busch, C. (1992). A new biodegradable patch for closure of atrial septal defect. An experimental study. *Scand J Thorac Cardiovasc Surg.; 26(1)*: 9–14 (a).
66. Malm, T., Bowald, S., Bylock, A., Saldeen, T., & Busch C (1992). Regeneration of pericardial tissue on absorbable polymer patches implanted into the pericardial sac. An immunohistochemical, ultrastructural and biochemical study in the sheep. Scandinavian Journal of Thoracic and Cardiovascular Surgery, 26(1): 15–21 (b).
67. Malm, T., Bowald, S., Bylock A & Busch C. Prevention of postoperative pericardial adhesions by closure of the pericardium with absorbable polymer patches. An experimental study. The Journal of Thoracic and Cardiovascular Surgery, (1992) 104: 600–607. (c)
68. Malm, T., Bowald, S., Bylock, A., Busch, C., & Saldeen T (1994). Enlargement of the right ventricular outflow tract and the pulmonary artery with a new biodegradable patch in transannular position. European Surgical Research, 26: 298–308.
69. Duvernoy, O., Malm, T., Ramström, J., & Bowald S. A biodegradable patch used as a pericardial substitute after cardiac surgery: 6– and 24–month evaluation with CT. Thorac Cardiovasc Surg. 1995 Oct;43(5):271–274
70. Unverdorben, M., Spielberger, A., Schywalsky, M., Labahn, D., Hartwig, S., Schneider, M., Lootz, D., Behrend, D., Schmitz, K., Degenhardt, R., Schaldach, M., & Vallbracht C. A polyhydroxybutyrate biodegradable stent: preliminary experience in the rabbit. Cardiovasc Intervent Radiol., (2002) 25(2), 127–132.
71. Kawaguchi, T., Tsugane, A., Higashide, K., Endoh, H., Hasegawa, T., Kanno, H., Seki, T., Juni, K., Fukushima, S., & Nakano M. Control of drug release with a combination of prodrug and polymer matrix: antitumor activity and release profiles of 2',3'–Diacyl–5–

fluoro–2'–deoxyuridine from poly(3–hydroxybutyrate) microspheres. Journal of Pharmaceutical Sciences, (1992) 87(6): 508–512.
72. Baptist, J. N., (Assignor to, W. R., Grace Et Co., New York), US Patent No. 3 225 766, 1965
73. Holmes P. Biologically produced (R)–3–hydroxy–alkanoate polymers and copolymers. In: Bassett DC (Ed.) Developments in crystalline polymers. London, Elsevier, (1988) Vol. 2: 1–65.
74. Saad, B., Ciardelli, G., Matter, S., Welti, M., Uhlschmid, G. K., Neuenschwander, P., & Suterl, U. W., Characterization of the cell response of cultured macrophages and fibroblasts td particles of short–chain poly[(R)–3–hydroxybutyric acid]. Journal of Biomedical Materials Research, (1996) 30, 429–439.
75. Fedorov, M., Vikhoreva G., Kildeeva N., Maslikova A., Bonartseva G., & Galbraikh L. [Modeling of surface modification process of surgical suture]. Chimicheskie volokna (2005) (6), 22–28. [Article in Russian]
76. Rebrov, A. V., Dubinskii, V. A., Nekrasov, Y. P., Bonartseva, G. A., Shtamm, M., & Antipov, E. M., [Structure phenomena at elastic deformation of highly oriented polyhydroxybutyrate]. Vysokomol. Soedin. (Russian) (2002) 44, 347–351 [Article in Russian].
77. Kostopoulos, L., & Karring T. Augmentation of the rat mandible using guided tissue regeneration. Clin Oral Implants Res. (1994) 5(2), 75–82.
78. Zharkova, I. I., Bonartsev, A. P., Boskhomdzhiev, A. P., Efremov, Iu. M., Bagrov, D. V., Makhina, T. K., Myshkina, V. L., Voinova, V. V., Iakovlev, S. G., Filatova, E. V., Zernov, A. L., Andreeva, N. V., Ivanov, E. A., Bonartseva, G. A., & Shaĭtan K. V., The effect of poly(3–hydroxybutyrate) modification by poly(ethylene glycol) on the viability of cells grown on the polymer films]. Biomed. Khim. (2012) 58(5), 579–591.
79. Kil'deeva, N. R., Vikhoreva, G. A., Gal'braikh, L. S., Mironov, A. V., Bonartseva, G. A., Perminov, P. A., & Romashova A. N., [Preparation of biodegradable porous films for use as wound coverings] Prikl. Biokhim. Mikrobiol. (2006) 42(6): 716–720. [Article in Russian].
80. Unverdorben, M., Spielberger, A., Schywalsky, M., Labahn, D., Hartwig, S., Schneider, M., Lootz, D., Behrend, D., Schmitz, K., Degenhardt, R., Schaldach, M., & Vallbracht C. A polyhydroxybutyrate biodegradable stent: preliminary experience in the rabbit. Cardiovasc. Intervent. Radiol., (2002) 25, 127–132.
81. Solheim, E., Sudmann, B., Bang, G., & Sudmann E. Biocompatibility and effect on osteogenesis of poly(ortho ester) compared to poly(DL–lactic acid). J. Biomed. Mater. Res. (2000) 49(2), 257–263.
82. Bostman, O., & Pihlajamaki H. Clinical biocompatibility of biodegradable orthopaedic implants for internal fixation: a review. Biomaterials, (2000) 21(24), 2615–2621.
83. Lickorish, D., Chan, J., Song, J., & Davies, J. E., An in–vivo model to interrogate the transition from acute to chronic inflammation. Eur. Cell. Mater., (2004) 8, 12–19.
84. Khouw, I. M., van Wachem, P. B., de Leij, L. F., & van Luyn, M. J., Inhibition of the tissue reaction to a biodegradable biomaterial by monoclonal antibodies to IFN–gamma. J. Biomed. Mater. Res., (1998) 41, 202–210.
85. Su, S. H., Nguyen, K. T., Satasiya, P., Greilich, P. E., Tang, L., Eberhart, R. C., & Curcumin impregnation improves the mechanical properties and reduces the inflammatory response associated with poly(L–lactic acid) fiber. J. Biomater. Sci. Polym. Ed. (2005) 16(3), 353–370.
86. Lobler, M., Sass, M., Schmitz, K. P., & Hopt U. T., Biomaterial implants induce the inflammation marker CRP at the site of implantation. J. Biomed. Mater. Res., (2003) 61, 165–167.

87. Novikov, L. N., Novikova, L. N., Mosahebi, A., Wiberg, M., Terenghi, G., & Kellerth, J. O., A novel biodegradable implant for neuronal rescue and regeneration after spinal cord injury. Biomaterials, (2002) 23, 3369–3376.
88. Cao, W., Wang, A., Jing, D., Gong, Y., Zhao, N., & Zhang X. Novel biodegradable films and scaffolds of chitosan blended with poly(3–hydroxybutyrate). J. Biomater. Sci. Polymer Edn., (2005) 16(11), 1379–1394.
89. Wang Y.–W., Yang, F., Wu, Q., Cheng, Y. C., Yu, P. H., Chen, J., & Chen, G. Q. Effect of composition of poly(3–hydroxybutyrate–co–3–hydroxyhexanoate) on growth of fibroblast and osteoblast. Biomaterials. (2005) 26(7), 755–761 (b).
90. Ostwald, J., Dommerich, S., Nischan, C., & Kramp B. [In vitro culture of cells from respiratory mucosa on foils of collagen, poly–L–lactide (PLLA) and poly–3–hydroxy–butyrate (PHB)]. Laryngorhinootologie, (2003) 82(10), 693–699 [Article in Germany].
91. Bonartsev, A. P., Yakovlev, S. G., Boskhomdzhiev, A. P., Zharkova, I. I., Bagrov, D. V., Myshkina, V. L., Mahina, T. K., Charitonova, E. P., Samsonova, O. V., Zernov, A. L., Zhuikov, V. A., Efremov, Yu. M., Voinova, V. V., Bonartseva, G. A., & Shaitan, K. V., The terpolymer produced by Azotobacter chroococcum 7B: effect of surface properties on cell attachment. PLoS ONE 8(2): e57200.
92. Bonartsev, A. P., Yakovlev, S. G., Zharkova, I. I., Boskhomdzhiev, A. P., Bagrov, D. V., Myshkina, V. L., Makhina, T. K., Kharitonova, E. P., Samsonova, O. V., Voinova, V. V., & Zernov, A. L., Efremov, Yu. M., Bonartseva, G. A., & Shaitan, K. V., Cell attachment on poly(3–hydroxybutyrate)–poly(ethylene glycol) copolymer produced by Azotobacter chroococcum 7B. BMC Biochemistry, 2013 (in press)
93. Wollenweber, M., Domaschke, H., Hanke, T., Boxberger, S., Schmack, G., Gliesche, K., Scharnweber, D., & Worch H. Mimicked bioartificial matrix containing chondroitin sulphate on a textile scaffold of poly(3–hydroxybutyrate) alters the differentiation of adult human mesenchymal stem cells. Tissue Eng. 2006 Feb; 12(2):345–359.
94. Wang Y.–W, Wu, Q., & Chen, G. Q. Attachment, proliferation and differentiation of osteoblasts on random biopolyester poly(3–hydroxybutyrate–co–3–hydroxyhexanoate) scaffolds. Biomaterials, (2004) 25(4), 669–675.
95. Nebe, B., Forster, C., Pommerenke, H., Fulda, G., Behrend, D., Bernewski, U., Schmitz, K. P., & Rychly J. Structural alterations of adhesion mediating components in cells cultured on poly–beta–hydroxy butyric acid. Biomaterials (2001) 22(17): 2425–2434.
96. Qu X.–H., Wu, Q., & Chen G. Q. In vitro study on hemocompatibility and cytocompatibility of poly(3–hydroxybutyrate–co–3–hydroxyhexanoate) *J. Biomater. Sci. Polymer Edn.*, (2006) 17(10), 1107–1121 (a).
97. Pompe, T., Keller, K., Mothes, G., Nitschke, M., Teese, M., Zimmermann, R., & Werner C. Surface modification of poly(hydroxybutyrate) films to control cell–matrix adhesion. Biomaterials. (2007) 28(1), 28–37.
98. Deng, Y., Lin, X. S., Zheng, Z., Deng, J. G., Chen, J. C., Ma, H., & Chen G. Q. Poly(hydroxybutyrate–co–hydroxyhexanoate) promoted production of extracellular matrix of articular cartilage chondrocytes in vitro. Biomaterials, (2003) 24(23), 4273–4281.
99. Zheng, Z., Bei F.–F., Tian H.–L., & Chen G. Q. Effects of crystallization of polyhydroxyalkanoate blend on surface physicochemical properties and interactions with rabbit articular cartilage chondrocytes, *Biomaterials*, (2005) 26, 3537–3548.
100. Qu, X. H., Wu, Q., Liang, J., Zou, B., & Chen, G. Q. Effect of 3–hydroxyhexanoate content in poly(3–hydroxybutyrate–co–3–hydroxyhexanoate) on in vitro growth and differentiation of smooth muscle cells. Biomaterials. 2006 May;27(15):2944–2950
101. Sangsanoh, P., Waleetorncheepsawat, S., Suwantong, O., Wutticharoenmongkol, P., Weeranantanapan, O., Chuenjitbuntaworn, B., Cheepsunthorn, P., Pavasant, P., & Supaphol P.

In vitro biocompatibility of schwann cells on surfaces of biocompatible polymeric electrospun fibrous and solution cast film scaffolds. Biomacromolecules, (2007) 8(5), 1587–1594.
102. Shishatskaya, E. I., & Volova, T. G., A comparative investigation of biodegradable polyhydroxyalkanoate films as matrices for in vitro cell cultures. J. Mater. Sci–Mater. M., (2004) 15, 915–923.
103. Iordanskii, A. L., Ol'khov, A. A., Pankova, Yu. N., Bonartsev, A. P., Bonartseva, G. A., & Popov, V. O. Hydrophilicity impact upon physical properties of the environmentally friendly poly(3–hydroxybutyrate) blends: modification via blending. Fillers, Filled Polymers and Polymer Blends, Willey–VCH, 2006 г., 233, p. 108–116.
104. Suwantong, O., Waleetorncheepsawat, S., Sanchavanakit, N., Pavasant, P., Cheepsunthorn, P., Bunaprasert, T., & Supaphol P. In vitro biocompatibility of electrospun poly(3–hydroxybutyrate) and poly(3–hydroxybutyrate–co–3–hydroxyvalerate) fiber mats. Int J Biol Macromol. (2007) 40(3), 217–23.
105. Heidarkhan Tehrani, A., Zadhoush, A., Karbasi, S., & Sadeghi–Aliabadi H. Scaffold percolative efficiency: in vitro evaluation of the structural criterion for electrospun mats. J. Mater. Sci. Mater. Med., (2010) 21(11), 2989–2998.
106. Masaeli, E., Morshed, M., Nasr–Esfahani, M. H., Sadri, S., Hilderink, J., van Apeldoorn, A., van Blitterswijk, C. A., & Moroni, L. (2013). Fabrication, characterization and cellular compatibility of poly(hydroxy alkanoate) composite nanofibrous scaffolds for nerve tissue engineering. PLoS One. 8(2), e57157.
107. Zhao, K., Deng, Y., Chun Chen, J., & Chen, G. Q. (2003). Polyhydroxyalkanoate (PHA) scaffolds with good mechanical properties and biocompatibility. *Biomaterials, 24(6),* 1041–1045.
108. Cheng, S. T., Chen, Z. F., & Chen, G. Q. (2008). The expression of cross–linked elastin by rabbit blood vessel smooth muscle cells cultured in polyhydroxyalkanoate scaffolds. *Biomaterials 29(31)*, 4187–4194.
109. Francis, L., Meng, D., Knowles, J. C., Roy, I., & Boccaccini, A. R. (2010). Multi–functional P (3HB) microsphere/45S5 Bio glass–based composite scaffolds for bone tissue engineering. *Acta Biomater 6 (7),* 2773–2786.
110. Misra, S. K., Ansari, T. I., Valappil, S. P., Mohn, D., Philip, S. E., Stark, W. J., Roy, I., Knowles, J. C., Salih, V., & Boccaccini, A. R. (2010). Poly (3–hydroxybutyrate) multifunctional composite scaffolds for tissue engineering applications. *Biomaterials, 31(10),* 2806–2815.
111. Lootz, D., Behrend, D., Kramer, S., Freier, T., Haubold, A., Benkiesser, G., Schmitz, K. P., & Becher, B. Laser cutting: influence on morphological and physicochemical properties of polyhydroxybutyrate. Biomaterials. (2001) 22(18), 2447–2452.
112. Fischer, D., Li, Y., Ahlemeyer, B., Kriglstein, J., & Kissel T. In vitro cytotoxicity testing of polycations: influence of polymer structure on cell viability and hemolysis. Biomaterials (2003) 24(7), 1121–1131.
113. Nitschke, M., Schmack, G., Janke, A., Simon, F., Pleul, D., & Werner C. Low pressure plasma treatment of poly(3–hydroxybutyrate): toward tailored polymer surfaces for tissue engineering scaffolds. J. Biomed. Mater. Res., (2002) 59(4), 632–638.
114. Chanvel–Lesrat, D. J., Pellen–Mussi, P., Auroy, P., & Bonnaure–Mallet, M. Evaluation of the in vitro biocompatibility of various elastomers. Biomaterials (1999) 20, 291–299.
115. Boyan, B. D., Hummert, T. W., Dean D. D., & Schwartz Z. Role of material surfaces in regulating bone and cartilage cell response. Biomaterials (1996) 17, 137–146.

116. Bowers, K. T., Keller, J. C., Randolph, B. A., Wick, D. G., & Michaels C. M., Optimization of surface micromorphology for enhanced osteoblasts responses in vitro. Int. J. Oral. Max. Impl., (1992) 7, 302–310.
117. Cochran, D., Simpson, J., Weber, H., & Buser D. Attachment and growth of periodontal cells on smooth and rough titanium. Int. J. Oral. Max. Impl., (1994) 9, 289–297.
118. Bagrov, D. V., Bonartsev, A. P., Zhuikov V.A , Myshkina, V. L., Makhina, T. K., Zharkova, I. I., Yakovlev, S. G., Voinova, V. V., Boskhomdzhiev, A. P., Bonartseva, G. A., & Shaitan K. V. Amorphous and semicrystalline phases in ultrathin films of poly(3–hydroxybutirate) Technical Proceedings of the 2012 NSTI Nanotechnology Conference and Expo, NSTI–Nanotech 2012 , pp. 602–605.
119. Sevastianov, V. I., Perova, N. V., Shishatskaya, E. I., Kalacheva, G. S., & Volova T. G., Production of purified polyhydroxyalkanoates (PHAs) for applications in contact with blood. J. Biomater. Sci. Polym. Ed., (2003) 14, 1029–1042.
120. Seebach, D., Brunner, A., Burger, H. M., Schneider, J., & Reusch, R. N., Isolation and 1H–NMR spectroscopic identification of poly(3–hydroxybutanoate) from prokaryotic and eukaryotic organisms. Determination of the absolute configuration (R) of the monomeric unit 3–hydroxybutanoic acid from Escherichia coli and spinach. Eur. J. Biochem., (1994) 224(2), 317–328.
121. Myshkina, V. L., Nikolaeva, D. A., Makhina, T. K., Bonartsev, A. P., & Bonartseva, G. A. (2008). Effect of growth conditions on the molecular weight of poly–3–hydroxybutyrate produced by Azotobacter chroococcum 7B. *Applied biochemistry and microbiology, 44(5)*, 482–486.
122. Myshkina, V. L., Ivanov, E. A., Nikolaeva, D. A., Makhina, T. K., Bonartsev, A. P., Filatova, E. V., Ruzhitskiĭ, A. O., & Bonartseva G. A. (2010). Biosynthesis of poly–3–hydroxybutyrate–3–hydroxyvalerate copolymer by Azotobacter chroococcum strain 7B. *Applied biochemistry and microbiology, 46(3)*, 289–296.
123. Reusch, R. N. (1989). Poly–B–Hydroxybutryate Calcium Polyphosphate Complexes in Eukaryotic Membranes Proc. Soc. Exp. *Biol. Med., 191*, 377–381.
124. Reusch, R. N. (1992). Biological Complexes of Poly–B–Hydroxybutyrate *FEMS Microbiol Rev., 103*, 119–130.
125. Reusch, R. N. (1995). Low Molecular Weight Complexed Poly (3–Hydroxybutyrate) A Dynamic and Versatile Molecule in Vivo. *Can. J. Microbiol., 41(Suppl. 1)*, 50–54.
126. Müller, H. M., & Seebach, D. (1994). Polyhydroxyalkanoates: a fifth class of physiologically important organic biopolymers? *Angew Chemie, 32*, 477–502.
127. Huang, R., & Reusch, R. N. (1996). Poly (3–hydroxybutyrate) is associated with specific proteins in the Cytoplasm and Membranes of Escherichia Coli. *J. Biol. Chem. 271*, 22196–22201.
128. Reusch, R. N., Bryant, E. M., & Henry D. N. (2003). Increased Poly (R) 3–Hydroxybutyrate Concentrations in Streptozotocin (STZ) Diabetic Rats *Acta Diabetol, 40(2),* 91–94.
129. Reusch, R. N., Sparrow, A. W., & Gardiner, J. (1992). Transport of Poly–β–hydroxybutyrate in Human Plasma, *Biochim Biophys Acta*, (1123) 33–40.
130. Reusch, R. N., Huang, R., & Kosk–Kosicka, D. (1997). *Novel components and enzymatic activities of the human erythrocyte plasma membrane calcium pump*. FEBS Lett. *412(3),* 592–596.
131. Pavlov, E., Zakharian, E., Bladen, C., Diao, C. T. M., Grimbly, C., Reusch, R. N., & French, R. J. (2005). A large, voltage–dependent channel, isolated from mitochondria by water–free chloroform extraction. Biophysical Journal, 88, 2614–2625.
132. Theodorou, M. C., Panagiotidis, C. A., Panagiotidis, C. H., Pantazaki, A. A., & Kyriakidis, D. A. (2006). Involvement of the AtoS–AtoC signal transduction system in poly–

(R)–3–hydroxybutyrate biosynthesis in Escherichia coli. *Biochim Biophys. Acta. 1760(6),* 896–906.

133. Wiggam, M. I., O'Kane, M. J., Harper, R., Atkinson, A. B., Hadden, D. R., Trimble, E. R., & Bell, P. M. (1997). Treatment of Diabetic Ketoacidosis Using Normalization of Blood 3 Hydroxy–Butyrate Concentration as the Endpoint of Emergency Management *Diabetes Care, 20*, 1347–1352.
134. Larsen, T., & Nielsen, N. I. (2005). Fluorometric Determination of Beta–Hydroxybutyrate in Milk and Blood Plasma *J. Dairy Sci., 88(6)*, 2004–2009.
135. Agrawal, C. M., & Athanasiou, K. A. (1997). Technique to control pH in vicinity of biodegrading PLA–PGA implants. *J. Biomed. Mater Res., 38(2)*, 105–114.
136. Ignatius, A. A., & Claes, L. E. (1996). In vitro biocompatibility of bioresorbable polymers: poly (l, dl–lactide) and poly(l–lactide–co–glycolide), *17(8*), 831–839.
137. Rihova, B. (1996). Biocompatibility of biomaterials: hemo compatibility, immune compatibility and biocompatibility of solid polymeric materials and soluble targetable polymeric carriers. *Adv Drug. Delivery Rev., 21*, 157–176.
138. Ceonzo, K., Gaynor, A., Shaffer, L., Kojima, K., Vacanti, C. A., & Stahl, G. L., Polyglycolic acid–induced inflammation: role of hydrolysis and resulting complement activation. Tissue Eng. (2006) 12(2), 301–308.
139. Chasin, M., Langer, R. (eds) (1990), Biodegradable Polymers as Drug Delivery Systems, New York, Marcel Dekker.
140. Johnson, O. L., & Tracy, M. A., Peptide and protein drug delivery. In: Mathiowitz, E., ed. Encyclopedia of Controlled Drug Delivery. Vol 2. Hoboken, NJ: John Wiley and Sons; (1999) 816–832.
141. Jain, R. A., The manufacturing techniques of various drug loaded biodegradable poly(lactide–co–glycolide) (PLGA) devices. Biomaterials. (2000) 21, 2475–2490.
142. Gursel, I., & Hasirci, V. (1995). Properties and drug release behaviour of poly(3–hydroxybutyric acid) and various poly(3–hydroxybutyrate–hydroxyvalerate) copolymer microcapsules. *J. Micro encapsul. 12(2),*185–193.
143. Li, J., Li, X., Ni, X., Wang, X., Li, H., & Leong, K. W. (2006). Self–assembled supramolecular hydrogels formed by biodegradable PEO–PHB–PEO triblock copolymers and a–cyclodextrin for controlled drug delivery. *Biomaterials 27*, 4132–4140.
144. Akhtar, S., Pouton, C. W., & Notarianni, L. J. (1992). Crystallization behaviour and drug release from bacterial polyhydroxyalkanoates *Polymer, 33(1),* 117–126.
145. Akhtar, S., Pouton, C. W., & Notarianni, L. J. (1991). The influence of crystalline morphology and copolymer composition on drug release from solution cast and melting processed P (HB–HV) copolymer matrices. *J. Controlled Release, 17*, 225–234.
146. Korsatko, W., Wabnegg, B., Tillian, H. M., Braunegg, G., & Lafferty, R. M. (1983). Poly–D–hydroxybutyric acid–a biologically degradable vehicle to regard release of a drug. Pharm. Ind. 45:1004–1007.
147. Korsatko, W., Wabnegg, B., Tillian, H. M., Egger, G., Pfragner, R., & Walser, V. (1984). Poly D (–) 3–hydroxybutyric acid (poly–HBA)–a biodegradable former for long–term medication dosage 3 *Studies on compatibility of poly–HBA implantation tablets in tissue culture and animals*. Pharm. Ind., *46*, 952–954.
148. Kassab, A. C., Piskin, E., Bilgic, S., Denkbas, E. B., & Xu, K. (1999). Embolization with Polyhydroxybutyrate (PHB) Micromerspheres: in vivo studies, *J. Bioact. Compat. Polym. 14*, 291–303.
149. Kassab, A. C., Xu, K., Denkbas, E. B., Dou, Y., Zhao, S., & Piskin, E. (1997). Rifampicin carrying Polyhydroxybutyrate Microspheres as a Potential Chemoembolization Agent J *Biomater Sci. Polym* Ed., *8,* 947–961.

150. Sendil, D., Gursel, I., Wise, D. L., & Hasirci, V. (1999). Antibiotic release from biodegradable PHBV microparticles. *J. Control Release 59*, 207–17.
151. Gursel, I., Yagmurlu, F., Korkusuz, F., & Hasirci, V. (2002). In Vitro Antibiotic release from poly (3–hydroxybutyrate–co–3–hydroxyvalerate) rods *J. Microencapsul. 19*, 153–164.
152. Turesin, F., Gursel, I., & Hasirci, V. (2001). Biodegradable Polyhydroxyalkanoate Implants for Osteomyelitis Therapy: *In Vitro Antibiotic Release J. Biomater. Sci. Polym.* Ed., *12*, 195–207.
153. Turesin, F., Gumusyazici, Z., Kok, F. M., Gursel, I., Alaeddinoglu, N. G., & Hasirci, V. (2000). Biosynthesis of Polyhydroxybutyrate and its copolymers and their use in controlled drug release. Turk *J. Med. Sci., 30*, 535–541.
154. Gursel, I., Korkusuz, F., Turesin, F., Alaeddinoglu, N. G., & Hasirci, V. (2001). In vivo application of biodegradable controlled antibiotic release systems for the treatment of implant related osteomyelitis *Biomaterials, 22(1)*, 73–80.
155. Korkusuz, F., Korkusuz, P., Eksioglu, F., Gursel, I., & Hasirci, V. (2001). In Vivo response to biodegradable controlled antibiotic release systems. *J. Biomed. Mater Res. 55(2)*, 217–228.
156. Yagmurlu, M. F., Korkusuz, F., Gursel, I., Korkusuz, P., Ors, U., & Hasirci, V. (1999). Sulbactam–cefoperazone polyhydroxybutyrate–co–hydroxyvalerate (PHBV) local antibiotic delivery system: In vivo effectiveness and biocompatibility in the treatment of implantrelated experimental osteomyelitis. *J. Biomed. Mater Res., 46*, 494–503.
157. Yang, C., Plackett, D., Needham, D., & Burt, H. M. (2009). PLGA and PHBV microsphere formulations and solid–state characterization: possible implications for local delivery of fusidic acid for the treatment and prevention of orthopaedic infections. *Pharm. Res. 26(7)*, 1644–1656.
158. Kosenko, R. Yu., Iordanskii, A. L., Markin, V. S., Arthanarivaran, G., Bonartsev, A. P., & Bonartseva, G. A. (2007). Controlled release of antiseptic drug from poly(3–hydroxybutyrate)–based membranes. Combination of Diffusion and Kinetic Mechanisms *Pharmaceutical Chemistry Journal, 41(12)*, 652–655.
159. Khang, G., Kim, S. W., Cho, J. C., Rhee, J. M., Yoon, S. C., & Lee, H. B. (2001). Preparation and characterization of poly(3–hydroxybutyrate–co–3–hydroxyvalerate) microspheres for the sustained release of 5–fluorouracil. *Biomed Mater Eng., 11*, 89–103.
160. Bonartsev, A. P., Yakovlev, S. G., Filatova, E. V., Soboleva, G. M., Makhina, T. K., Bonartseva, G. A., Shaĭtan, K. V., Popov, V. O., & Kirpichnikov, M. P. (2012). Sustained release of the antitumor drug paclitaxel from poly (3–hydroxybutyrate) based microspheres. Biochemistry (Moscow) Supplement Series B: *Biomedical Chemistry, 6(1)*, 42–47.
161. Yakovlev, S. G., Bonartsev, A. P., Boskhomdzhiev, A. P.,, Bagrov, D. V.,, Efremov, Yu. M., Filatova, E. V., Ivanov, P. V., Mahina, T. K., & Bonartseva, G. A. (2012). In vitro cytotoxic activity of poly(3–hydroxybutyrate) nanoparticles loaded with antitumor drug paclitaxel. Technical Proceedings of the 2012 NSTI Nanotechnology Conference and Expo, NSTI–Nanotech, 190–193.
162. Shishatskaya, E. I., Goreva, A. V., Voinova, O. N., Inzhevatkin, E. V., Khlebopros, R. G., & Volova, T. G. (2008). Evaluation of antitumor activity of rubomycin deposited in absorbable polymeric microparticles *Bull. Exp. Biol. Med., 145(3)*, 358–361.
163. Filatova, E. V., Yakovlev, S. G., Bonartsev, A. P., Mahina, T. K., Myshkina, V. L., & Bonartseva, G. A. (2012). Prolonged release of chlorambucil and etoposide from poly–3–oxybutyrate–based microspheres *Applied Biochemistry and Microbiology, 48(6)*, 598–602.
164. Bonartsev, A. P., Bonartseva, G. A., Makhina, T. K., Mashkina, V. L., Luchinina, E. S., Livshits, V. A., Boskhomdzhiev, A. P., Markin, V. S., & Iordanskiĭ, A. L. (2006). New

poly–(3–hydroxybutyrate)–based systems for controlled release of dipyridamole and indomethacin. *Applied biochemistry and microbiology, 42(6)*, 625–630.
165. Coimbra, P. A., De Sousa, H. C., & Gil, M. H. (2008). Preparation and characterization of flurbiprofen–loaded poly(3–hydroxybutyrate–co–3–hydroxyvalerate) microspheres. *J. Microencapsul. 25(3)*, 170–178.
166. Wang, C., Ye, W., Zheng, Y., Liu, X., & Tong, Z. (2007). Fabrication of drug–loaded biodegradable microcapsules for controlled release by combination of solvent evaporation and layer–by–layer self–assembly. *Int. J. Pharm. 338(1–2)*, 165–173.
167. Salman, M. A., Sahin, A., Onur, M. A., Oge, K., Kassab, A., & Aypar, U. (2003). Tramadol encapsulated into polyhydroxybutyrate microspheres: in vitro release and epidural analgesic effect in rats. *Acta Anaesthesiol Scand., 47*, 1006–1012.
168. Bonartsev, A. P., Livshits, V. A., Makhina, T. A., Myshkina, V. L., Bonartseva, G. A., & Iordanskii, A. L. (2007). Controlled release profiles of dipyridamole from biodegradable microspheres on the base of poly (3–hydroxybutyrate). *Express Polymer Letters, 1(12)*, 797–803.
169. Livshits, V. A., Bonartsev, A. P., Iordanskii, A. L., Ivanov, E. A., Makhina, T. A., Myshkina, V. L., & Bonartseva, G. A. (2009). Microspheres based on poly(3–hydroxy)butyrate for prolonged drug release. *Polymer Science Series B, 51(7–8)*, 256–263.
170. Bonartsev, A. P., Postnikov, A. B., Myshkina, V. L., Artemieva, M. M., & Medvedeva. N. A. (2005). A new system of nitric oxide donor prolonged delivery on basis of controlled–release polymer, polyhydroxybutyrate. *American Journal of Hypertension, 18(5A)*, A
171. Bonartsev, A. P., Postnikov, A. B., Mahina, T. K., Myshkina, V. L., Voinova, V. V., Boskhomdzhiev, A. P., Livshits, V. A., Bonartseva, G. A., & Iorganskii, A. L. (2007). A new in vivo model of prolonged local nitric oxide action on arteries on basis of biocompatible polymer *The Journal of Clinical Hypertension, Suppl. A., 9(5)*, A152 (c).
172. Stefanescu, E. A., Stefanescu, C., & Negulescu, I. I. (2011). Biodegradable polymeric capsules obtained via room temperature spray drying: preparation and characterization. *J. Biomater. Appl., 25(8)*, 825–849.
173. Costa, M. S., Duarte, A. R., Cardoso, M. M., & Duarte, C. M. (2007). Super critical Anti solvent Precipitation of PHBV Micro particles *Int. J. Pharm 328(1)*, 72–77.
174. Riekes, M. K., Junior, L. R., Pereira, R. N., Borba, P. A., Fernandes, D., & Stulzer, H. K. (2013). Development and Evaluation of Poly (3–hydroxybutyrate co 3–hydroxyvalerate) and polycaprolactone microparticles of nimodipine *Curr Pharm Des* Mar 12. [Epub ahead of print].
175. Bazzo, G. C., Caetano, D. B., Boch, M. L., Mosca, M., Branco, L. C., Zétola, M., Pereira, E. M., & Pezzini, B. R. (2012). Enhancement of felodipine dissolution rate through its incorporation into Eudragit® E–PHB polymeric microparticles: in vitro characterization and investigation of absorption in rats. *J. Pharm. Sci., 101(4)*, 1518–1523.
176. Zhu, X. H., Wang, C. H., & Tong, Y. W. (2009). In vitro characterization of hepatocyte growth factor release from PHBV/PLGA microsphere scaffold. J. Biomed. Mater. Res, A., 89(2), 411–423.
177. Parlane, N. A., Grage, K., Mifune, J., Basaraba, R. J., Wedlock, D. N., Rehm, B. H., & Buddle, B. M. (2012). Vaccines displaying mycobacterial proteins on biopolyester beads stimulate cellular immunity and induce protection against tuberculosis. *Clin. Vaccine Immunol 19(1)*, 37–44.
178. Yilgor, P., Tuzlakoglu, K., Reis, R. L., Hasirci, N., & Hasirci, V. (2009). Incorporation of a sequential BMP–2/BMP–7 delivery system into chitosan based scaffolds for bone tissue engineering *Biomaterials 30(21)*, 3551–3559.

179. Errico, C., Bartoli, C., Chiellini, F., & Chiellini, E. (2009). Poly (hydroxyalkanoates)–based polymeric nanoparticles for drug delivery. *J. Biomed. Biotechnol 2009,* 571702.
180. Althuri, A., Mathew, J., Sindhu, R., Banerjee, R., Pandey, A., & Binod, P. (2013). Microbial synthesis of poly–3–hydroxybutyrate and its application as targeted drug delivery vehicle. *Bioresour Technol*. pii: S0960–8524(13)00129–6.
181. Pouton, C. W., & Akhtar, S. 1996. Biosynthetic polyhydroxyalkanoates and their potential in drug delivery *Adv. Drug Deliver Rev 18*, 133–162.

CHAPTER 2

THE EFFECT OF ANTIOXIDANT DRUG MEXIDOL ON BIOENERGETIC PROCESSES AND NITRIC OXIDE FORMATION IN THE ANIMAL TISSUES

Z. V. KUROPTEVA, O. L. BELAYA, L. M. BAIDER, and T. N. BOGATIRENKO

CONTENTS

ABSTRACT

The method of electron paramagnetic resonance was used to study the influence of mexidol (2-ethyl-6-methyl-3-hydropyridine succinate) on the heart and liver tissues of experimental animals. It has been shown that mexidol protects the iron–sulfur centers of the respiratory chain of the heart mitochondria in the composition of integral animal tissues from oxidizing damage and increases nitric oxide formation in cells and the degree of hemoglobin oxygenation. It has been suggested that the maintenance of the mitochondrial iron–sulfur centers in the reduced working state (under oxidation conditions) is most likely to be due to the presence of succinate, a metabolic component of the cycle of tricarboxylicacids, in the composition of mexidol.

2.1 INTRODUCTION

It is known that atherosclerosis and associated dyslipidemia and coronary heart disease (CHD) are accompanied by the intensification of the processes of lipid peroxidation (LPO) in membranes, deficiency of endogenous antioxidant system (AOS), reduced antioxidant activity of blood [1–3].

After the experimental studies have found that the decisive role in the prevention and elimination of the processes LPO play antioxidants [4–7], began an intensive search of compounds able to adjust the damaging effects of free radicals, which led to the creation of synthetic antioxidants, which inhibiting properties often superior to the famous natural analogs [8, 9]. One of these antioxidants is 2-ethyl-6-methyl-3-hydropyridine succinate, on the basis of which the created products Mexidol and Mexicor (Russia), which in clinical use have shown positive therapeutic effect in complex therapy of acute and chronic forms of ischemic heart disease, hypertension, improve the clinical course of disease in patients with myocardial infarction. Study of the mechanisms of action of Mexidol and Mexicor was devoted to a series of studies showing that drugs have a significant impact on the level of LPO products and state of the antioxidant system, support activity of FAD-dependent succinate link of Krebs cycle under hypoxia [9–11]. Nevertheless, molecular mechanisms of action of Mexidol at the organ tissues remain insufficiently studied. Moreover, in complex treatment of cardiovascular diseases antioxidant pills are usually used in the background of nitrate therapy. Therefore in the present as a direct effect of Mexidol on the tissue of the heart and liver of the

experimental animals, and joint action with nitroglycerine (NG) on the state of the mitochondrial iron–sulfur centers (ISC) and heme-containing proteins in the composition of integrated animal tissue were studied using the method of electron paramagnetic resonance (EPR).

2.2 EXPERIMENTAL PART

Drugs The experiments used mexidol (Farm soft, Russia) (the concentration of 2-ethyl-6-methyl-3-hydropyridine succinate is 2.5×10^{-3} M), succinic acid (2.5×10^{-3} M), and nitroglycerin (2.5×10^{-6} M) (Institute of New Technologies, Russian Academy of Medical Sciences).

The *study subjects* were murine heart and liver tissues. The SHK male mice weighing 20–22 g and maintained under the standard vivarium conditions were used. In the experiment, the hearts and livers of the animals were isolated immediately after decapitation. The isolated organs were washed in saline, cut into small pieces, and incubated at room temperature (18–20°C) with: (*1*) mexidol; (*2*) NG; (*3*) mexidol and NG; (*4*) succinic acid; and (5) saline (control). At different time intervals after the beginning of incubation, the tissues were sampled, prepared in the form of 30×3 mm columns, and frozen at 77K.

The *EPR spectra* of the prepared samples were recorded on an X EPR-300 spectrometer (Brucker-Analitische-Messtechnik, Germany). The spectrum recording conditions: power 20 mW; the amplitude of magnetic field modulation 5 G; the spectrum measurement temperature 77K.

2.3 RESULTS AND DISCUSSION

Earlier, we obtained the data testifying to the ability of mexidol to protect the ISC of the respiratory chain of hepatic mitochondria from oxidation, including NG-induced oxidation [12]. We showed in this work that a considerable effect of defense of the ISC against oxidation was also observed for the heart and brain tissue mitochondria. Figure 2.1 shows the EPR spectra of the animal heart tissues in health (Fig. 2.1) and those incubated for 21 h at room temperature in the presence of mexidol (Fig. 2.2) and without it (Fig. 2.3, the control samples with the addition of saline alone).

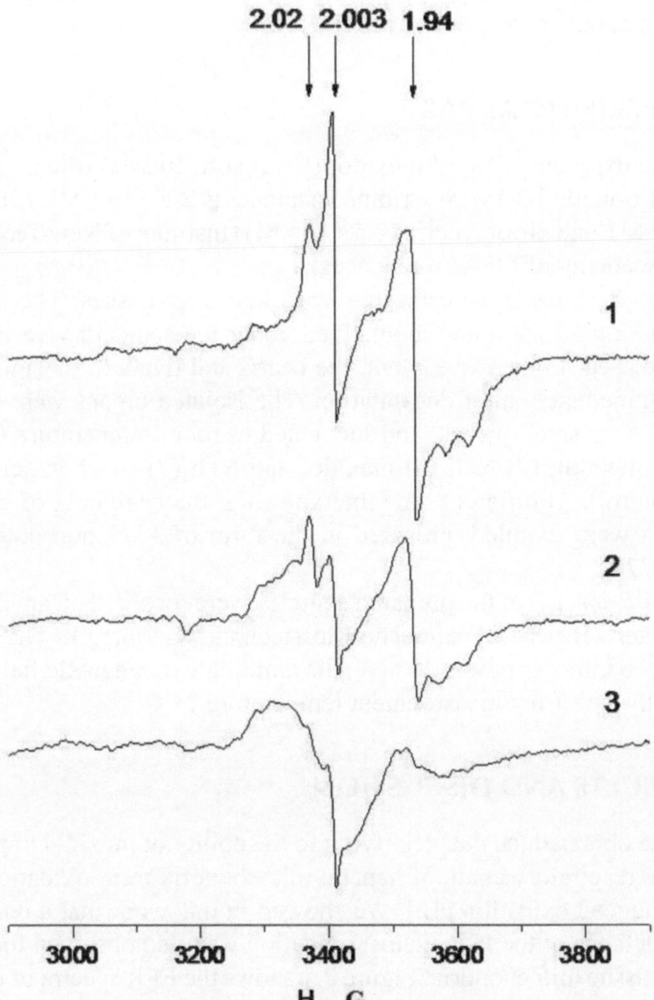

Fig. 1.

FIGURE 2.1 EPR spectra of the heart tissues in health (*1*) and after incubation for 21 h at room temperature in the presence of mexidol (*2*) and saline (the control sample) (*3*). The spectrum recording conditions: microwave power 20 mW; the amplitude of magnetic field modulation 5 G; temperature 77K.

The EPR signals determined by the mitochondrial respiratory chain components are observed in the EPR spectra of the heart tissues in health: the signal of N-1b centers of the NADH–dehydrogenase complex (iron–sulfur tetra-nuclear clusters [4Fe-4S]) with g-factors: $g_1 = 2.02$; $g_2 = 1.94$; as well as the signal in the free-radical region with g = 2.003 determined by flavo- and ubisemiquinones. The ISC N-1b are recorded with EPR in the reduced state and are the most vulnerable sites of the respiratory chain in any pathological process in a cell, including oxidative processes. Note that the intensity of the ISC signal decreases; therefore, the level of disorders in the NADH-dehydrogenase complex of the mitochondrial respiratory chain can be judged by the decrease in the intensity of the ISC signal. The ISC are destroyed on complete oxidation, and the EPR signal disappears. In freshly isolated animal tissues, the ISC are in the reduced state.

In the control heart tissue samples, up to 70–80% of the ISC were destroyed during the incubation in different experiments (the signal with g = 1.94, Fig. 2.3). At the same time, the intensity of the ISC signal in the presence of mexidol remained significant over the time of incubation (Fig. 2.2).

The active substance of mexidol (ethyl-methyl-hydropyridine succinate molecule, EMHPS) dissociates in water into two constituents – succinate and hydroxypyridine, and each of them is able to function independently in a cell. Therefore, in this work we also studied the influence of individual mexidol (EMHPS) constituents – succinate and hydroxypyridine on the ISC of the mitochondrial respiratory chain of the heart.

For this purpose, the tissues were incubated separately with succinate and hydroxypyridine under the same conditions as with mexidol. As in the experiments with mexidol, the changes in the intensity of the ISC signals in the EPR spectra of the heart tissue samples were studied under the action of the constituents after incubation at room temperature.

Figure 2.2 shows the data on changes in the intensity of the ISC signals in the heart tissues after incubation for 8 and 21 h with mexidol and succinate. Both after 8 and 21 h of incubation, the intensity of the ISC signals in the control samples was lower than in the samples with mexidol and succinate. Note that the protective effect of succinic acid was approximately 20% higher than that of mexidol. A similar protective effect of mexidol and succinic acid was also observed for the ISC of the brain mitochondria in the integral brain tissues (the authors' findings).

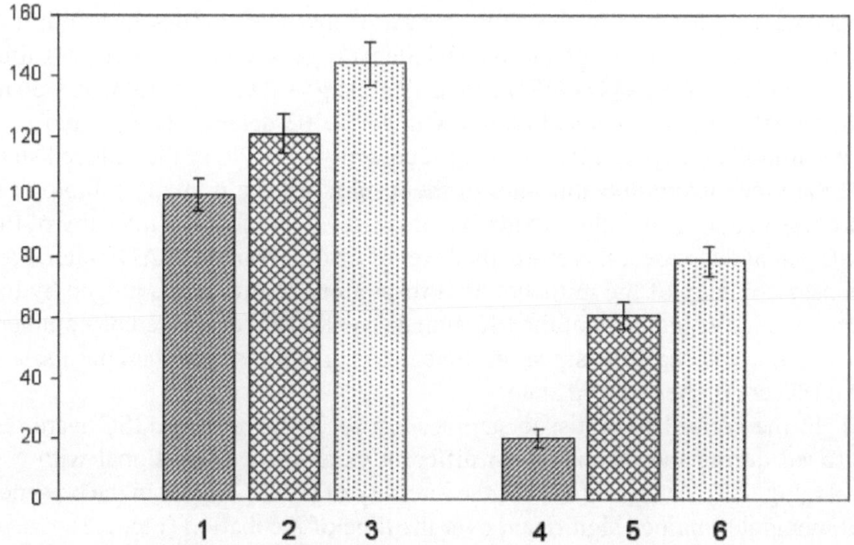

FIGURE 2.2 Change in the intensity of the EPR signal of the iron–sulfur centers of the respiratory chain of the heart mitochondria after 8 h (*1, 2, 3*) and 21 h (*4, 5, 6*) of incubation at room temperature: *1* and *4*, the control samples (incubation with saline); *2* and *5*, in the presence of mexidol; *3* and *6*, in the presence of succinic acid.

The vertical scale shows a change in the intensity of the EPR signals of the ISC in arbitrary units.

The first stage of oxidation of tetra-nuclear clusters ISC [4Fe-4S] is typically associated with the oxidation of one of the iron atoms, release it from the cluster, and loss of activity of the protein. According to Ref. [13], upon exposure to O_2, cluster $[4Fe\text{-}4S]^{2+}$ conversion occurs in two stages: 1 – with the release of Fe2+ and 2 – with further oxidation to Fe3+ and sulfide ions S^2:

$$[4Fe\text{-}4S]^{2+} + O_2 + 2H^+ \rightarrow [3Fe\text{-}4S]^{1+} + Fe^{2+} + HO_2^{\cdot} \rightarrow [2Fe\text{-}2S]^{2+} + 2Fe^{3+} + H_2O_2 + 2S^{2-}$$

Thus, the reaction of disintegration iron-sulfur tetra-nuclear cluster under the influence of O_2 occurs through the oxidation of iron in the cluster. Mexidol can, firstly, to protect IS centers, accepting oxygen at the initial stage of oxidation. Secondly, mexidol can reduce oxidized iron to the state of Fe^{2+} and restore tetra-nuclear cluster.

For hydroxypyridine, no significant changes in the intensity of the ISC were observed, compared to the control.

Antioxidant drugs are known to be used in the clinical setting against the background of the basic cardiac therapy that includes nitroglycerin-based drugs. It was shown that the physiological action of nitro compounds is determined by their ability to release nitric oxide in the process of biotransformation in the cells and tissues [14, 15]. That was the reason why we also studied the influence of the combined action of mexidol and NG on the process of nitric oxide formation, which can be controlled by the appearance of Heme–NO nitrosyl complexes. Figure 2.3 shows the EPR spectra of the liver tissue sampled 24 h after the beginning of incubation with nitroglycerin and mexidol (Fig. 2.1) and nitroglycerin alone (Fig. 2.2).

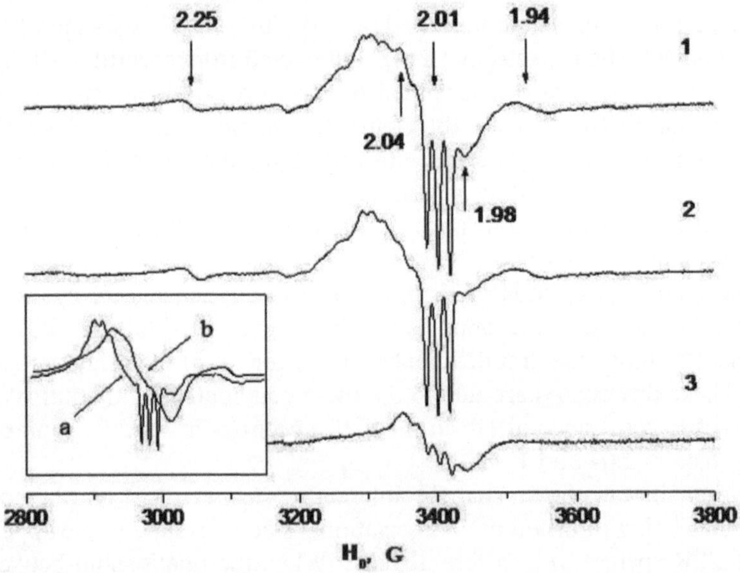

FIGURE 2.3 EPR spectra of the liver tissue sampled 24 h after the beginning of incubation with (*1*) nitroglycerin and mexidol and (*2*) nitroglycerin alone. Figure 2.3 shows the EPR spectrum obtained when spectrum 3.2 was subtracted from spectrum 3.1. Spectrum recording conditions: microwave power, 20 mW; the amplitude of magnetic field modulation, 5 G, temperature, 77 K.

Insertion below, on the left: (a) the blood EPR spectra in decreased oxygen content where Hb T-conformers predominate and (b) the heart tissue EPR

spectra under the usual oxygenation conditions where no Hb T-conformers are observed.

The EPR signal of IS centers with the g-factor 1.94 (the N-1b centers of the NADH–dehydrogenase complex of the respiratory chain of mitochondria), the free-radical signal, and the cytochrome P-450 signal with $g_1 = 2.42$ and $g_2 = 2.25$ are usually observed in the liver samples of intact animals. An intensive signal with a characteristic triplet splitting at g=2.01 was recorded in the liver tissue samples after incubation with NG. It is a well-known signal, which is determined by the nitrosyl (Heme–NO) complexes [16, 17]. In the liver EPR spectra, this signal is due to the nitrosyl complexes of two heme-containing proteins: Hb-NO and cytochrome P-450–NO. The appearance of the Heme–NO complexes gives evidence of the now well-known effect of formation of a large amount of NO during reduction of NG and other nitro compounds in the body and also recorded in tissue homogenates that was described for the first time in Refs. [14, 15]. Figure 2.3 shows the EPR spectrum obtained when spectrum 3.2 was subtracted from spectrum 3.1. The difference spectrum is also determined by the nitrosyl Heme–NO complexes. It means that the intensity of the Heme–NO signal in the samples incubated simultaneously with NG and mexidol was higher than with NG alone, that is, mexidol contributes to additional nitric oxide formation.

The differences in the form of the EPR spectra of the nitrosyl Heme–NO complexes in Figs. 2.1 and 2.2 are also worth noting. The level of Heme-NO complexes belonging to Hb R-conformers (Hb-oxygenated form) was higher in the EPR spectra of the samples incubated with mexidol and NG. Figures 2.1 and 2.2 shows spectral differences in the region of the g-factors 2.04 and 1.98. These differences are due to the increased content of the nitrosyl complexes of R-conformer of Hb in the EPR spectrum in Fig. 2.1 (components with g-factors 2.04 and 1.98).

Earlier, it was shown that the interaction between NO and hemoglobin molecules being in different conformational states resulted in the appearance of the EPR spectra of a different shape. When the relationship between the EPR spectra of the nitrosyl Hb-NO complexes and the degree of NO saturation was studied, it was established that the Hb-NO α-chains are sensitive to the R- and T-states of the quaternary hemoglobin structure [17–19]. The insertion in Fig. 2.3 (below, on the left) shows the nitrosyl Heme–NO complexes: (a) in the blood EPR spectra when the oxygen content was decreased and (b) in the heart EPR spectra under the usual oxygenation conditions (the author's data).

As shown in Ref. [19], the appearance of a well-resolved triplet structure from the nitrogen atom in the EPR spectrum is determined by the hemoglo-

bin α-chains in T-conformation (a). The α-chains in R- conformation do not have a markedly pronounced triplet structure (b). The Hb-NO EPR spectra for β-subunits do not depend on the quaternary hemoglobin structure and have a form of a wide signal with a g-factor of 2.02 and a half width of 70 Gs. Note that the affinity to NO for the α-chains is approximately four times as large as for the β-chains; therefore, the nascent NO will preferably bind to the α-chains of the hemoglobin molecule.

To put it differently, the resultant form of the EPR spectra of the nitrosyl Hb-NO complexes will be determined by the hemoglobin oxygenated R- to deoxygenated T-conformer ratio. The registration of more intense components with the g-factor 1.98 and in the region of the g-factor 2.04in the liver tissue EPR spectra in the presence of mexidol in Fig. 2.1 gives evidence of the higher content of the complexes belonging to the oxygenated form of hemoglobin. These data show that the presence of mexidol contributes to an increase in the degree of hemoglobin oxygenation.

Thus, the results of the studies have shown that mexidol protects the ISC of the respiratory chain of the heart and liver mitochondria from oxidative damage, increases nitric oxide production in the tissues and the degree of hemoglobin oxygenation.

The maintenance of the mitochondrial ISC in the reduced working state (under oxidation conditions) is most likely to be determined by the presence of succinate (succinic acid), a metabolite of the tricarboxylic acid cycle, in the composition of mexidol. An important role of succinate in the maintenance of the work of mitochondria was repeatedly indicated in the Refs. [20, 21]. Their investigations showed that in a number of experimental effects in animals and in human pathologies, a reaction is developed in response to the growing impact: "progressive activation of succinate-dependent cell energy supply goes further to a progressive inhibition of this process" [21]. The reaction is most probably due to the depletion of both the succinate reserves and the substrates for its production. The data that Krebs cycle is depressed in hypoxia and ischemia much later compared to NAD-dependent oxidases and that it is capable of sustaining energy production in a cell long enough provided the succinate is present in the mitochondria are also available [22]. That is why the presence of succinate in the composition of mexidol is probably very important for the maintenance of supply of cardiac cells with energy in hypoxia (ischemia).

As shown in [12], the protective action of mexidol on the mitochondrial ISC manifested itself most markedly on oxidation of liver tissues in the presence of NG (the introduction of NG contributed to ISC oxidation). The use of mexidol in combination with NG decreased ISC oxidation, which is indicative

of the antioxidant effect of mexidol and its positive influence on the energy status of mitochondria.

The increase in NO production influenced by mexidol may be determined by the maintenance of the arginine-dependent pathway of nitric oxide formation with the participation of inducible NO-synthase (iNOS). There are data available in the literature that the compounds capable of maintaining the iNOS cofactors tetrahydrobiopterin and NADPH in a reduced working state increase NO production with the participation of iNOS [23]. Taking into account the data we may suggest that the mexidol-mediated increase in NO production (Fig. 2.3) may be due to reducing properties of constituents of mexidol – hydropyridine and succinate.

In the liver tissues, virtually all the cells are capable of expressing iNOS: hepatocytes, Kupffer's cells, endothelial cells, Ito cells; therefore, the increase in NO production will be higher and easier to observe in the liver tissues than in the heart tissues. Mexidol is also able to influence the activity of iNOS in the cells of vascular endothelium, because iNOS also functions in these cells.

Thus, mexidol (the active principle is 2-ethyl-6-methyl-3-hydropyridine succinate) exerts a multiple positive effect.

- IT increases NO production, which allows the prescribed dose of NG to be lowered;
- It decreases the negative action of NG on the mitochondria protecting the ISC from oxidation; and
- It increases the degree of hemoglobin oxygenation. Apparently, the latter is connected with the maintenance of the work of the mitochondrial respiratory chain and, consequently, with the maintenance of energy metabolism of the heart.

KEYWORDS

- **EPR**
- **Iron–sulfur centers**
- **Mexidol (2-ethyl-6-methyl-3-hydropyridine succinate)**
- **Nitric oxide**

REFERENCES

1. Heitzer, T., Schlinzig, T., Krohn, K., Meinertz, T., & Munzel, T. (2001). Endothelial Dysfunctions Oxidative Stress and Risk of Cardiovascular Events in Patients with Coronary Artery Disease *Circulation, 104, 2673.*

2. Munzell, T., Goril, T., Bruno, R., & Taddei, S. (2010). Is Oxidative Stress A Therapeutic Target In Cardiovascular Disease? *Eur Heart J, 31 (22), 2741.*
3. Belaya, O. L., Sulimov, V. A., Fomina, I. G., Kuropteva, Z. V., & Bayder, L. M. (2006). Antioxidant Status and Lipid Peroxidation In Stable Coronary Heart Disease Patients with Dyslipidemia. *Cardiovascular Therapy and Prevention, 35(5)*, 21.
4. Burlakova, E. B. (2007). Bioantioxidants *Russ J Gen Chem, 77(11)*, 3
5. Artamoshina, N. E., Bondar, K. Yu., Belaya, O. L., Bayder, L. M., & Kuropteva, Z. V. (2012). Antioxidant Effect of Simvastatin In Patients With Coronary Heart Disease and Dyslipidemia. *Cardiovascular Therapy and Prevention, 11(6)*, 16.
6. Lankin, V. Z., Tikhaze, A. K., Kapel'ko, V. I., Shepel'kova, G. S., Shumaev, K. B., Panasenko, O. M., Konovalova, G. G., & Belenkov, Y. N. (2007). Mechanisms of Oxidative Modification of Low Density Lipoproteins under Conditions Of Oxidative And Carbonyl Stress. *Biochemistry (Rus), 72(10)*, 1081.
7. Belaia, O. L., Artamoshina, N. E., Kalmykova, V. I., Kuropteva, Z. V., & Bayder, L. M. (2009). Lipid Peroxidation and Anti oxidative Protection in Patients with Coronary Heart Disease Klin Med. (Rus), *87(5)*, 21
8. Burlakova, E. B., Molochkina, E. M., & N,ikiforov, G. A. (2008). Hybrid Antioxidants *Oxid Commun. 31(4)*, 739.
9. Balashov, V. P., Smirnov, L. D., Balykova, L. A., Gerasimova, N. G., Markelova, I. A., Kruglyakov, P. P., & Talanova, E. V. (2007). The Role of Inducible No Synthase Isoform In Stress Protective Activity of Antioxidant Ethyl Methyl Hydroxypyridine Succinate Rus. *Cardiol J (2)*, 95.
10. Gatsura, V. V., Pichugin, V. V., Sernov, L. N., &Smirnov, L. D. (1996). Antiischemic Cardio protective Effect of Mexidol. *Kardiologia (Rus), 36(11)*, 59.
11. Golikov, A. P., Davydov, B. V., Rudnev, D. V., Klychnikova, E. V., Bykova, N. S., Riabinin, V. A., V. Iu. Polumiskov, N. Iu. Nikolaeva, & Golicov, P. P. (2005). Effect of Mexicor on Oxidative Stress in Acute Myocardial Infarction. Kardiologia (Rus), *45(7)*, 21
12. Belaia, O. L., Bayder, L. M., & Kuropteva, Z. V. (2006). Effect Of Mexidol And Nitroglycerine On Iron–Sulfur Centers, Cytochrome P–450, And Nitric Oxide Formation In Liver Tissue Of Experimental Animals *Bull Exp Biol Med., 142 (4)*, 422.
13. Crack, J. C., Green, J., Le Brun, N. E., & Thomson, A. J. (2006). Detection of Sulfide Release From The Oxygen–Sensing [4fe–4s] Cluster of Fnr. *J. Biol. Chem., 281 (28)*, 18909.
14. Kuropteva, Z. V., & Pastushenko, O. N. (1985). Changes in the Paramagnetic Complexes of the Blood and Liver of Animals Exposed to Nitroglycerin. Dokl. Akad. Nauk Sssr, *281 (1)*, 189.
15. Shubin, V. E., & Kuropteva, Z. V. (1983). Esr Study of No Generation During Reduction of Nitrofuranes And Nitroimidazoles. 1. Hemoglobin–Solutions. *Studia Biophysical, 97 (2)*, 157.
16. Szabo, A., & Perutz, M. F. (1976). Equilibrium between Six and Five Coordinated Hemes In Nitrosyl Hemoglobin: Interpretation of Esr Spectra. *Biochemistry 15,* 4427.
17. Perutz, M. F., Kilmartin, J. V., & Nagai, K. (1976). A. Szabo: Influence of Globin Structures on the State of Heme. Ferrous Low Spin Derivatives *Biochemistry, 15,* 378.
18. Nagai, K., Hori, H., Yoshida, S., Sakamoto, Y., & Morimoto, H. (1978). The Effect of Quaternary Structure on the State of the Alfa and Beta Sub units within Nitrosyl Hemoglobin *Biohim Biophys. Acta*, 532, 17.
19. Louro, S. R. W., Ribeiro, P. C., & Bemski, G. (1981). Epr Spectral Changes of Nitrosyl Hemes and their Relation to the Hemoglobin T R Transition Biohim. *Biophys. Acta, 670,* 56.

20. Aevskii, E. I. M., Grishina, E. V., Rozenfeld, A. S., Ziakun, A. M., Vereshchagina, V. M., & Kondrashova, M. N. (2000). An aerobic Formation of Succinate and Facilitation of its Oxidation Possible Mechanisms of Cell Adaptation to Oxygen Deficiency. Biofizika (Rus), *45(3)*, 509.
21. Saakyan, I. R., Saakyan, S. G., Kondrashova, M. N. (2001). Activation and Inhibition of Succinate–Dependent Ca2+ Transport in Liver Mitochondria during Adaptation *Biochemistry (Rus), 66 (7)*, 795.
22. Sakamoto, M., Takeshige, K., Yasui, H., Tokunaga, K. (1998). Cardio protective Effect of Succinate against Ischemia/Reperfusion Injury *Surg. Today, 28*, 522.
23. Shi, W. J., Meininger, C. J., Haynes, T. E., Hatakeyama, K., & Wu, G. Y. (2004). Regulation of Tetrahydrobiopterin Synthesis and Bioavailability in Endothelial Cells *Cell Biochem Biophys. 41*, 415.

CHAPTER 3

CALCIUM SOAP LUBRICANTS

ALAZ IZER, TUGCE NEFISE KAHYAOGLU, and DEVRIM BALKÖSE

CONTENTS

ABSTRACT

The reparation and characterization of calcium stearate ($CaSt_2$) and a lubricant by using calcium stearate were aimed at in this study. Calcium stearate powder was prepared from sodium stearate and calcium chloride by precipitation from aqueous solutions. $CaSt_2$ and the Light Neutral Base oil were mixed together to obtain lubricating oil. It was found that $CaSt_2$ had a melting temperature of 142.8 °C and in base oil it had a lower melting point, above 128 °C. It was dispersed as lamellar micelles as the optical micrographs had shown. From rate of settling the size of dispersed particles were found to be 1.88 μm and 0.11 μm for lubricants having 1% and 2% $CaSt_2$, respectively. The friction coefficient and wear scar diameter of base oil 0.099 and 1402 nm were reduced to 0.0730 and 627.61 nm respectively for the lubricant having 1% $CaSt_2$. Lower wear scar diameter (540 nm) was obtained for lubricant with 2% $CaSt_2$. $CaSt_2$ improved the lubricating property of the base oil but did not improve its oxidative and thermal stability.

3.1 INTRODUCTION

A lubricant provides a protective film, which allows for two touching surfaces to be separated, thus lessening the friction between them. Lubricating oil is a liquid lubricant that reduces friction, protects against corrosion, reduce electric currents and cool machinery temperature. It is most often used in the automobile industry and is applied to bearings, dies, chains, cables, spindles, pumps, rails and gears to make them run smoother and more reliably. Lubricating oil is a substance introduced between two moving surfaces to reduce the friction and wear between them. Lubricating oils consist of a liquid paraffinic or vegetable oil and surface-active agents, antioxidants and anticorrosive additives. Metal soaps in pure form or dispersed in paraffinic oils are used as lubricants. Felder et al. [1] used sodium and calcium soap coatings on steel wires for drawing the wires. Calcium stearate ($CaSt_2$) had good lubricating efficiency at low wire drawing rates. The possibility for the production of a motor oil with improved operating characteristics and a higher stability by applying of composite additives has been studied by Palichev et al. [2]. For this purpose two multifunctional additives, synthesized by them have been used. They used additives containing calcium stearate and calcium salts of nitrated polypropylene and oxidized paraffin, urea, ethylene diammine, stearic acid. The additives improved the anticorrosion, viscosity-temperature, anti wear and anti sludge properties of the lubricant [2]. The optimum concentration of the additive, which enables the production of a high-quality motor lubricant,

has been found to be 5% [2]. Cutting oils were obtained by adding $CaSt_2$ to dry paraffin oil up to 5% together with other additives [3]. Thus the gelation was prevented and an easily flowing cutting oil was obtained. Savrik et al. [4] prepared lubricants using base oil, surface-active agent Span 60 and zinc borate particles. They used 1% Span 60 and 1% zinc borate. Surface-active agent Span 60 was found to be very effective in reducing the friction coefficient and wear scar diameter in four ball tests. As surface-active agents metal soaps are also used. Metal soaps are transition metal salts of the fatty acids and the alkaline earth elements. Although, the alkali salts of the fatty acids such as sodium and potassium are water soluble, metal soap is water insoluble but more soluble in nonpolar organic solvents. Calcium stearate Ca $(C_{17}H_{35}COO)_2$, in short form $CaSt_2$ is the one of the important ionic surfactants of metal soaps. Calcium Stearate is a nontoxic, white powdery substance. It is a calcium salt derived from stearic acid and is widely used in cosmetics, plastics, pharmaceuticals and lubricants [5].

Metal soaps can be obtained by neutralization of long chain organic acids with bases or by precipitation process. Moreria et al. [6] investigated formation of $CaSt_2$ from stearic acid and calcium hydroxide in different solvents and a complete conversion to $CaSt_2$ was obtained in ethanol medium [6]. The precipitation process generally produces metal soap in powder form by the reaction of aqueous solutions of a water-soluble metal salt and a fatty acid alkali metal salt at a temperature below the boiling point of water at atmospheric pressure. Filtering, washing, drying are the important steps in this method. $CaSt_2$ is produced in pure form by using this process [5].

Production of a lubricant by using a neutral base oil and calcium stearate is the aim of this study. The lubricating effects were tested by a four-ball tester for this purpose.

3.2 MATERIALS AND THE METHOD

3.2.1 MATERIALS

Calcium chloride, $CaCl_2 \cdot 2H_2O$ (98%, Aldrich), and sodium stearate, (NaSt) $C_{17}H_{35}COONa$ (commercial product, Dalan Kimya, Turkey), were used in the synthesis of $CaSt_2$. The acid value of stearic acid, used in the NaSt synthesis, was 208.2 mg of KOH/g of stearic acid and it consists of a C16–C18 alkyl chain and with 47.7% and 52.3% by weight, respectively [5].

Spindle Oil from TUPRAS Izmir was used as base oil in the preparation of the lubricants.

3.2.2 PREPARATION OF CALCIUM STEARATE POWDER

Calcium stearate powder was prepared from sodium stearate and calcium chloride by precipitation from aqueous solutions according to Eq. (1).

$$2C_{17}H_{35}COO - {}^{+}Na_{(aq)} + Ca^{2+}{}_{(aq)} \rightarrow (C_{17}H_{35}COO)_2Ca_{(s)} + 2Na^{+}{}_{(aq)} \quad (1)$$

About 5 g (0.016 mol) of sodium stearate, (NaSt) was dissolved in 200 cm^3 of deionized water in a stainless steel reactor at 75 °C. About 1.7984 g (0.012 mol) of calcium chloride (50% excess) was dissolved in 100 cm^3 of deionized water at 30 °C and added to sodium soap solution at 75 °C. The mixture was stirred at a rate of 500 rpm at 75 °C by a mechanical stirrer for 30 min. Since the by-product, NaCl, is soluble in water the reaction media was filtered by using Büchner funnel and flask under 600-mmHg vacuum level. To remove the NaCl completely, wet $CaSt_2$ was washed by 200 cm^3 deionized water once and then, wet $CaSt_2$ cake was dried in a vacuum oven under 2×10^{-4} Pa pressure. The KBr disc spectrum of the powder was taken with Shimadzu FTIR spectrophotometer. The SEM micrograph of the dried powder was taken with scanning electron microscopy (Philips XL30 SFEG).

3.2.3 LUBRICANT PREPARATION

About 1 g of $CaSt_2$ and 100 cm^3 spindle oil were mixed together at 160 °C at 880 revolution min^{-1} rate for 30 min and then cooled to 25 °C by continuously stirring. At the mixing experiments, a heater and magnetic stirrer (Ika Rh Digital KT/C) and a thermocouple (IKAWerke) were used. The experiment was repeated with 2 g of $CaSt_2$ in 100 cm^3 oil.

3.2.4 LUBRICANT CHARACTERIZATION

The dispersion of $CaSt_2$ in base oil was observed by optical microscopy. The phase change behavior of $CaSt_2$ and lubricants with increasing temperature was observed with an optical microscope equipped with a hot plate. The stabilities of the lubricants having different calcium stearate contents were determined by measuring the rate of settling of calcium stearate particles in base oils. The chemical structures of calcium stearate and the prepared lubricants were investigated by FTIR spectroscopy. The tribologic behavior of the lubricants was tested with a four ball tester. Four ball tests were done using the four ball tester from DUCOM Corporation (Fig. 3.1) to determine the friction coefficient and wear scar diameter of the lubricants. The test was performed

according to ASTM D 4172–94 at 392 N and 1200 rpm and the test duration was 1 h. The wear scar diameter was reported as the average of the wear scar diameter of the three fixed balls.

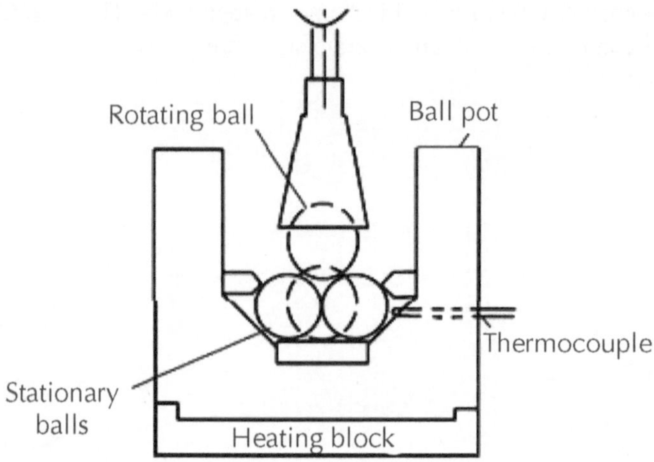

FIGURE 3.1 Four ball tester.

The visible spectrum of base oil separated by centrifugation from base oil was taken by using Perkin Elmer UV-Vis spectrophotometer by using base oil without any additive as the reference.

3.2.5 OPTICAL MICROSCOPE

Melting behavior of $CaSt_2$ in powder form and in dispersed form in the base oil on a microscope slide was observed by using the transmission optical microscope (Olympus, CH40) with a heated hot stage controlled by a temperature controller (Instec, STC 200 °C). The samples were heated at 5 °C/min rate from room temperature up to 190 °C. The photographs were taken with Camedia Master Olympus Digital camera.

3.3 RESULTS AND DISCUSSION

3.3.1 $CAST_2$ POWDERS

FTIR spectrum of calcium stearate powder obtained by precipitation process is shown in Fig. 3.2. The characteristic peaks of calcium stearate at 1542 cm^{-1}

and 1575 cm^{-1} were observed. These bands are due to anti symmetric stretching bands for unidendate and bidendate association of carboxylate groups with calcium ions [5, 7]. Anti symmetric and symmetric methylene stretching, and methylene scissoring bands ($\nu_a CH_2$, $\nu_s CH_2$, and $\delta_s CH_2$) were observed at about 2914 cm^{-1}, 2850 $cm^{-1}$and 1472 cm^{-1}, respectively. These bands are due to the alkyl chain in the calcium stearate structure [5, 7].

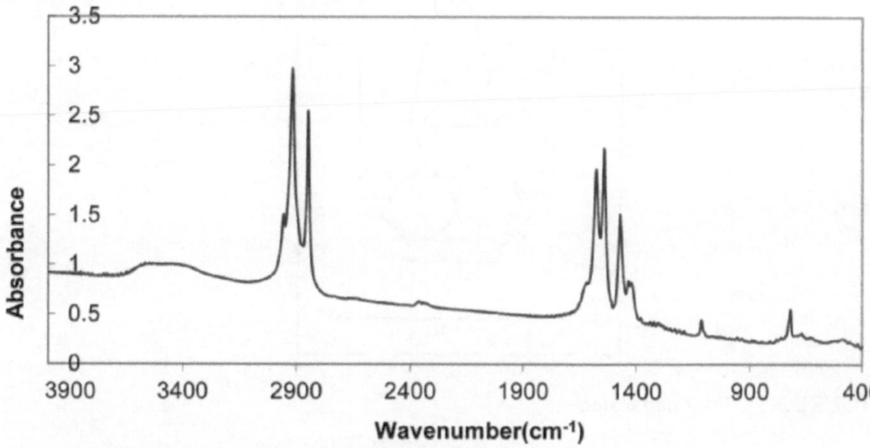

FIGURE 3.2 FTIR spectrum of bulk $CaSt_2$.

The SEM micrograph of the $CaSt_2$ powder shown in Fig. 3.3 indicated that the particles were flat in shape and had a broad size distribution ranging from 200 nm to 1 μm. The average diameter of particles was 600 nm.

FIGURE 3.3 SEM micrograph of $CaSt_2$ powder.

3.3.2 FTIR SPECTRA OF LUBRICANTS

The prepared lubricants were also examined by FTIR spectroscopy. Their FTIR spectra are shown in Fig. 3.4. The peaks at 2918 and 2848 cm^{-1}, 1454 $cm^{-1}$are due to stretching and bending vibrations of the methylene groups in base oil structure. The stretching and bending vibrations of the methyl group are observed at 2951 and 1385 cm^{-1}. At 3414 cm^{-1} a broad peak related to hydrogen bonded OH groups are present. The anti symmetric stretching bands for unidendate and bidendate association of carboxylate groups with calcium ions at 1542 cm^{-1} and 1575 cm^{-1} are observed as small peaks in the spectra.

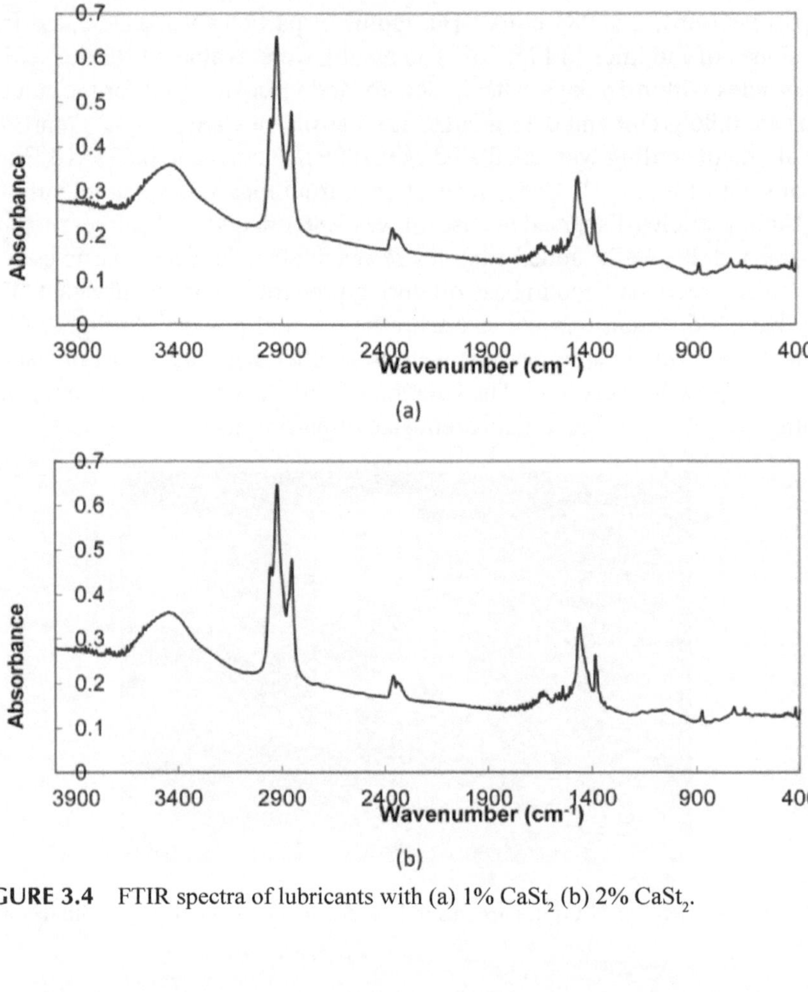

FIGURE 3.4 FTIR spectra of lubricants with (a) 1% $CaSt_2$ (b) 2% $CaSt_2$.

3.3.3 STABILITY OF LUBRICANTS AND THE PARTICLE SIZE OF THE $CAST_2$ DISPERSED IN BASE OIL

The stability of the lubricant suspensions was determined by recording the height of the line separating the oil phase and the suspension phase. Due to gravity settling of the particles the level of this line decreases continuously with time as seen in Fig. 3.5. The settling velocity is directly proportional to the radius of the particle as shown in Eq. (2) [8].

$$dx/dt = 2r^2(\rho-\rho_o)\, g/ 9\eta \qquad (2)$$

where; dx/dt is rate of settling (cm/s); ρ_o is the density of medium (g/cm^3), ρ is the density of particle (g/cm^3), η is viscosity of medium (g/(cm.s)), r is radius of particle (cm), g is 981 cm/s^2. The radius of particles was calculated from the slopes of the lines in Fig. 3.6. The results were evaluated for the settling of particles within 15 days. The oil density and viscosity used for the calculations are 0.86 g/cm^3 and 0.35 g/cm.s. The density of $CaSt_2$ is 1.12 g/cm^3. The initial rate of settling was calculated as 0.188×10^{-7} cm/s and 0.635×10^{-12} cm/s for oils with 1% and 2% $CaSt_2$, respectively, from Fig. 3.6. Apparent radius of the $CaSt_2$ particles dispersed in base oil was 1.88 μm and 0.11 μm respectively for 1% and 2% $CaSt_2$ added samples respectively. The $CaSt_2$ particles were molten and recrystallized in base oil during preparation of the lubricant. Thus they have a different particle size than the original powder. At higher $CaSt_2$ content the formed $CaSt_2$ crystals were in smaller size due to fast nucleation and slow growth of crystals. The gelation of $CaSt_2$ and base oil system is also another possibility affecting apparent size of particles.

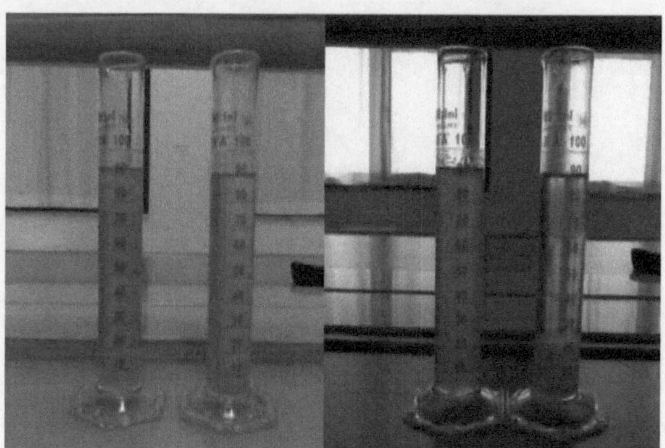

(a) (b)

FIGURE 3.5 Settling of $CaSt_2$ particles in base oil on the (a) 1st day, (b) 15th day after mixing.

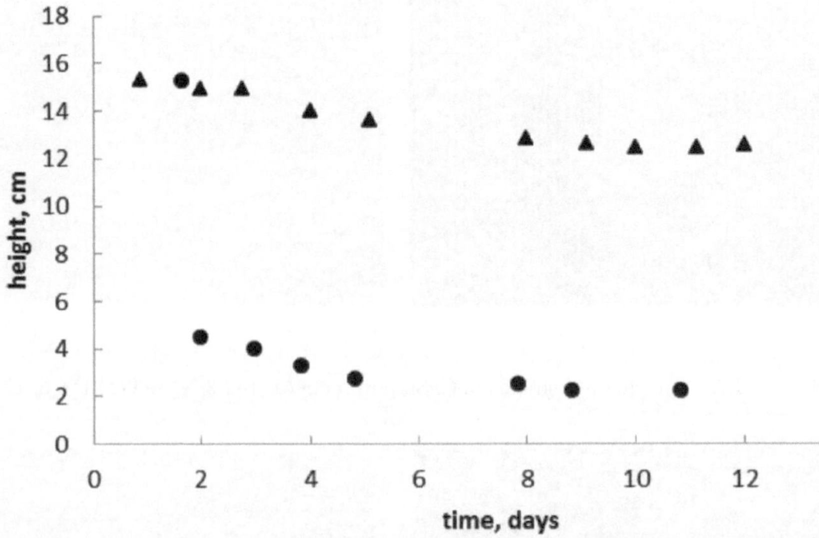

FIGURE 3.6 The height of the boundary between clear base oil and $CaSt_2$ particles settling in base oil.

3.3.4 MELTING BEHAVIOR OF PURE $CAST_2$ AND $CAST_2$ IN MINERAL OIL

$CaSt_2$ powder melts at 120 °C as determined by DSC and at 148 °C by optical microscopy in a previous study [3]. In Fig. 3.7, micrographs of the $CaSt_2$ powders before and after melting are seen. Before melting $CaSt_2$ appears as a white powder and on melting it is transformed into a transparent liquid. It was found that $CaSt_2$ had a melting temperature of 142.8 °C by optical microscopy in the present study. $CaSt_2$ particles in base oil also showed a similar phase transition behavior as bulk $CaSt_2$. They were dispersed as particles in base oil at room temperature. The particles kept their shape up to 113 °C and they melted and mixed with mineral oil homogeneously at 128 °C as seen in Fig. 3.8.

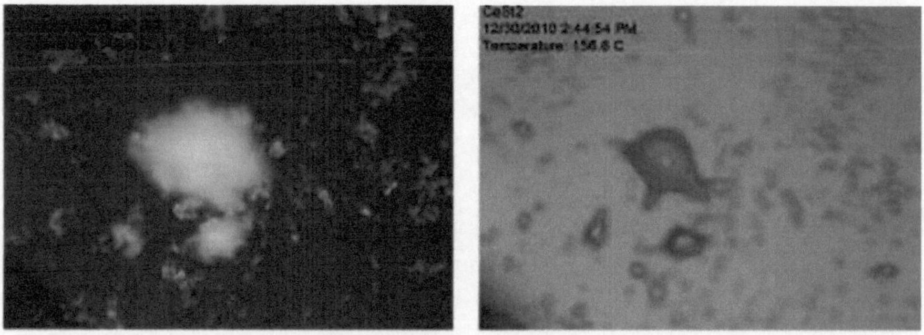

(a) (b)

FIGURE 3.7 Optical micrographs of $CaSt_2$ powder at (a) 142.8 °C and (b) 156.6 °C.

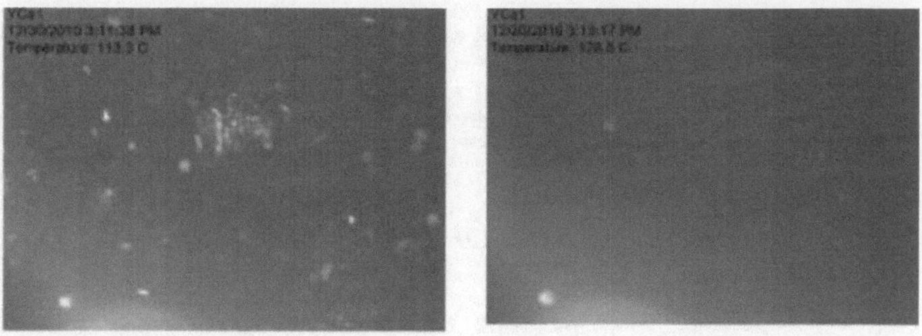

(a) (b)

FIGURE 3.8 Optical micrographs of $CaSt_2$ (1%) dispersed in mineral oil at (a) 113 °C (b) 128 °C.

3.3.5 FRICTION AND WEAR BEHAVIOR OF THE LUBRICANTS

The lubricants with $CaSt_2$ efficiently decreased the friction and wear between metal surfaces. The four ball test results are shown in Table 3.1, wear scar's optical micrographs are seen in Fig. 3.9 and the change of friction coefficient during 1 h test duration is seen in Fig. 3.10. The friction coefficient and wear scar diameter of base oil 0.099 and 1402 nm were reduced to 0.0730 and 627.61 nm, respectively for the lubricant having 1% $CaSt_2$. For 2% $CaSt_2$ containing lubricant the friction coefficient and the wear scar diameter were 0.815 and 0.540, respectively. As the $CaSt_2$ content increased better lubricating efficiency were observed. The four ball tests are done at 75 °C. At this

temperature $CaSt_2$ is in solid form in base oil. However, by the kinetic energy of the rotating ball over fixed balls the temperature of the oil should have been increased to melt the $CaSt_2$ crystals in base oil and to cover the surface of the balls by a smooth lubricating layer. The solid $CaSt_2$ particles similar to other nano particles can also fill the crevices and holes on the steel surface reducing the friction and wear.

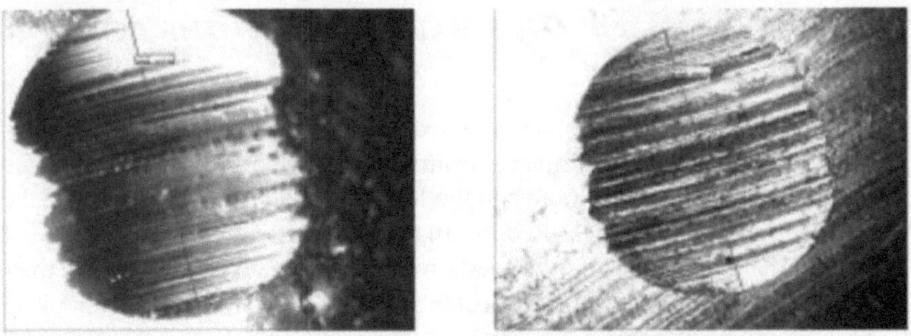

(a) (b)

FIGURE 3.9 Optical micrographs of the wear scar diameters of the one of the fixed balls of four ball tests for (a) 1% $CaSt_2$ (b) 2% $CaSt_2$ containing lubricant.

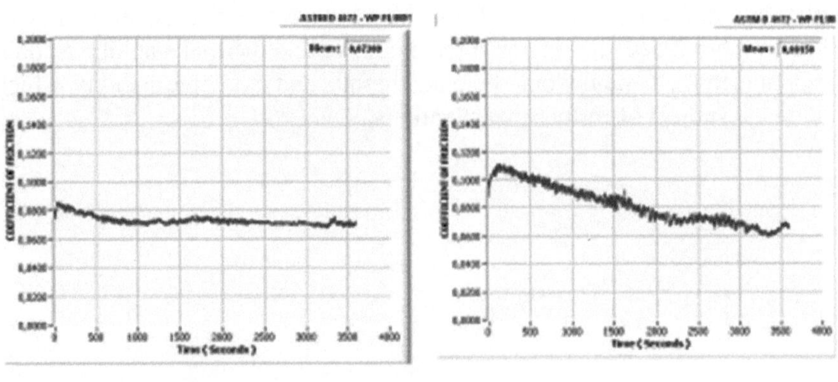

(a) (b)

FIGURE 3.10 Change of the friction coefficient with time for (a) 1% $CaSt_2$ (b) 2% $CaSt_2$ containing lubricants.

TABLE 3.1 Friction Coefficient and Wear Scar Diameter of Base Oil and Lubricants with 1% and 2% $CaSt_2$

Property	Base oil [2]	Base oil with 1% $CaSt_2$	Base oil with 2% $CaSt_2$
Friction coefficient	0.099	0.0730	0.8150
Wear Scar Diameter, nm	1402	627.61	540.88

3.3.6 THE EFFECT OF FOUR BALL TESTS ON THE COLOR OF THE BASE OIL

The lubricants change their color due to oxidation, hydrolysis and thermal degradation during its use. Contaminants from the eroding surfaces also change the color of the oil. The solid colorants in the lubricating oil can be filtered and the filter surface color can be measured [9]. In the present study the color change of the lubricating oils during four ball tests were investigated by visible spectroscopy. The visible spectra of the lubricants shown in Fig. 3.11 were taken using the base oil as the reference. The base oil with 1% $CaSt_2$ were lighter in color than the reference base oil as indicated by the negative absorbance values in Fig. 3.11. $CaSt_2$ adsorbed the coloring material initially existing in the base oil. After four ball test the base oil become dark yellow due to oxidation and cross linking reactions in base oil. The base oil having 2% $CaSt_2$ had higher absorbance values at all wavelengths and the absorbance was maximum at 420 nm. It also had a darker color after the test. Thus $CaSt_2$ improves the lubricating efficiency of the base oil, but it does not increase the oxidative and thermal stability. Adding antioxidants to the system would help the thermal and oxidative stability, which could be the subject of further investigations.

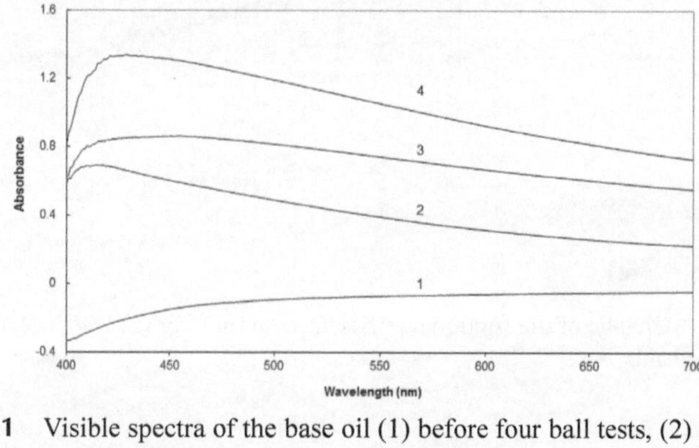

FIGURE 3.11 Visible spectra of the base oil (1) before four ball tests, (2) after four ball tests for 1% $CaSt_2$, (3) before four ball tests, and (4) after four ball tests for 2% $CaSt_2$.

3.4 CONCLUSION

Calcium stearate powder prepared from sodium stearate and calcium chloride by precipitation from aqueous solutions and Light Neutral Base oil were mixed together to obtain lubricating oils. It was found that $CaSt_2$ powder had a melting temperature of 142.8 °C and in the base oil it melted above 128 °C. From rate of settling of the particles in base oil the size of dispersed particles were found to be 1.88 μm and 0.11 μm, respectively, for lubricants having 1% and 2% $CaSt_2$. The friction coefficient (0.099) and wear scar diameter of base oil (1402 nm) were reduced to 0.0730 and 627.61 nm respectively for the lubricant having 1% $CaSt_2$. Lower wear scar diameter (540 nm) was obtained for lubricant with 2% $CaSt_2$. Calcium stearate when added to base oil reduces the friction and wear of metal surfaces sliding on each other. It covers the cracks and groves of the metal surface with a smooth film. Thus it is an efficient lubricant additive. However, $CaSt_2$ did not increased oxidative and thermal stability of the base oil. Thus further studies for the antioxidant selection should be made.

ACKNOWLEDGEMENTS

The authors thank to Opet Fuchs Turkey for the four ball tests.

KEYWORDS

- **Aqueous solutions**
- **Characterization of calcium stearate**
- **Friction coefficient**
- **Lubricating oil**
- **Oxidative**
- **Thermal stability**

REFERENCES

1. Felder, E., Levrau, C., Mantel, M., & Dinh, N. G. T. (2011). "Experimental study of the lubrication by soaps in stainless steel wire drawing" Proceedings of the Institution of Mechanical Engineers. *Part J–Journal of Engineering Tribology, 225(19, J9)*, 915–923.
2. Palichev, T., Kramis, F. S., & Petkov, P. (2008). "Operating–Conservation Motor Oil with Composite Additives" *Oxidation Communications 31(1)* 223–230.
3. Hughes, E. C., & Harrison, M. S. (1956). US 2995516 A, 1956.

4. Savrık, A. S., Balköse, D., & Ülkü, S. (2011). "Synthesis of zinc borate by inverse emulsion technique for lubrication" *J. Therm. Anal. Calorim. 104*, 605–612.
5. Gönen, M., **Öztürk**, S., Balköse, D., & **Ülkü**, S. (2010). "Preparation and Characterization of Calcium Stearate Powders and Films Prepared by Precipitation and Langmuir–Blodgett Techniques" *Ind. and Eng. Chem. Res. 49,* **1732–1736.**
6. Moreira, A. P. D., Souza, B. S., & TeixeiraA. M. R. F. (2009). "Monitoring of calcium stearate formation by thermogravimetry" *J Therm Anal Calorim. 97,* 647–652.
7. Lu, Y., & Miller, J. D. (2002). Carboxyl Stretching Vibrations of Spontaneously Adsorbed and LB Transferred Calcium Carboxylates as Determined by FTIR Internal Reflection Spectroscopy" *J. Colloid and Interface Sci. 256,* 41–52.
8. Alberty, R. (1997). Physical Chemistry McGraw Hill New York.
9. Akira Sasaki, Hideo Aoyama, Tomomi Honda, Yoshiro Iwai & Yong, C. K. (2014). A Study of the Colors of Contamination in Used Oils, *Tribology Transactions, 57(1)*, 1–10.

CHAPTER 4

RADICAL SCAVENGING CAPACITY OF N-(2-MERCAPTO-2-METHYLPROPIONYL)-L-CYSTEINE: DESIGN AND SYNTHESIS OF ITS DERIVATIVE WITH ENHANCED POTENTIAL TO SCAVENGE HYPOCHLORITE

MÁRIA BAŇASOVÁ, LUKÁŠ KERNER, IVO JURÁNEK, MARTIN PUTALA, KATARÍNA VALACHOVÁ, and LADISLAV ŠOLTÉS

CONTENTS

ABSTRACT

High-molar-mass hyaluronan (HA) – glycosaminoglycan composed of repeating disaccharide units of *N*-acetyl-D-glucosamine and D-glucuronic acid linked by β-(1→4) and β-(1→3) glycoside bonds – is an essential component of the extracellular matrix in many tissues. It is rather susceptible to free-radical mediated degradation resulting in various physiological and pathological consequences. In the present study, we investigated the effect of *N*-(2-mercapto-2-methylpropionyl)-L-cysteine against free-radical degradation of high-molar-mass HA. We applied rotational viscometry to record time-dependent changes of the dynamic viscosity of HA solutions, which reflected a decrease of HA molar-mass and thereby its fragmentation. The oxidative degradation of high-molar-mass hyaluronan was evoked by the Weissberger biogenic oxidative system comprising Cu(II) and ascorbate. Previously, we proved that stage I of oxidative HA degradation was mediated predominantly by hydroxyl radicals while stage II by peroxyl-type radicals. To determine the scavenging activity of the compound tested, ABTS and DPPH assays were used. Oximetry was used to monitor the consumption of oxygen by the treated HA solutions.

In this study, rotational viscometry demonstrated a significant ability of *N*-(2-mercapto-2-methylpropionyl)-L-cysteine to protect high-molar-mass HA from oxidative degradation in both stage I and II. This implies that this substance may possess a certain scavenging activity towards hydroxyl- and peroxyl-type radicals. Furthermore, the IC_{50} values obtained by the ABTS and DPPH assays revealed a high free-radical scavenging activity of this substance. Oxygen consumption in the reaction mixture containing the Weissberger biogenic oxidative system was found to occur along the very similar trajectory as the curve recorded by rotational viscometry. After addition of *N*-(2-mercapto-2-methylpropionyl)-L-cysteine into the reaction mixture, consumption of oxygen was significantly inhibited. These results may indicate that the possible mechanism of action of this compound, revealed as an effective remedy in treating arthritic joints, could be its free-radical scavenging activity and hence the ability to protect high-molar-mass HA of synovial fluid from its oxidative degradation. In the last part, based on existing literature, we discuss the design and synthetic routes towards the preparation of a new derivative of *N*-(2-mercapto-2-methylpropionyl)-L-cysteine. This novel compound should react with another reactive species formed during oxidative stress, the hypochlorite, plausibly damaging HA, while the beneficial activity of the parent molecule, i.e. the ability to scavenge hydroxyl- and peroxyl-type radicals, would be maintained. Preliminary synthetic results are also described.

4.1 INTRODUCTION

Many human diseases are associated with harmful actions of reactive oxygen species [11]. These species are involved in oxidative modification and damage of essential biomacromolecules in-vivo. Of them, high-molar-mass hyaluronan (HA) is of particular interest [17]. The reported reduction of HA molar-mass in the synovial fluid of patients suffering from arthritic diseases led to in-vitro studies on HA degradation by reactive oxygen species [26]. One of the earliest investigations was carried out by Pigman et al. [27] and since then numerous studies have investigated the action of various reactive oxygen species (ROS) on HA.

Hyaluronan (Fig. 4.1) is a glycosaminoglycan composed of repeating disaccharide units formed by *N*-acetyl-D-glucosamine and D-glucuronic acid. These disaccharide units are linked together by β-(1–3) and β-(1–4) glycosidic bonds [8, 17, 19, 23, 25]. HA is a main component of extracellular matrix in many tissues and organs, for example, the skin, vitreous humor, trachea, heart valves, and synovial joints. The extracellular matrix represents a foundation of tissue microenvironments, which provide biophysical and biochemical cues that maintain tissue architecture integrity [39].

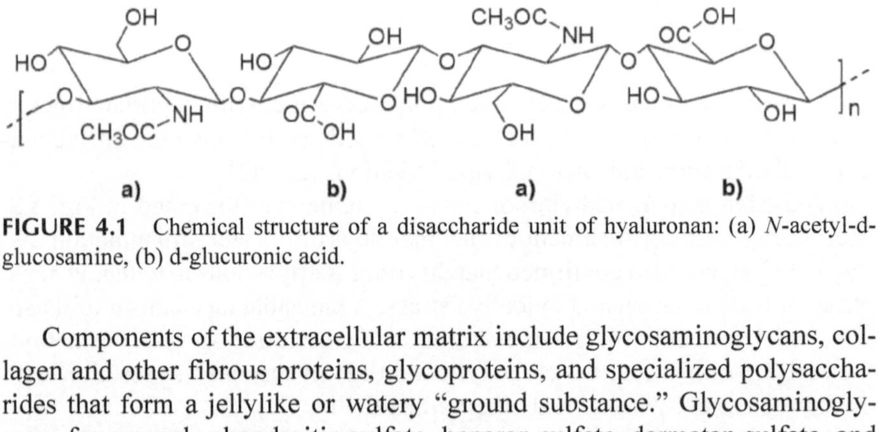

FIGURE 4.1 Chemical structure of a disaccharide unit of hyaluronan: (a) *N*-acetyl-d-glucosamine, (b) d-glucuronic acid.

Components of the extracellular matrix include glycosaminoglycans, collagen and other fibrous proteins, glycoproteins, and specialized polysaccharides that form a jellylike or watery "ground substance." Glycosaminoglycans, for example chondroitin sulfate, heparan sulfate, dermatan sulfate, and also hyaluronan, are major components of the extracellular matrix molecules (Table 4.1) [5, 9, 16].

TABLE 4.1 Types of Glycosaminoglycans

NAME	Approx. M_w (Da)	COMPONENTS		
		SACCHARIDE	SULFATE	PROTEOGLY-CANS
Chondroitin sulfate	$(10–50) \times 10^3$	Glucuronic acid Galactosamine	✓	✓
Heparan sulfate	$(10–50) \times 10^3$	Glucuronic acid Iduronic acid Glucosamine	✓	✓
Dermatan sulfate	$(10–50) \times 10^3$	Iduronic acid Galactosamine	✓	✓
Keratan sulfate	$(5–15) \times 10^3$	Galactose Glucosamine	✓	✓
Hyaluronan	$10^5–10^7$	Glucuronic acid Glucosamine	-	-

✓ — present
- — absent

Glycosaminoglycans are complex polysaccharides with important roles in various physiological events, such as cell growth, morphogenesis, differentiation, cell migration and bacterial/viral infections [20, 43].

N-(2-Mercapto-2-methylpropionyl)-L-cysteine (**1**), illustrated in Fig. 4.2, is clinically used for treatment of rheumatoid arthritis and inflammation diseases [13]. It has been confirmed that this drug is efficacious in reducing acute phase of inflammation and oxidative stress. A plausible mechanism of its action is the formation of a 7-membered cyclic disulfide - (*R*)-7,7-dimethyl-6-oxo-1,2,5-dithiazepane-4-carboxylic acid (**2**) in response to reaction with oxidants such as hydroxyl and other radicals (Fig. 4.2).

N-(2-mercapto-2-methylpropionyl)-L-cysteine is metabolized in-vivo to cyclic disulfide 2 (Fig. 4.2) and *S*-methyl- and *S*, *S'*-dimethyl derivatives 3 and 4, respectively (Fig. 4.3).

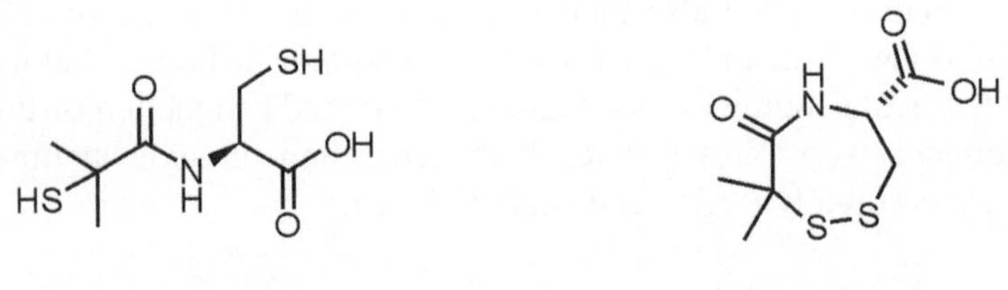

FIGURE 4.2 Chemical structure of *N*-(2-mercapto-2-methylpropionyl)-l-cysteine (1) and corresponding 7-membered cyclic disulfide (2).

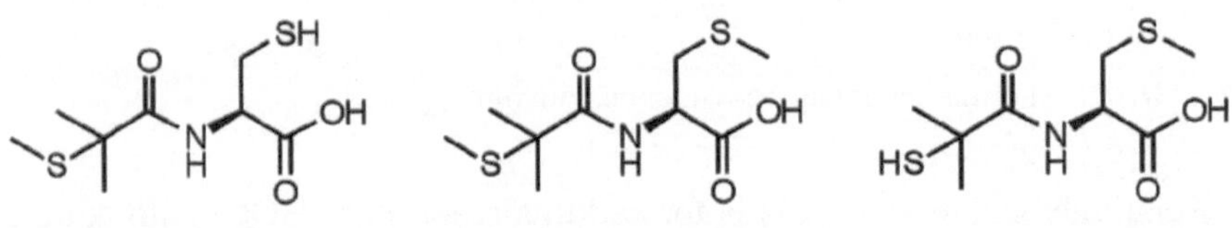

FIGURE 4.3 Chemical structures of *S*-methylated *N*-(2-mercapto-2-methylpropionyl)-l-cysteine derivatives.

Together with the parent structure, these compounds have been investigated in connection with their immunomodulating effects [22] as well as in respect to their potential to improve impaired angiogenesis *via* inducing synthesis of vascular endothelial growth factor (VEGF) [6]. Interaction of *N*-(2-mercapto-2-methylpropionyl)-L-cysteine (1) and its derivatives 3, 4, 5 (Fig. 4.3) with human serum albumin (HSA) has been examined in a study by Narazaki et al. [24], aimed to elucidate the reaction mechanism of covalent binding between HSA and thiol containing compounds. The ability of *N*-(2-mercapto-2-methylpropionyl)-L-cysteine and its oxidized derivative 2 to induce glutathione biosynthesis *via* activation of the antioxidant-response element (ARE) has also been studied and possible chemopreventive properties of these compounds against cancer have been suggested [41].

Together with reactive oxygen species produced by neutrophilic polymorphonuclear leukocytes, HOCl and taurine-*N*-monochloramine (TauCl, 7, Scheme 4.1) are also end-products of the neutrophilic polymorphonuclear leukocyte respiratory burst [21].

In human neutrophils, HOCl is generated from hydrogen peroxide by the action of myeloperoxidase in the presence of chloride anion [40]. Hypochlorous acid activates tyrosine kinase signaling cascades, leading to increased production of extracellular matrix components, growth factors, and inflam-

matory mediators. HOCl also increases the capacity of α_2-macroglobulin to bind TNF-alpha, IL-2, and IL-6. As the oxygen-chlorine bond is highly polarized, HOCl is the source of electrophilic Cl atom. The reaction of this bond with endogenic substrates can result in generation of toxic intermediates, which may eventually lead to cell death [10, 33].

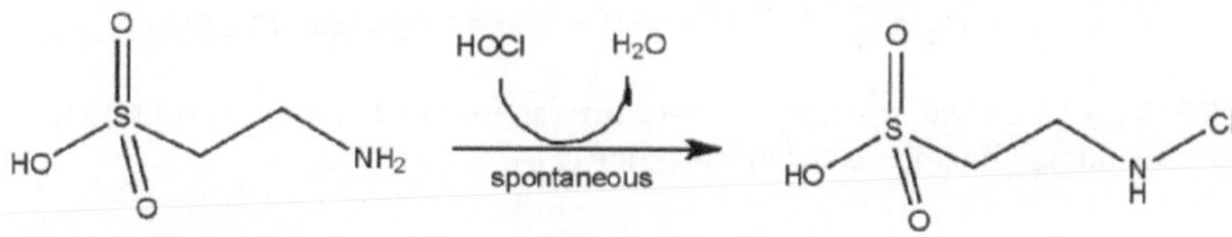

6 (M = 125.15)

7 (M = 159.6)

SCHEME 4.1 Formation of taurine-*N*-monochloramine.

Relatively stable TauCl (7) is formed in a reaction of HOCl with taurine (6), which is also generated by neutrophils. Compared to HOCl, TauCl is less toxic and inhibits the production of inflammatory mediators, prostaglandins, and nitric oxide [21].

For targeting oxygen radicals and hypochlorous acid at the same time, the two compounds – *N*-(2-mercapto-2-methylpropionyl)-L-cysteine and taurine might be administered simultaneously. However, such application has certain limitations, mainly due to different distribution and hence bioavailability of these distinct compounds within the body compartments. It is desirable to have both functions in one single molecule, which would be then able to deliver both effects to the target site at once. Therefore, we turned our focus to extend the activity of *N*-(2-mercapto-2-methylpropionyl)-L-cysteine by the ability of taurine to scavenge HOCl.

Accordingly, the aims of the present study were (i) to establish the scavenging activity of *N*-(2-mercapto-2-methylpropionyl)-L-cysteine by the ABTS and DPPH decolorization assays; (ii) to study the mode of action of *N*-(2-mercapto-2-methylpropionyl)-L-cysteine in-vitro by rotational viscometry; (iii) to assess the effect of *N*-(2-mercapto-2-methylpropionyl)-L-cysteine on the degradation of HA by means of oximetry, i.e. to compare the consumption of oxygen in the HA solutions containing the Weissberger's biogenic oxidative system in the presence and absence of *N*-(2-mercapto-2-methylpropionyl)-L-cysteine; and (iv) to design and synthesize a *N*-(2-mercapto-2-methylpropionyl)-L-cysteine derivative capable of scavenging HOCl by combining the key structural features of the parent molecule and taurine. The original activity of the parent drug has to be retained and molar-mass of the resulting derivative should not exceed 500 Da.

4.2 RESULTS AND DISCUSSION

Oxygen is an indispensable element for aerobic organisms. It is required especially for energy metabolism. Sufficient tissue oxygenation is therefore a prerequisite for adequate energy production that is essential for the maintenance of key cell functions [30]. On the other hand, utilization of the electronegative potential of molecular oxygen results also in a certain production of reactive oxygen species and nitrogen species (RNS). Overproduction of ROS, RNS and reactive chlorous species may lead to oxidative stress. In the inflammatory reaction, ROS are among the products released by activated neutrophils, monocytes, macrophages, endothelial cells and fibroblasts into the affected tissue [30]. In the acute phase of inflammation, the targeted treatment with antioxidants may be expected to modulate the increase of ROS and/or RNS towards their baseline levels. Our present study suggests that compound **1** may act in this manner.

4.2.1 *WEISSBERGER BIOGENIC OXIDATIVE SYSTEM*

As an established method for the initiation of radical degradation, we decided to employ the Weissberger oxidative system under biogenic conditions (Scheme 4.2). Reaction between $AscH^-$ and Cu(II) in the presence of oxygen gives hydrogen peroxide and since ascorbate reduces Cu(II) to cuprous ions it is feasible to assume that Cu(I) and hydrogen peroxide generates hydroxyl radicals [37].

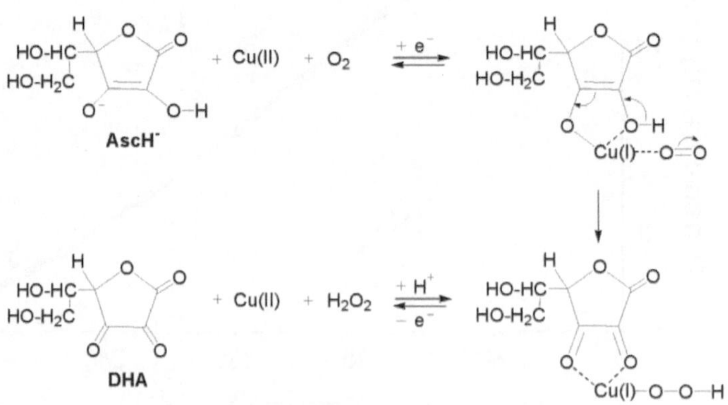

SCHEME 4.2 The Weissberger biogenic oxidative system. $AscH^-$ and DHA denote ascorbate anion and dehydroascorbate, respectively.

$$2\ Cu(I) + H_2O_2 \rightarrow 2\ Cu(II) + {}^{\bullet}OH + OH^-\ \text{(Fenton-like reaction)}$$

The OH radicals abstract proton from the HA macromolecule. The resulting HA (macro)radical (A·) reacts under aerobic conditions with molecular oxygen yielding a secondary peroxyl (macro)radical (A-O-O·). The latter participates in the phase of propagation of free-radical-mediated oxidative degradation of HA [9, 12, 18, 29, 42].

4.2.2 ROTATIONAL VISCOMETRY

Figure 4.4 illustrates changes of the dynamic viscosity of the HA solution due to a prooxidative action of the two reactants, namely Cu(II) and ascorbic acid. After a few-minute induction period the initial value of the HA solution dynamic viscosity equaling 11.48 mPa·s started to gradually decrease. After a 5-h measurement, dynamic viscosity of the HA solution reached the value 7.35 mPa·s. As already proved, the decrease of the HA solution viscosity is a result of the degradation of high-molar-mass HA, whose initial M_w value equals 970 kDa. The rate of decrease of the dynamic viscosity value in the basic experimental set is sufficiently high to be used as the probe reference. This reference curve (Fig. 4.4, curve 0) is depicted also in the next (Figs. 4.5 and 4.6, curve 0).

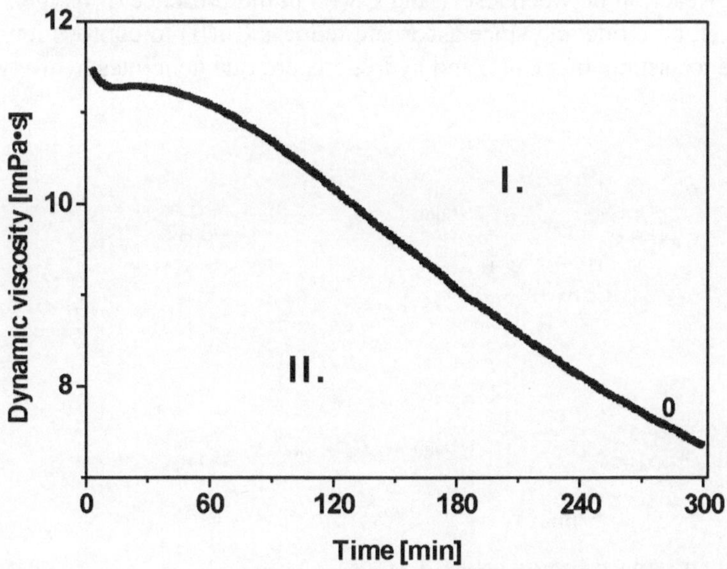

FIGURE 4.4 Free-radical degradation of high-molar-mass HA by the action of the prooxidative system containing 1 μM Cu(II) *plus* 100 μM ascorbate (curve 0).

It can be expected that lowering of the concentration ratio of cupric chloride/ascorbate, e.g. by applying higher concentrations of Cu(II) solutions, would lead to increasing the rate of HA degradation, reflected by the probe reference. Indeed, any prooxidative action of a substance added into the reference oxidative system would results in a greater decline of the curves situated within area II (Fig. 4.4). On the contrary, any antioxidative effect of a compound added into the same system would cause a retardation of the dynamic viscosity drop. In a limiting situation providing that the high-molar-mass HA is completely protected against any degradation, we can speak about a total inhibitory action of the admixed antioxidative compound with resulting curves situated in area I (Fig. 4.4). Hence, area II represents the action of the compounds functioning in a prooxidative mode, while area I. is represented by the action of the compounds acting in an antioxidative mode.

Our results suggest that addition of *N*-(2-mercapto-2-methylpropionyl)-L-cysteine (1, 10 and 100 μM) at the beginning of free-radical degradation of high-molar-mass HA led to a complete inhibition of HA degradation (Fig. 4.5).

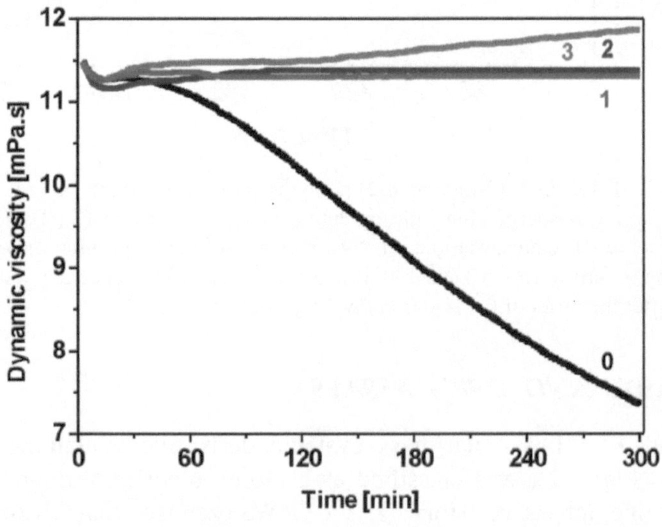

FIGURE 4.5 Effect of *N*-(2-mercapto-2-methylpropionyl)-L-cysteine on HA degradation induced by the Weissberger biogenic oxidative system 1.0 μM Cu(II) and 100 μM ascorbate (curve 0). Concentrations of *N*-(2-mercapto-2-methylpropionyl)-L-cysteine in the reaction mixture were 1, 10 and 100 μM, curves 1, 2 and 3, respectively. The drug was added before the onset of the HA oxidative degradation.

Similar effects of this substance were observed when it was added to the reaction mixture 1 h after the onset of HA oxidative degradation (Fig. 4.6). The compound tested acted in an antioxidative mode. These results demonstrated the ability of *N*-(2-mercapto-2-methylpropionyl)-L-cysteine to act as an effective scavenger of reactive oxygen species, such as $^{\bullet}OH$ and peroxyl-type radicals (Figs. 4.5 and 4.6, respectively).

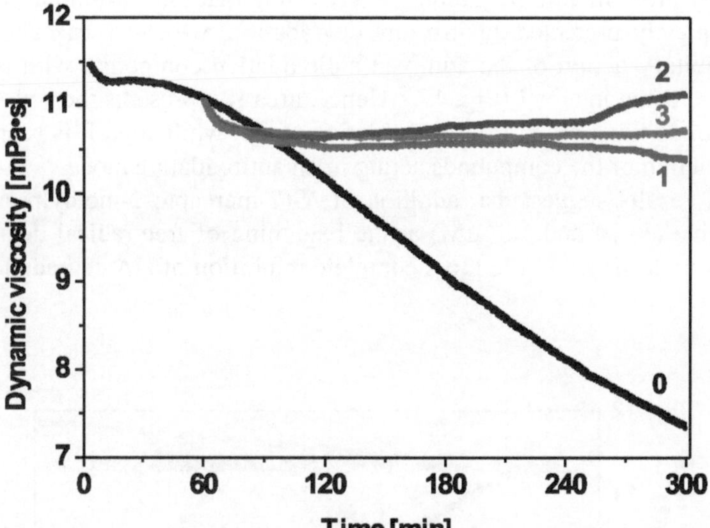

FIGURE 4.6 Effect of *N*-(2-mercapto-2-methylpropionyl)-L-cysteine on HA degradation induced by the Weissberger biogenic oxidative system – 1.0 μM Cu(II) and 100 μM ascorbate (curve 0). Concentrations of *N*-(2-mercapto-2-methylpropionyl)-L-cysteine in the reaction mixture were 1, 10 and 100 μM, curves 1, 2 and 3, respectively. The drug was added 1 h after the onset of the HA oxidative degradation.

4.2.3 ABTS AND DPPH ASSAYS

Compound **1**, a low-molar-mass cysteine derivative containing two thiol groups (Fig. 4.2), was classified as a more potent thiol donor than any other cysteine derivative (Horwitz, 2003). We proposed that *N*-(2-mercapto-2-methylpropionyl)-L-cysteine potentially acted as an H donor. The chemical structure of *N*-(2-mercapto-2-methylpropionyl)-L-cysteine predetermines this drug to be an effective reducing agent as it contains two thiol groups, potential donors of two electrons. Indeed, under our experimental conditions, it was found to be a very potent antioxidant (see Table 4.2).

Table 4.2 depicts the IC_{50} values of the ABTS and DPPH assays for *N*-(2-mercapto-2-methylpropionyl)-L-cysteine and quercetin. The IC_{50} values obtained for quercetin by these assays were 2.9 and 4.4 μM, respectively [35]. Based on its IC_{50} values (4.0 and 9.0 μM, respectively), *N*-(2-mercapto-2-methylpropionyl)-L-cysteine may be classified as an effective scavenger of $ABTS^{\cdot+}$ and $DPPH^{\cdot}$. Quercetin is a substance classified as the standard natural antioxidant. The IC_{50} values were calculated from the respective dose-response inhibition curves.

TABLE 4.2 IC_{50} Values of the Compounds Tested Determined by ABTS and DPPH Decolorization Assays (n=4; data expressed as mean ± s.e.m.; significance of differences was determined by the One-Way ANOVA test: * $p<0.05$, ** $p<0.01$, *** $p<0.001$ compared to the quercetin value)

Substance	**ABTS IC_{50}[μM]**	**DPPH IC_{50}[μM]**
N-(2-mercapto-2-methylpropionyl)-L-cysteine	4.0 ± 0.4 n.s.	9.0 ± 0.4 ***
Quercetin	2.9 ± 0.2	4.4 ± 0.2

n.s. – not significant.

The decolorization reaction of the $ABTS^{\cdot+}$ radical cation solution or that of the $DPPH^{\cdot}$ radical can be simply described by the chemical reaction during which a proper reductant, in our case *N*-(2-mercapto-2-methylpropionyl)-L-cysteine, provides an electron to the acceptor – the $ABTS^{\cdot+}$ radical cation or $DPPH^{\cdot}$ radical. The reaction of a thiol-derived compound (R-SH), which itself undergoes oxidation, can be described as $R\text{–}SH - e^- \rightarrow R\text{–}S^{\cdot} + H^+$.

4.2.4 OXIMETRY

Oxygen consumption in the reaction mixture containing the Weissberger biogenic oxidative system, that is, 1.0 μM Cu(II) and 100 μM ascorbate (curve 0, Fig. 4.7), was found to occur along the very similar trajectory as the curve recorded by rotational viscometry (compare with the curve 0 in Fig. 4.5). After addition of *N*-(2-mercapto-2-methylpropionyl)-L-cysteine in 100 μM concentration into the reaction mixture (curve 1, Fig. 4.7), consumption of oxygen was significantly inhibited. This finding indicated that the substance analyzed effectively reduced consumption of oxygen by the Weissberger biogenic oxidative system.

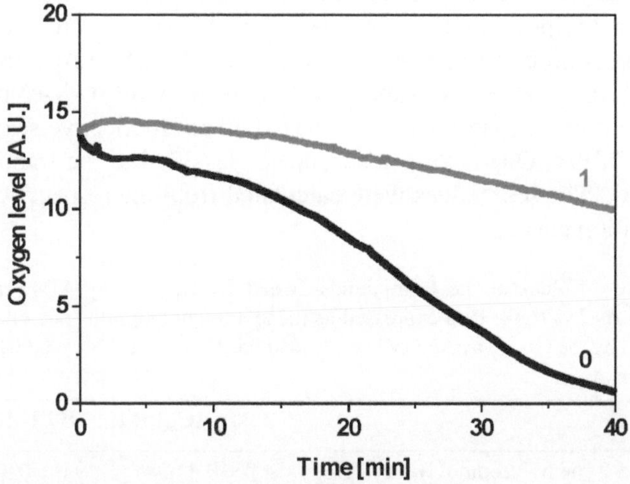

FIGURE 4.7 Oxygen concentration decrease in the reaction mixture with the Weissberger biogenic oxidative system – 1.0 μM Cu(II) and 100 mM ascorbate (curve 0). Addition of *N*-(2-mercapto-2-methylpropionyl)-l-cysteine (100 μM) significantly inhibited the oxygen consumption by the Weissberger biogenic oxidative system (curve 1).

We speculate that this effect of substance **1** may occur due to the effective inhibition of the Weissberger biogenic oxidative system mediated free-radical generation resulting in the reduction of its cycling rate and thus decreasing oxygen consumption (Scheme 4.2).

4.2.5 SYNTHESIS OF N-(2-MERCAPTO-2-METHYLPROPIONYL)-L-CYSTEINE DERIVATIVE

When a molecular structure is to be modified, a detailed investigation of its features is required. Taking into account the mode of action of *N*-(2-mercapto-2-methylpropionyl)-L-cysteine (**1**) and the resulting disulfide metabolite (*R*)-7,7-dimethyl-6-oxo-1,2,5-dithiazepane-4-carboxylic acid (**2**), it is evident that the two thiol functional groups are inevitable for the original activity of the compound. These have to be retained. Of the remaining groups, only carboxylate is amenable to simple modification. From the synthetic point of view, however, free thiol groups complicate synthetic strategy, as once deprotonated, they are fairly nucleophilic and could readily compete with -COO^- or form cyclic thioester under dehydrating conditions. We therefore thought it would be elegant to choose the available compound **2** as the starting material

for the transformation followed by the reduction of the disulfide bond. In this way, the –S-S– bond acts as a protecting group for the thiols (Scheme 4.3).

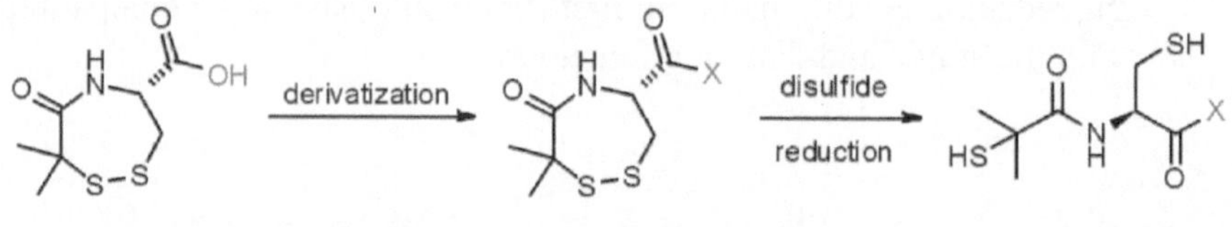

SCHEME 4.3 The synthetic strategy for derivatization of 1. Cyclic disulfide 2 is first transformed to an intermediate compound 8. Subsequently, the free thiols are regenerated by –S-S– reduction to give the target structure 9.

HOCl scavenging activity of taurine is due to the primary amino group in its structure. Direct reaction of **2** with taurine is not feasible because the NH_2 group would preferably form an amide with the carboxylic function. It is conceivable to first protect the amino group of taurine, but still the resulting mixed carboxylic-sulfonic anhydride would be prone to hydrolysis and at least two more reaction steps would be necessary. Introduction of -CH_2-CH_2-NH_2 moiety to (*R*)-7,7-dimethyl-6-oxo-1,2,5-dithiazepane-4-carboxylic acid (**2**) could also be carried out by means of:

a) esterification with commercially available *N*-protected aminoethanol **10** in the presence of a dehydrating agent (Scheme 4.4). However, conditions for the removal of the protecting group (PG) would have to be compatible with the rest of the molecule (ester and amide functions).

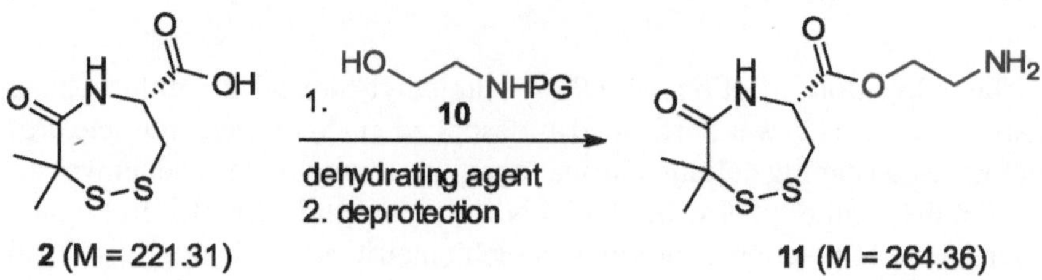

SCHEME 4.4 Esterification approach with *N*-protected aminoethanol 10.

b) reduction of carboxylate to carbinol 12 and subsequent reaction with *N*-protected taurine **13** (Scheme 4.5). Yet, this approach would present a significant alteration to the parent structure **1**. Moreover and again, the reducing conditions in the first step would have to be compatible with the amide and disulfide functions.

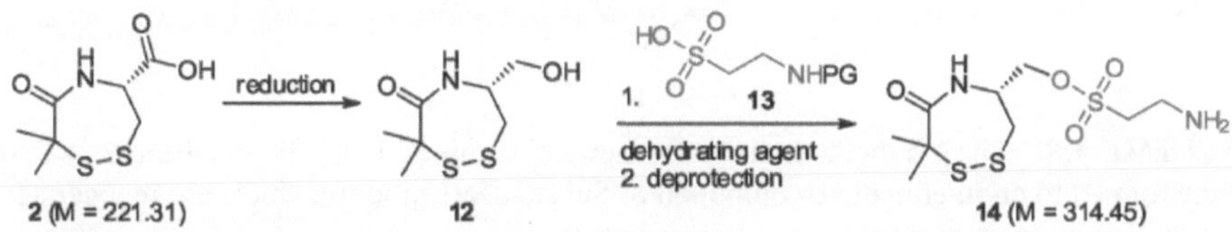

SCHEME 4.5 Carboxylate reduction in 2 followed by esterification with *N*-protected taurine 13.

c) direct alkylation of carboxylate in 2 by 2-bromoethylamine salt 15 in the presence of a base. The reaction should proceed under mild conditions and besides 2-bromoethylamine, a number of analogous reagents is commercially available as hydrochlorides or -bromides for further investigation. Moreover, to increase the solubility of target compound 16 in water, its primary amino group could be eventually transformed into hydrochloride. We decided to favor this approach.

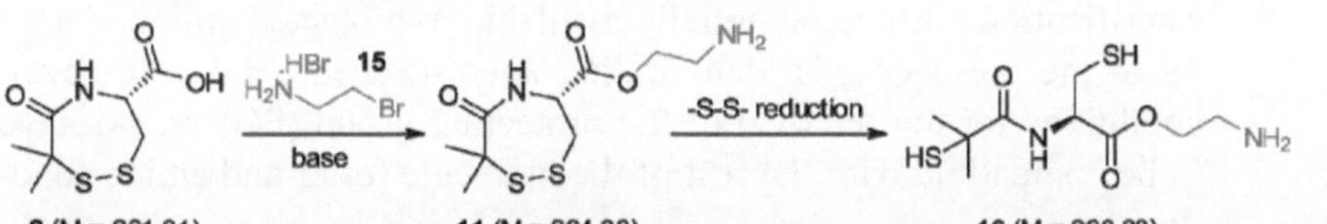

SCHEME 4.6 Direct alkylation of carboxylate in 2 with 2-bromoethylamine salt 15.

The alkylation reaction of (*R*)-7,7-dimethyl-6-oxo-1,2,5-dithiazepane-4-carboxylic acid (**2**) was performed as described in *Experimental procedures*. After time-consuming column chromatography, 10 mg of the product was obtained in the form of yellow oil. Its ^{1}H NMR spectrum is depicted in Fig. 4.8. Under identical reaction conditions, stoichiometric addition of silver(I) salt ($AgNO_3$), contrary to the previously described procedure in the literature, did not lead to any improvement [1].

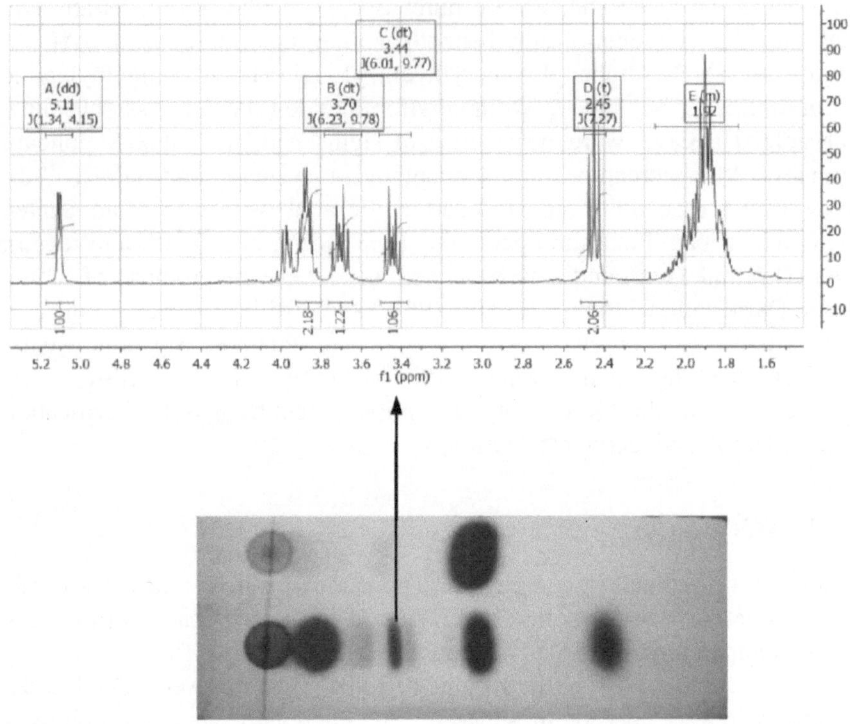

FIGURE 4.8 ^{1}H NMR spectrum of a (*R*)-7,7-dimethyl-6-oxo-1,2,5-dithiazepane-4-carboxylic acid alkylation product and reaction TLC plate. The arrow indicates which fraction was used to obtain the spectrum.

Based on these preliminary results, the first reaction step did indeed afford a new compound. Nevertheless, its structure has not been fully elucidated. In its ^{1}H NMR spectrum (Fig. 4.8), the triplet at 2.45 ppm with doubled integrated intensity (and therefore corresponding to 2 hydrogen atoms) clearly suggests incorporation of -CH_2-CH_2- moiety into the structure. The triplet arising from the nearby -CH_2- is presumably hidden in the multiplet at 1.76–2.11 ppm. Detailed analysis of the spectrum would be possible after additional purification. Further investigations are underway.

4.3 EXPERIMENTAL PROCEDURES

4.3.1 BIOPOLYMER AND CHEMICALS

High-molar-mass hyaluronan sample P0207–1A (M_w = 970 kDa) was purchased from Lifecore Biomedical Inc., Chaska, MN, USA. Analytical purity

grade NaCl and $CuCl_2 \cdot 2H_2O$ were purchased from Slavus Ltd., Bratislava, Slovakia. 2,2'-azinobis-(3-ethylbenzothiazoline)-6-sulfonic acid (ABTS; *purum*, >99%), 2,2-diphenyl-1-picrylhydrazyl (DPPH), 2-bromoethylamine hydrobromide (99%) and *N*, *N*-dimethylformamide (DMF, puriss., absolute, over molecular sieve) were the products of Sigma-Aldrich GmbH, Steinheim, Germany. *N*-(2-mercapto-2-methylpropionyl)-L-cysteine and (*R*)-7, 7-dimethyl-6-oxo-1,2,5-dithiazepane-4-carboxylic acid were gifts from Santen Co., Osaka, Japan. Potassium persulfate (*p.a.* purity; max 0.001% nitrogen), silver (I) nitrate (*p.a.*) and L-ascorbic acid used were the products of Merck KGaA, Darmstadt, Germany. Ethanol (96%) and distilled methanol, both *p.a.* purity grades, were purchased from Mikrochem, Pezinok, Slovakia. Redistilled deionized high quality water, with conductivity of <0.055 μS/cm, was produced by using the TKA water purification system from Water Purification Systems GmbH, Niederelbert, Germany.

4.3.2 SOLUTIONS

Solutions for rotational viscometry were prepared as follows: the high-molar-mass hyaluronan sample solutions (2.5 mg/mL) were prepared in the dark at room temperature in 0.15 M aqueous NaCl in two steps: first, 4.0 mL and after 6 h the same solvent in the volume of 3.90 or 3.85 mL were added, when working in the absence or presence of *N*-(2-mercapto-2-methylpropionyl)-L-cysteine, respectively. The stock solutions of ascorbate, *N*-(2-mercapto-2-methylpropionyl)-L-cysteine (16, 1.6, 0.16 mM), and cupric chloride (16 mM diluted to a 160 μM solution) were also prepared in 0.15 M aqueous NaCl.

Solutions for oximetry were prepared as follows: the HA sample solutions (2.5 mg/mL) were prepared in the dark at room temperature in 0.15 M aqueous NaCl in two steps: first, 1.0 mL of the solvent was added to 5 mg of HA and after 6 h another 1.0 mL of the solvent was added. The stock solution of ascorbate (16 mM) was prepared in deionized water, while solutions of *N*-(2-mercapto-2-methylpropionyl)-L-cysteine (16 mM), and cupric chloride (16 mM diluted to a 160 μM solution) were prepared in 0.15 M aqueous NaCl.

4.3.3 ROTATIONAL VISCOMETRY

The dynamic viscosity of the reaction mixture (8 mL; 0.15 M aqueous NaCl) containing high-molar-mass HA,) Cu(II) ions (1 μM) and ascorbate (100 μM in the absence and presence of *N*-(2-mercapto-2-methylpropionyl)-L-cysteine (1, 10 and 100 μM) was monitored by a Brookfield LVDV-II+PRO digital

rotational viscometer (Brookfield Engineering Labs., Inc., Middleboro, MA, USA) at 25.0 ± 0.1 °C at a shear rate of 237.6 s^{-1} for 5 h in a reservoir-spindle couple made of Teflon [14, 15, 31, 32]. The drug was introduced into the reservoir vessel before initiating HA oxidative degradation or 1 h after the reaction onset [2, 3, 34, 36, 38].

4.3.4 ABTS AND DPPH ASSAYS

The standard ABTS decolorization assay was used as reported previously [4, 28]. Briefly, the aqueous solution of $ABTS^{\cdot+}$ cation radical was prepared 24 h before the measurements at room temperature as follows: ABTS aqueous stock solution (7 mmol/L) was mixed with $K_2S_2O_8$ aqueous solution (2.45 mmol/L) in equivolume ratio. The next day, 1.1 mL of the resulting solution was diluted with 96% ethanol to the final volume of 50 mL. The ethanol-aqueous reagent in the volume of 250 μL was added to 2.5 μL of the ethanolic solutions of *N*-(2-mercapto-2-methylpropionyl)-L-cysteine. The concentration of stock solutions of *N*-(2-mercapto-2-methylpropionyl)-L-cysteine was 101−0.808 mmol/L. The light absorbance of the sample mixtures was recorded at 734 nm 6th minute after mixing the reactants.

For the DPPH decolorization assay, 2,2-diphenyl-1-picrylhydrazyl (1.1 mg) was dissolved in 50 mL of distilled methanol to generate $DPPH^{\cdot}$. The $DPPH^{\cdot}$ radical solution in the volume of 225 μL was added to 25 μL of the methanolic solutions of *N*-(2-mercapto-2-methylpropionyl)-L-cysteine (in the concentration range of 10–0.078 mmol/L) and in the 30th min the absorbance of the samples was measured at 517 nm.

All measurements of both assays were performed quadruplicately in 96-well Greiner UV-Star microplates (Greiner-Bio-One GmbH, Germany) by using the Tecan Infinite M 200 reader (Tecan AG, Austria).

4.3.5 OXIMETRY

The oxygen consumption of the reaction mixture (1600 μL; 0.15 M aqueous NaCl) containing high-molar-mass HA, Cu(II) ions (1 μM) and ascorbate (100 μM) in the absence and presence of *N*-(2-mercapto-2-methylpropionyl)-L-cysteine (100 μM) was monitored by a Strathkelvin 782–2-Chanel Oxygen System version 1.0 (Strathkelvin Instruments, Ltd., UK). Oximetric experiments were carried out at 37 °C and oxygen consumption was recorded for 40 min. The oxygen electrodes used had the advantage of small size, high precision, and the ability to operate in either stirred or unstirred media.

4.3.6 SYNTHESIS OF N-(2-MERCAPTO-2-METHYLPROPIONYL)-L-CYSTEINE DERIVATIVE

The alkylation of (*R*)-7,7-dimethyl-6-oxo-1,2,5-dithiazepane-4-carboxylic acid (**2**) was carried out as follows: **2** (50 mg, 0.23 mmol) was dissolved in 1 mL of DMF, and triethylamine (0.11 mL, 0.79 mmol) was added drop-wise *via* syringe followed by 2-bromoethylamine hydrobromide (70 mg, 0.34 mmol). The mixture was heated to 50 °C for 1 h and then left to reach room temperature. Reaction progress was monitored by thin-layer chromatography (TLC) on Merck silica gel F-254 plates and visualization was performed using UV light (254 nm). After 24 h, the reaction mixture was diluted with distilled water, extracted with ethyl acetate (3×), dried over Na_2SO_4 and concentrated under reduced pressure. The crude mixture was separated by column chromatography ($CHCl_3$:CH_3OH = 30:1) on Merck silica gel 60. Structure of the compounds was analyzed by ^{1}H nuclear magnetic resonance (NMR) spectroscopy on a Varian Mercury Plus Instrument (300 MHz) at 20 °C in $CDCl_3$ with $(CH_3)_4Si$ (0.00 ppm) as an internal standard.

4.4 CONCLUSIONS

In conclusion, our experimental findings suggest that in-vivo the beneficial effect of the disease-modifying antirheumatic drug *N*-(2-mercapto-2-methylpropionyl)-L-cysteine may be partly due to its ability to protect the synovial joint high-molar-mass HA from oxidative degradation. Particularly the free-radical scavenging ability of this drug is likely to be involved.

By combining the beneficial features of the compound studied and those reported for taurine, we have designed a synthetic route towards a new *N*-(2-mercapto-2-methylpropionyl)-L-cysteine derivative with an increased potential to act as a scavenger of hypochlorite. However, further elaboration of the synthetic procedure employed (replacement of base, solvent) and evaluation of the structure of the compound obtained is still required.

ACKNOWLEDGMENTS

The work was supported by the VEGA grants 2/0011/11, 2/0149/12, APVV 0351–10.

We thank Dr. Fumio Tsuji from Santen Pharmaceutical Co., Ltd., Osaka, Japan for providing *N*-(2-mercapto-2-methylpropionyl)-L-cysteine and (*R*)-7,7-dimethyl-6-oxo-1,2,5-dithiazepane-4-carboxylic acid.

KEYWORDS

- **ABTS**
- **DPPH**
- **Extracellular matrix**
- **Glycosaminoglycans**
- **Oximetry**
- **Rotational viscometry**
- **Thiols**

REFERENCES

1. Aron, Z. D., & Overman, L. E. (2005). Total synthesis and properties of the crambescidin core zwitterionic acid and crambescidin 359. *Journal of American Chemical Society* 127: 3380–3390.
2. Baňasová, M., Sasinková, V., Mendichi, R., Perečko, T., Valachová, K., Juránek, I., & Šoltés, L. (2012). Free–radical degradation of high–molar–mass hyaluronan induced by Weissberger´s oxidative system: potential anti oxidative effect of bucillamine. *Neuroendocrinology Letters 33*, 151–154.
3. Baňasová, M., Valachová, K., Juránek, I., & Šoltés, L. (2013). Aloe Vera and methylsulfonylmethane as dietary supplements: their potential benefits for arthritic patients with diabetic complications. *Journal of Information, Intelligence and Knowledge 5*, 51–68.
4. Cheng, Z., Moore, J., & Yu, L. (**2006**). High throughput relative DPPH radical scavenging capacity assay *Journal of Agricultural and Food Chemistry 54*, 7429–7436.
5. DeAngelis, P. L. (2012). Microbial glycosaminoglycan glycosyltransferases. *Glycobiology 12*, 9–16.
6. Distler, J. H., Hagen, C., Hirth, A., Müller–Ladner, U., Lorenz, H. M., del Rosso, A., Michel, B. A., Gay, R. E., Nanagara, R., Nishioka, K., Matucci–Cerinic, M., Kalden, J. R., Gay, S., & Distler, O. (2004). Bucillamine induces the synthesis of vascular endothelial growth factor dose–dependently in systemic sclerosis fibroblasts via nuclear factor–kappa B and simian virus 40 promoter factor 1 pathways. *Molecular Pharmacology 65*, 389–399.
7. Dreyfuss, J. L., Regatieri, C. V., Jarrouge, T. R., Cavalheiro, R. P., Sampaio, L. O., Nader, H. B. (2009). Heparan sulfate proteoglycans: structure, protein interactions and cell signaling. *Annals of the Brazilian Academy of Sciences 81*, 409–429.
8. Erickson, M., & Stern, R. (2012). Chain gangs: new aspects of hyaluronan metabolism. *Biochemistry Research International 2012*, 1–9.
9. Greenwald, R. A., & Moy, W. W. (1980). *Effect of oxygen derived free radicals on hyaluronic acid.* Arthritis & Rheumatism 23, 455-463.
10. Grisham, M. B., Jefferson, M. M., Melton, D. F., & Thomas, E. L. (1984). Chlorination of Endogenous Amines by Isolated Neutrophils Journal of Biological Chemistry 259, 10404–10413.
11. Halliwell, B., & Gutteridge, J. M. C., (1989). Free Radicals in Biology and Medicine (2nd ed). In: *Free Radicals in Biology and Medicine*, Oxford: *Clarendon Press, 152*–156.

12. Hawkins, C. L., & Davies, M. J. (1996). *Direct detection and identification of radicals generated during the hydroxyl radical–induced degradation of hyaluronic acid and related materials* Free Radical Biology and Medicine, 21, 275−290.
13. Horwitz, L. D. (2003). Bucillamine: A Potent Thiol Donor with Multiple Clinical Applications. *Cardiovascular Drug Reviews 21*, 77–90.
14. Hrabárová, E., Valachová, K., Rychlý, J., Rapta, P., Sasinková, V., Malíková, M., & **Šoltés**, L. (2009). High molar mass hyaluronan degradation by Weissberger›s system: pro– and anti–oxidative effects of some thiol compounds *Polymer Degradation and Stability*, *94*, 1867−1875.
15. Hrabárová, E., Valachová, K., Juránek, I., & **Šoltés**, L. (2012). Free–radical degradation of high–molar–mass hyaluronan induced by Ascorbate *Plus* cupric ions. Anti–oxidative properties of the Piešťany–spa curative waters from healing peloid and maturation pool, **In:** *Kinetics, Catalysis and Mechanism of Chemical Reactions: From Pure to Applied Science. 2. Tomorrow and Perspectives*, Islamova, R. M., & Kolesov, S. V., Zaikov, G. E. (Eds.) New York: *Nova Science Publishers*, 29−36.
16. Hull, R. L., Johnson, P. Y., Braun, K. R., Day, A. J., & Wight, T. N. (2012). Hyaluronan and hyaluronan binding proteins are normal components of mouse pancreatic islets and are differentially expressed by islet endocrine cell types. *Journal of Histochemistry & Cytochemistry 60*, 749–760.
17. Jiang, D., Liang, L., & Noble, P. W. (2011). Hyaluronan as an Immune Regulator in Human Diseases *Physiological Reviews 91,* 221–264.
18. Kvam, C., Granese, D., Flaibani, A., Pollesello, P., & Paoletti, S. (1993). *Hyaluronan can be protected from free–radical depolymerization by 2, 6–diisopropylphenol, a novel radical scavenger* Biochemical and Biophysical Research Communications 193, 927-933.
19. Laurent, T. C., Laurent, U. B. G., Fraser, J. R. E. (1995). Functions of Hyaluronan *Annals of the Rheumatic Diseases 54,* 429-432.
20. Lee, J. Y., & Spicer, A. P. (2000). Hyaluronan: a multifunctional, mega Dalton, stealth molecule. *Current Opinion in Cell Biology 12,* 581–586.
21. Mainnemare, A., Mégarbane, B., Soueidan, A., Daniel, A., & Chapple, I. L. C. (2004). Hypochlorous acid and taurine–N–nonochloramine in periodontal diseases *Journal of Dental Research* 83: 823–831.
22. Matsuno, H., Sugiyama, E., Muraguchi, A., Nezuka, T., Kubo, T., Matsuura, K., & Tsuji, H. (1998). Pharmacological effects of SA96 (bucillamine) and its metabolites as immune modulating drugs–the disulfide structure of SA–96 metabolites plays a critical role in the pharmacological action of the drug. *International Journal of Immunopharmacology 20,* 295–304.
23. Meyer, K., Palmer, J. W. (1934). The Polysaccharide of the Vitreous Humor *Journal of Biological Chemistry* 107, 629–634.
24. Narazaki, R., Hamada, M., Harada, K, & Otagiri, M. (1996). Covalent Binding between Bucillamine Derivatives and Human Serum Albumin *Pharmaceutical Research 13*, 1317–1321.
25. Noble, P. W. (2002). Hyaluronan and its Catabolic Products in Tissue Injury and Repair *Matrix Biology 21*, 25–29.
26. Parsons, B. J., Al Assaf, S., Navaratnam, S., & Phillips, G. O. (2002). Comparison of the reactivity of different oxidative species (ROS) towards hyaluronan In: *Hyaluronan: Chemical, Biochemical and Biological Aspects. 1.* Kennedy, J. F., Phillips, G. O., Williams, P. A., Hascall, V. C. (Eds.). Cambridge: *Wood head Publishing Ltd.* 141−150.
27. Pigman, W., Rizvi, S., & Holley, H. L. (1961). Depolymerization of hyaluronic acid by the ORD reaction. *Arthritis & Rheumatology 4*, 240−252.

28. Re, R., Pellegrini, N., Proteggente, A., Pannala, A., Yang, M., & Rice–Evans, C. (1999). Antioxidant activity applying an improved ABTS radical cation decolorization assay *Free Radical Biology and Medicine 26*, 1231–1237.
29. Saari, H., Kontinen, Y. T., Friman, C., & Sorsa, T. (1993). *Differential Effects of Reactive Oxygen Species on Native Synovial Fluid and Purified Human Umbilical Cord hyaluronate* Inflammation 17, 403-415.
30. Schreml, S., Szeimies, R. M., Prantl, L., Karrer, S., Landthaler, M., & Babilas, P. (2010). Oxygen in Acute and Chronic Wound Healing *British Journal of Dermatology 163*, 257–268.
31. Stankovská, M., & **Šoltés**, L. (2010). Oxidative Degradation of Hyaluronan: Is Melatonin an Antioxidant or Prooxidant. In: *Monomers, Oligomers, Polymers, Composites, and Nanocomposites 23*, Polymer Yearbook Pethrick, R. A., Petkov, P., Zlatarov, A., Zaikov, G. E., Rakovský, S. K. (Eds.) *Nova Science Publishers,* New York 59-67.
32. Surovčíková–Machová, Ľ., Valachová, K., Baňasová, M., **Šnirc**, V., Priesolová, E., Nagy, M., Juránek, I., & **Šoltés**, L. (2012). Free–radical degradation of high–molar–mass hyaluronan induced by ascorbate *plus* cupric ions: testing of stobadine and its two derivatives in function as antioxidants *General Physiology and Biophysics*, *31*, 57-64.
33. Thomas, E. L., Grisham, M. B., & Jefferson, M. M. (1983). Myeloperoxidase–Dependent Effect of Amines on Functions of Isolated Neutrophils *Journal of Clinical Investigation 72*, 441–454.
34. Valachová, K., Kogan, G., Gemeiner, P., & **Šoltés**, L., (2010a). Protective Effects of Manganese(II) Chloride on Hyaluronan Degradation by Oxidative System Ascorbate *Plus* Cupric Chloride *Interdisciplinary Toxicology 3*, 26–34.
35. Valachová, K., Hrabárová, E., Dráfi, F., Juránek, I., Bauerová, K., Priesolová, E., Nagy, M., & **Šoltés**, L., (2010b). Acrobat and Cu (II) induced Oxidative Degradation of High Molar Mass hyaluronan, Pro and anti oxidative effects of some thiols. *Neuroendocrinology Letters 31*, 101–104.
36. Valachová, K., Hrabárová, E., Priesolová, E., Nagy, M., Baňasová, M., Juránek, I., & **Šoltés**, L. (2011). Free Radical Degradation of High Molecular Weight Hyaluronan Induced by Ascorbate *Plus* Cupric Ions. Testing of bucillamine and its SA981–metabolite as antioxidants *Journal of Pharmaceutical and Biomedical Analysis 56*, 664-670.
37. Valachová, K., Rapta, P., Slováková, M., Priesolová, E., Nagy, M., Mislovičová, D., Dráfi, F., Bauerová, K., & **Šoltés**, L., Radical degradation of high molar mass hyaluronan induced by acrobat *plus* cupric ions. Testing of arbutin in the function of antioxidant *Kinetics, Catalysis and Mechanism of Chemical Reactions: From Pure to Applied Science 2* Tomorrow and Perspectives, Islamova, R. M., Kolesov, S. V., Zaikov. G. E. (Eds.) *Nova Science Publishers*, New York, (2012) 11–28.
38. Valachová, K., Baňasová, M., Machová, Ľ., Juránek, I., Bezek, Š., & Šoltés, L. (2013). Antioxidant Activity of Various Hexahydropyridoindoles *Journal of Information Intelligence and Knowledge 5*, 15–31.
39. Veiseh, M., & Turley, E. A. (2011). Hyaluronan metabolism in remodeling extracellular matrix: probes for imaging and therapy of breast cancer. *Integrative Biology 3,* 304–315.
40. Yamada, S., Sugahara, K., & Özbek, S. (2011). Evolution of glycosaminoglycans, Comparative biochemical study. *Communicative and Integrative Biology 4*, 150–158.
41. Weiss, S. J., Klein, R., Slivka, A., & Wei, M. (1982). Chlorination of Taurine by Human Neutrophils Evidence for Hypochlorous Acid Generation *Journal of Clinical Investigation.* 70, 598–607.

42. Wielandt, A. M., Vollrath, V., Farias, M., & Chianale, J. (2006). Bucillamine induces glutathione biosynthesis via activation of the transcription factor Nrf2. *Biochemical Pharmacology 72*, 455–462.
43. Wong, S. F., Halliwell, B., Richmond, R., & Skowroneck, W. R. (1981). *The Role of Superoxide and Hydroxyl Radicals in the Degradation of Hyaluronic Acid induced by Metal Ions and by Ascorbic Acid. Journal of Inorganic Biochemistry, 14*, 127−34.

CHAPTER 5

MAGNETIC PROPERTIES OF ORGANIC PARAMAGNETS

M. D. GOLDFEIN, E. G. ROZANTSEV, and N. V. KOZHEVNIKOV

CONTENTS

ABSTRACT

Communication between the phenomenon of a magnetism and paramagnetism, which stable radicals of different type possess is probed. It is shown that under the influence of an outside magnetic field there is a change of physical and chemical properties of some organic free radicals to the localized and nonlocalized unpaired electrons. Changes of properties of low-molecular and high-molecular organic paramagnets are caused by magnetic interactions occurring in them. Some iminoxyl polyradicals, possessing high value of a magnetic susceptibility, can weaken or increase strength of the enclosed magnetic field. One of the most important applications of stable paramagnets is their use as components to polarized proton targets in high-energy physics. It allowed to create nuclear precession magnetometers for geophysics and astronautics.

5.1 GENERAL STATEMENTS

When a magnetic field *H* is imposed, all substances show a macroscopic magnetic moment *M*. The value *M* relates to the imposed field *H* with a coefficient of proportionality c (the magnetic susceptibility):

$$M = \chi H.$$

In diamagnetic substances with completely filled orbitals, the induced moment is oriented *against* the external field; their magnetic susceptibility is negative and temperature-independent. In paramagnetic substances with half-filled orbitals, the induced moment vector under the influence of the imposed magnetic field is directed *parallel* to the latter. For noninteracting (independent) spins, the value of the magnetic moment is inversely proportional to temperature and their susceptibility can be approximated by Curie's expression:

$$\chi = C/T,$$

where *C* is Curie's constant, *T* is absolute temperature.

The value of magnetic susceptibility is usually recalculated to the effective magnetic moment μ_{eff} defined as

$$\mu_{eff}=[(3k/Na)\chi T]^{0.5} = \mu_s g[S(S+1)]^{0.5},$$

where k – Boltzmann's constant, N_a – Avogadro's number, $\mu_Б$ – Bohr's magneton, and S – spin.

For the case of interacting spins, numerous deviations from Curie's law are known. As a first approximation, such behavior is described by the Curie–Weiss law (Fig. 5.1):

$$\chi = C/(T - \theta).$$

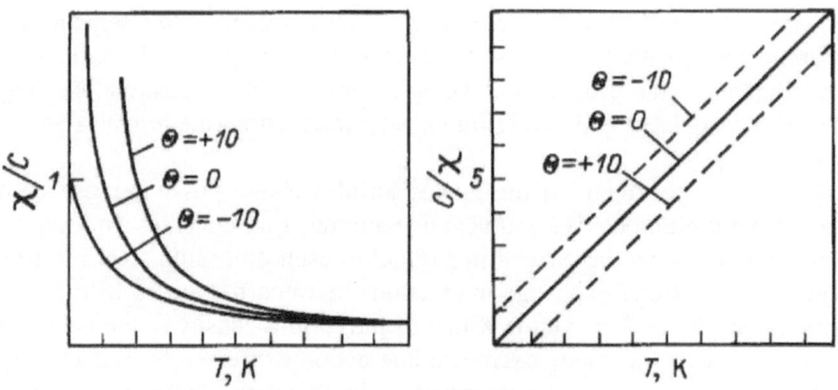

FIGURE 5.1 Temperature dependence of magnetic susceptibility (left) and that of inverse magnetic susceptibility (right) for noninteracting ferromagnetically coupled spins and antiferromagnetically coupled spins, respectively.

Here, the "characteristic temperature" θ is determined by the crystal field and may be either positive, corresponding to ferromagnetic interactions (with parallel spin orientation), or negative, corresponding to antiferromagnetic interactions (with antiparallel spin orientation).

The interradical interactions of uncoupled electrons are classified into two types, namely: dipole–dipole ones and exchange ones; the latter is determined by the overlap of the wave functions of uncoupled electrons and quickly decreases with increasing distance. The exchange interaction averages both the dipolar interaction between uncoupled electrons and the intraradical superfine interaction of uncoupled electrons with atomic nuclei. When there are a couple of electrons on neighboring centers with pronounced overlapping wave functions, some interaction between the spins S_1 and S_2 arises. It leads to the formation of singlet and triplet states. According to Heitler–London's description of chemical bonds this interaction is expressed by the Hamiltonian:

$$\hat{H} = -2J\hat{S}_1\hat{S}_2 \tag{1}$$

Extension of Eq. (1) to a multielectron system is described by Heisenberg's exchange Hamiltonian as:

$$\hat{H} = -\sum_{i,j} J_{i,j}\hat{S}_i\hat{S}_j, \tag{2}$$

where $J_{i,j}$ is the exchange integral between atoms i and j having total spins S_i and S_j.

The exchange integral *J* characterizes the exchange interaction degree and is expressed in energy units. Negative *J* values correspond to interactions of the antiferromagnetic type (the state of the lowest energy with the antiparallel spin orientation, the ground state veing a spin singlet). A positive exchange integral is associated with the ferromagnetic interaction (the ground state is a spin triplet) [1].

In 1963, McConnell [2] formulated an idea of the possible presence of particles with high positive and negative atomic spin densities. In a crystal, such compounds can be packed in parallel to each other into stacks to form conditions for strong exchange interactions between the atoms with a positive spin density and the atoms with a negative spin density in the neighboring radicals. Ferromagnetic exchange interaction expressed by Heisenberg's exchange integral is a consequence of the incomplete compensation of the antiferromagnetic coupled spins,

$$H^{AB} = -\sum_{i,j} J_{ij}^{AB} S_i^A S_j^B = -S^A S^B \sum_{i,j} J_{ij}^{AB} \rho_i^A \rho_j^B \tag{3}$$

where S^A and S^B are the total spins of radicals A and B; ρ_i^A and ρ_j^B are the -spin densities on atoms i and j in radicals A and B, J_{ij}^{AB} the interradical exchange integral for i and j.

Buchachenko [3] stated that "it would be almost impossible to realize this way since it is impossible to construct such a crystal lattice of the radical 'to turn-on' the intermolecular exchange interaction among the atoms with opposite spin densities only and to 'turn-out' it among the atoms with spin densities of an identical sign." McConnell's model, nevertheless, was involved to interpret complex interradical interactions found in the crystals of stable organic paramagnets. Rather recently [4], direct experimental evidence on *bis*-phenylmethylenyl-[2,2]-*p*-cyclophanes has been obtained that ferromagnetic exchange can be reached within McConnell's model.

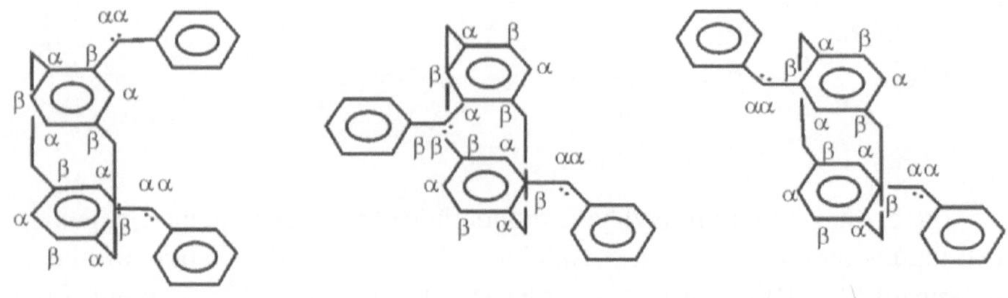

Pseudo-*o*-, pseudo-*m*-, and pseudo-*p*-*bis*-phenylmethylenyl-[2.2]-*p*-cyclophanes respectively, were obtained through photolysis in a vitrified matrix at low temperatures. The spin-spin interaction between the two triplet diphenylcarben fragments built into the [2, 2]-*p*-cyclophane frame, was explored by the EPR technique.

For the pseudo-*o*-dicarben a quintet state has been revealed, and within a temperature range of 11–50 K the EPR signal intensity obeys the Curie law. When $T > 20$K, another signal caused by a changed population of the triplet level was observed. Therefore, the pseudo-*o*-isomer is in its ground quintet state with $D = 0.0624$ and $E = 0.0190$ cm^{-1}, and the triplet state lies 63 cm^{-1} higher by energy.

Pseudo-*m*-*bis*-phenylmethylenyl-[2.2]-*n*-cyclophane gives no resonance signal at 11K. But a triplet state with $D = 0.1840$ and $E = 0.0023$ cm^{-1} was recorded with increasing temperature. The pseudo-*m*-isomer is in its ground singlet state, and the value of singlet–triplet splitting is 98 cm^{-1}. At 15 K for the pseudo-*n*-isomer the quintet nature of the ground state with $D = 0.1215$ and $E = 0.085$ cm^{-1} has been established, but it is not stable chemically.

5.2 MAGNETIC INTERACTIONS IN STABLE ORGANIC PARAMAGNETS

As our work deals with stable radicals only, it seems expedient to analyze literature data, having limited ourselves to stable organic paramagnets. According to our goal, it is worthwhile focusing attention mainly on measurements of magnetic susceptibility and, in particular, on clarification of the dependence of the magnetic properties of substances on their chemical structure. One of the most studied stable aroxyls, the so-called galvinoxyl, possesses a highly delocalized uncoupled electron. The formula of galvinoxyl is

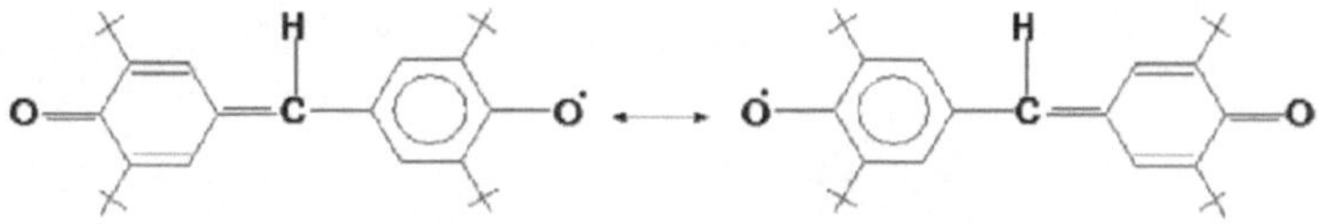

The crystals of galvinoxyl have monoclinic symmetry with the elementary cell parameters $a = 23.78$, $b = 10.87$, $c = 10.69$ nm and the angle of nonorthogonality $\beta= 106.6^{\circ}$; a second-order symmetry axis; a 12° deviation from coplanarity, and a 134° angle formed by the C–C bonds at the central carbon atom [5]. The crystal structure of galvinoxyl allows the possibility of the formation of a magnetic linear chain structure extended along the *c* axis.

The temperature dependence of the paramagnetic susceptibility of galvinoxyl obeys the Curie–Weiss law with a positive Weiss constant $\theta= +19$ K above 85 K, which allows one to assume ferromagnetic interactions between neighboring particles. However, at 85 K a phase transition is observed, upon which the paramagnetic susceptibility sharply decreases, and at 55 K its value corresponds to the content of free radicals 1.1% [6].

It is interesting that galvinoxyl radicals form couples in a diluted crystal, which have a ground triplet state; and a thermally achievable excited singlet state lies 2J higher [7]. Therefore, a ferromagnetic interradical exchange interaction with $2J_F = 1.5 \pm 0.7$ meV is realized in every radical couple. In other words, within a temperature range 10–100 K a diluted galvinoxyl crystal shows no phase transition since it retains ferromagnetic interactions. On the contrary, antiferromagnetic-type interactions with $2J_{AF} = -45 \pm 2$ meV prevail in chemically pure galvinoxyl below 85 K. Apparently, the phase transition in this case is caused by radical dimerization.

This is also confirmed by data on the temperature dependence of the magnetic susceptibility of mixed galvinoxyl crystals. From magnetization curves it follows that the spin multiplicity is almost proportional to the radical concentration in a mixed crystal. As calculations show [8], the ferromagnetic intermolecular interactions in galvinoxyl can be explained by superposition of the effects of intraradical spin polarization and charge transfer between free radicals.

Hydrazyl and hydrazidyl radicals are inclined to the formation of various complexes with solvents. This circumstance slightly influences the value of the *g* factor but strongly changes the EPR linewidth. The discordance in the magnetic data of different researchers is probably caused by the presence of impurities in the samples studied, owing to experimental difficulties in purification of organic paramagnets.

The magnetic susceptibility of 1,3,5-triphenyl verdazyl [8]

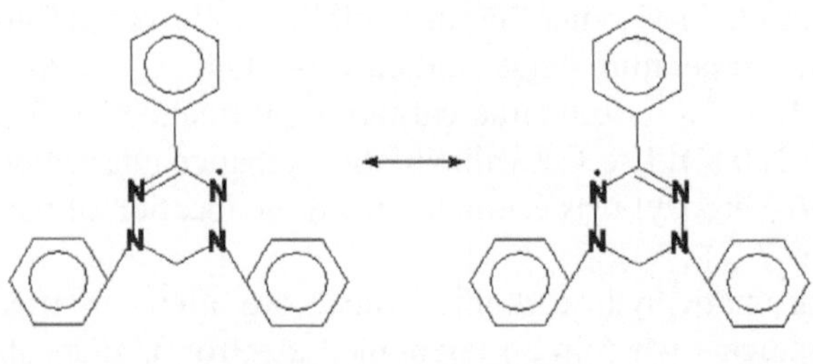

was measured in a temperature range of 1.6–300K.

In the high-temperature range the magnetic susceptibility obeys the Curie–Weiss law with a negative Weiss constant θ= −8 K. The susceptibility deviates from the Curie–Weiss law at lower temperatures and shows a wide maximum near 6.9 K.

The usage of Heisenberg's linear model with isotropic exchange interaction with J/k = −5.4 K above 6 K provides satisfactory agreement with experiment. The distant order of interactions caused by ferromagnetic-type interchain interactions arises at 1.7 K. The crystals of 1,3,5-triphenylverdazine have orthorhombic symmetry with the elementary cell parameters: a = 18.467, b = 9.854, c = 8.965 nm. All the four nitrogen atoms and the substituent at position 3 are almost coplanar, the two other phenyl groups turned relative to the C–N bond by 23 and 13°, respectively. The radicals in a possible magnetic chain are shown [9] to be bound with each other by a second-order screw axis parallel to the c axis so that interchain ferromagnetic exchange interactions are formed between these antiferromagnetically ordered chains.

In this regard, verdazyl biradicals with strongly delocalized uncoupled electrons are of interest, namely: *n*-di-1, 5-diphenyl-3-verdazyl benzene and *m*-di-1, 5-diphenyl-3-verdazyl benzene:

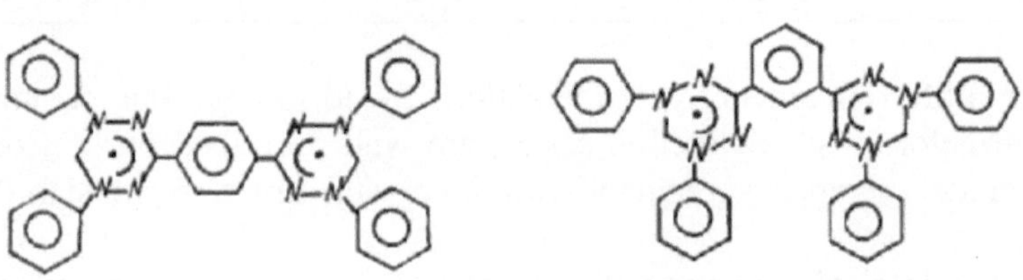

The susceptibility of the *n*-isomer obeys the Curie–Weiss law above 100 K with a Weiss constant θ= −100 ± 20 K and a Curie constant C = 1.0 ± 0.01 K·emu/mol, and the χ *vs.* T curve passes through a maximum at 19 ± 1 K when temperature reduces.

In the case of the *m*-isomer the susceptibility follows the Curie–Weiss law over the whole temperature range studied 1.8–300 K ($C = 0.90 \pm 0.05$ K·emu/mol and $\theta = -12 \pm 3$ K). Both biradicals are supposed to exist in a ground triplet state (J/k > 300 K). The J'/k value of the exchange interaction between the triplets in *n-bis*-verdazyl was estimated from the location of the maximum, it was negative (–7 K).

Classical aromatic hydrocarbonic radicals are often classified as so-called π-electronic radicals wherein an uncoupled electron is delocalized over the whole aromatic bond system. In their majority, arylmethyl radicals in solution exist in thermodynamic equilibrium with their dimer.

Ballester's perchloro-triphenylmethyl radicals sharply differ from classical hydrocarbonic ones by properties: they are rather stable in the absence of light, completely monomeric in both solution and their solid state.

The perchloro-triphenylmethyls studied in Ref. [10] within the range 293–77 K obey the Curie–Weiss law (Table 5.1).

TABLE 5.1 Characteristic Temperature θ (K) of Some Perchloro-triphenylmethyl Radicals: Ar, Ar[1], Ar[2] C

Ar	Ar[1]	Ar[2]	θ, K	μ_{eff}
4H-C_6HCl_4	C_6Cl_5	C_6Cl_5	-4.8	1.76
4H-C_6HCl_4	4H-C_6HCl_4	4H-C_6HCl_4	+1.9	1.73
3H,5H-$C_6H_2Cl_3$	C_6Cl_5	C_6Cl_5	–10.4	1.76
3H,5H-$C_6H_2Cl_3$	3H,5H-$C_6H_2C_3$	3H,5H-$C_6H_2Cl_3$	–10.1	1.74
2H-C_6HCl_4	C_6Cl_5	C_6Cl_5	–12.0	1.71
2H-C_6HCl_4	2H-C_6HCl_4	C_6Cl_5	–3.3	1.69

The antiferromagnetic-type interactions found in the stable paramagnets of the trichloro-triphenylmethyl series, are well described by McConnell's above model, being in agreement with the crystal structure and spin density values.

Unlike classical aromatic radicals, the NO group in the iminoxyl radicals takes no part in the formation of a conjugated bond system; the uncoupled electron in such radicals is therefore mainly localized on the nitrogen–oxygen bond. The rather reliable steric shielding of the uncoupled electron (due to the effects of the voluminous methyl groups and the σ-bond system interfering uncoupled electron delocalization) provides conditions for nonradical reactions to proceed in the row of functionalized radicals of this class. This

allows synthesizing many chemically pure paramagnets of various chemical structures (Fig. 5.2) [11].

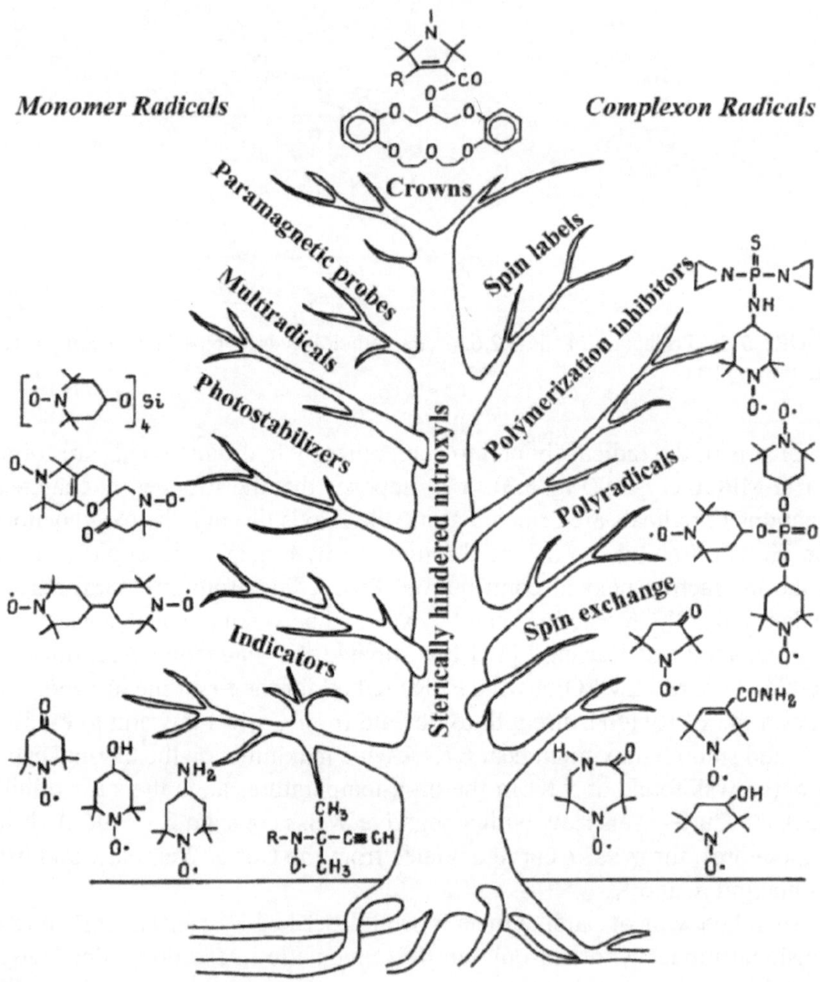

FIGURE 5.2 Genealogic tree of stable iminoxyl (nitroxyl) radicals, whose synthesis and application were promoted by the discovery of nonradical reactions of radicals.

One can easily see that the majority of works is devoted to 2,2,6,6-tetramethyl-4-hydroxypiperidino-1-oxyl (TEMPOL) derivatives obtained by Rozantsev [12]. TEMPOL crystallizes in a monoclinic cell with the axis param-

eters $a = 0.705$; $b = 1.408$; $c = 0.578$ nm; β=118°40 and belongs to the spatial group C (Fig. 5.3).

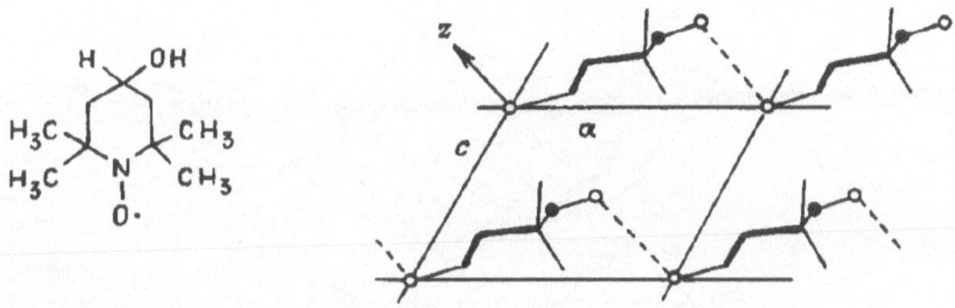

FIGURE 5.3 Projection of the 2,2,6,6-tetramethyl-4-hydroxypiperidyl-1-oxyl structure onto the *ac* plane.

Chains of the radicals bound to each other by hydrogen bonds are formed in a TEMPOL crystal (Fig. 5.3). It is supposed that the strongest exchange interactions of radicals are oriented along the Z axis through the oxygen atoms. The direction along the *a* axis through the hydrogen bond could probably be the interaction next in contribution. Proceeding from structural reasons, weaker magnetic interactions can be expected between the *ac* planes.

Rozantsev and Karimov [13] investigated the magnetic susceptibility of chemically pure TEMPOL by the EPR method for the first time in 1966. They showed the EPR signal strength to deviate from Curie's law and to exhibit a wide and smooth maximum near 6 K. A wide maximum on the thermal capacity curve was found at 5 K. In the high-temperature range the susceptibility obeys the Curie–Weiss law with a negative Weiss constant $\theta = -6$K. At lower temperatures, the χ vs. T curve deviates from the Curie–Weiss law and has a flat maximum at 6 K.

Such behavior of paramagnets is well described by Heisenberg's one-dimensional model with isotropic antiferromagnetic interactions. For Heisenberg's linear system with $S = 1/2$ the magnetic susceptibility should have a flat maximum determined by $\chi_{max}/(N_a g^2 \mu^2_B/J)$ 0.07346 at $kT_{max}/J \approx 1.282$.

The value of the exchange J/k parameter is estimated as −5 K. Therefore, independent studies of the magnetic susceptibility of TEMPOL evidence strong exchange interactions experienced by radicals in one direction, which results in the near order of interactions and the formation of linear antiferromagnetic chains near 6K.

As nonzero interaction always exists between the chains in one-dimensional magnetic systems, it could be expected that below some critical temperature it would get rather expressed to cause transition to a distant order of interactions. In the case of TEMPOL the distant order caused by interchain interactions arises at $T_N = 0.34$ K, the interchain to intrachain interaction ratio (J/J') estimated as 0.003, $J/k = 0.013$ K.

An alternating linear chain arises if $\gamma < 1$. The case $\gamma = 0$ corresponds to a simple dimer where paired interactions act only.

The stable di-2,2,6,6-tetramethyl-1-oxyl-4-piperidyl sulfite biradical exemplifies the alternating chain (Fig. 5.4). In the high-temperature range its magnetic susceptibility is described by the Curie–Weiss law with $\theta = -9$ K. When temperature falls, the susceptibility of this biradical deviates from the Curie–Weiss law near 25 K, and then sharply drops down to 2 K.

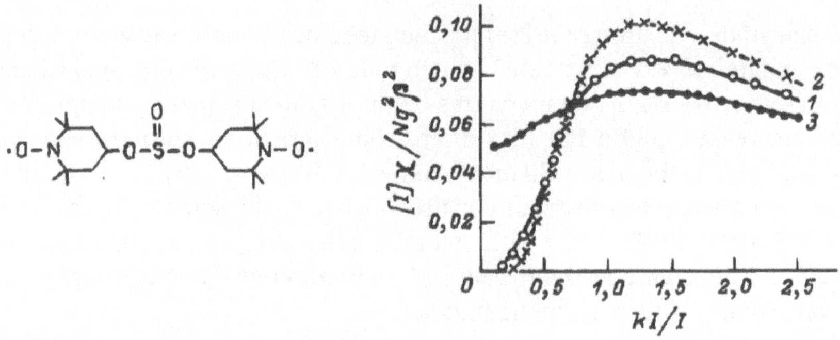

FIGURE 5.4 (a) A low-temperature fragment of the curve of magnetic susceptibility of the biradical; (b) calculated curve for Heisenberg's paired interaction with J/k = 9.6 K; (c) calculated curve for a regular spin chain with J/k = 9.6 K.

As is seen from Fig. 5.4, the susceptibility maximum of this biradical is less flat than it would be for a regular spin chain with $\gamma = 1$. The use of the model of an alternating spin chain with $J/k = 9.6$ K and $\gamma = 0.55$ provides satisfactory agreement with experiment, and the maximum exchange between interacting spins $J/k = 9.6$ K is a result of structural exchange interactions in the crystal.

The adduct of copper hexafluoro acetylacetonate with the iminoxyl radical of 2,2,6,6-tetramethyl-4-hydroxypiperidyl-1-oxyl is another example of the alternating linear chain [14].

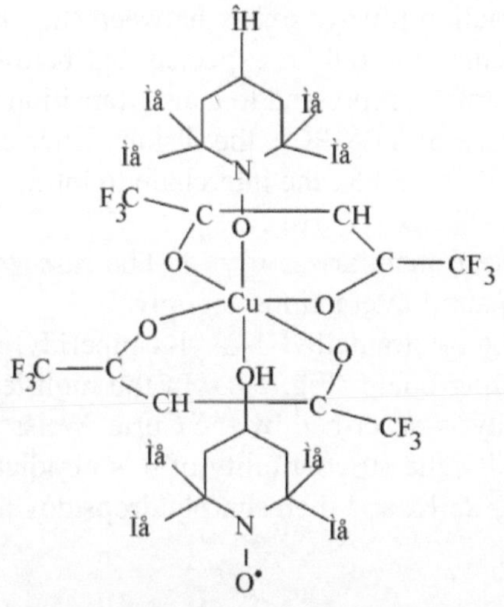

When studying, strong (19K) ferromagnetic interaction between the copper ion and iminoxyl was revealed, which followed from data obtained at temperatures above 4.2K. At temperatures below 1K, the magnetic susceptibility sharply increased, and a flat maximum, characteristic of antiferromagnetic linear chains, was found at $\sim$ 80 meV. Analysis of these data was carried out in the assumption that the substance consists of chains with spins $S = 1$ and weak ($2J = -78$ meV) antiferromagnetic interaction between spins. Therefore, the alternation in this case arises because of alternation of the strong ferromagnetic and weak antiferromagnetic interactions.

For the silicon-organic iminoxyl polyradicals

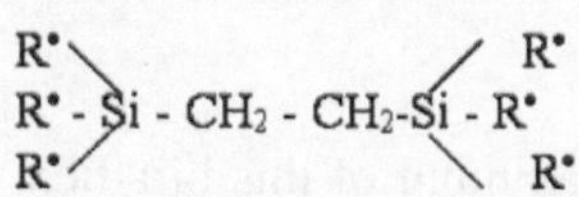

I

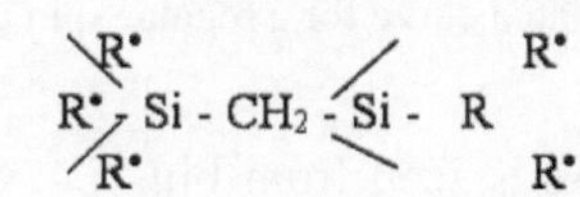

II

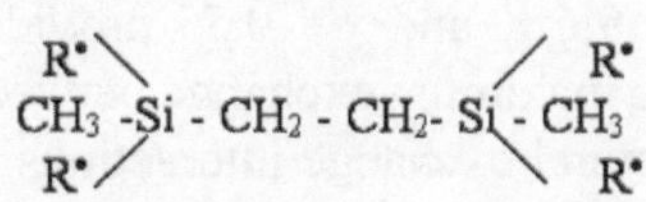

III

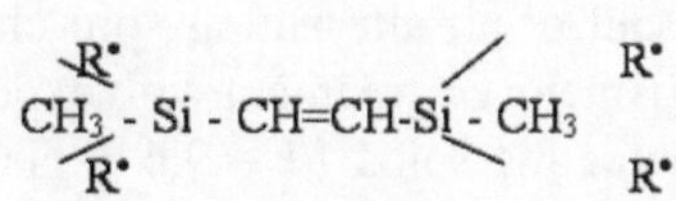

IV

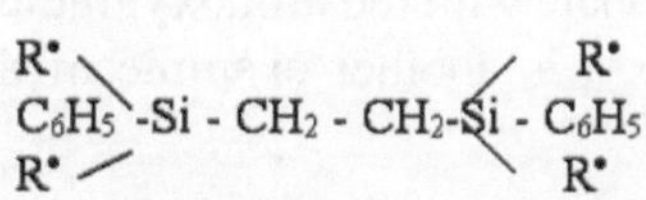

V

(R˙≡ 2,2,6,6-tetramethyl-1-oxyl-4-piperidyl fragment) paired spin-spin interactions are characteristic. All the studied paramagnets (I–V) exhibit low-temperature deviations of the course of their magnetic susceptibility from Curie's law χ= const/T (Fig. 5.5a), which are due to the existence of correlation between uncoupled electrons. E.g., the susceptibility of tetraradical V passes through a maximum at 8 K and decreases by 10 times when temperature falls down to 2 K (Fig. 5.5b). Such a course of susceptibility is well described by a model offered for paired exchange interactions of uncoupled electrons χ= $CT^{-1}[3 + \exp(J/kT)]^{-1}$.

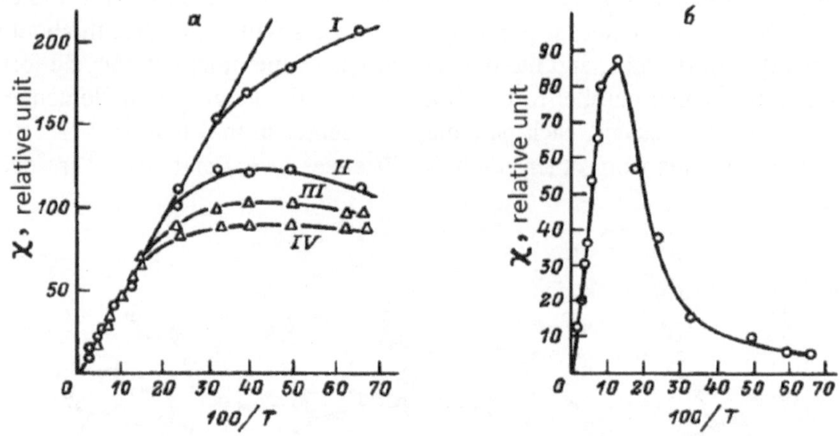

FIGURE 5.5 Magnetic susceptibility of polyradicals I–IV and V, respectively, graphs *a* and *b*.

If the ground state of such a couple is a singlet and the thermally excited state is a triplet (or a triplet magnetic exciton), it mainly contribute to the magnetic susceptibility. Excitons get energy for their excitation from thermal energy. Therefore, when kT becomes less than the exchange value J between electrons, the number of triplet states sharply falls and, hence, the susceptibility sharply decreases. From analysis of the course of susceptibility the exchange parameters of strongly bound spins (see Table 5.2) were estimated.

TABLE 5.2 Exchange Interaction Parameters of Silicon-Organic Polyradicals

Radical	I	II	III	IV	V
J/k, K	2.2±0.5	3.2±0.5	4.6±0.5	5.20.5	14.6±0.5

It is interesting that for tetraradical V the exchange interaction parameter J'/k between spin couples is 0.1 K.

The crystal structure of organic radicals, because of the asymmetry of the majority of chemical particles, as a rule, allows one to resolve topological linear chains of most strongly interacting spins. A study of the structure of, for example, radical V (Fig. 5.6) has shown that the nitrogen atoms of one radical heterocycle form a chain of paramagnetic centers with a link length about 6 nm parallel to the axis, and the nitrogen atoms of the other heterocycle form another spin chain parallel to the first one with the length of an elementary link of 6.6 nm. Besides, each paramagnetic center in the chain thus has two neighboring spins from other chains at distances of 6.4 and 6.6 nm, respectively.

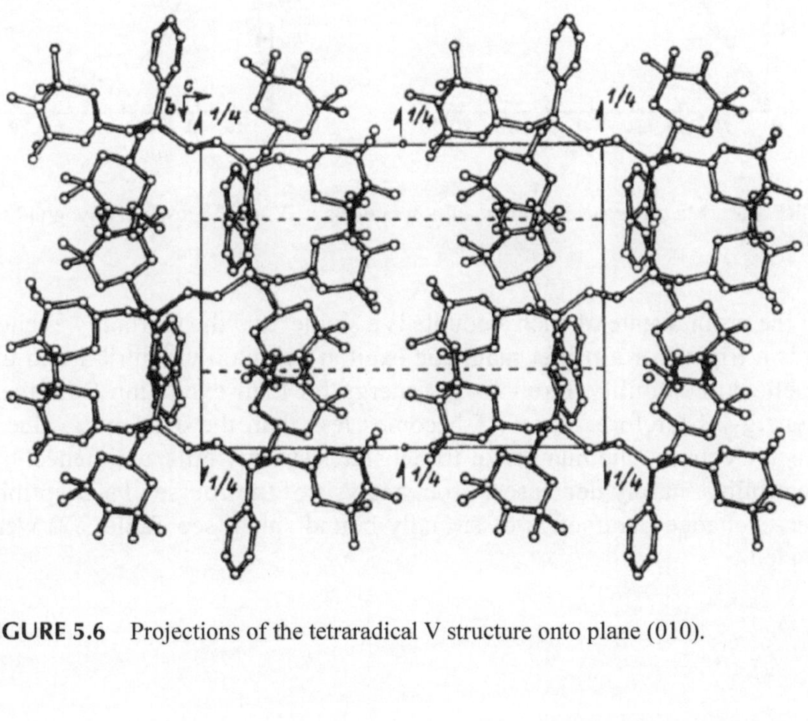

FIGURE 5.6 Projections of the tetraradical V structure onto plane (010).

In Ref. [15], the paired intermolecular interaction of the basic spin system in 1,4-*bis*-2,2,6,6-tetramethyl-1-oxyl-4-piperidyl-butane crystals was reported. This interaction is distinctly seen on the temperature dependence curve within a range of 10–300 K as the presence of a characteristic maximum near 40 K:

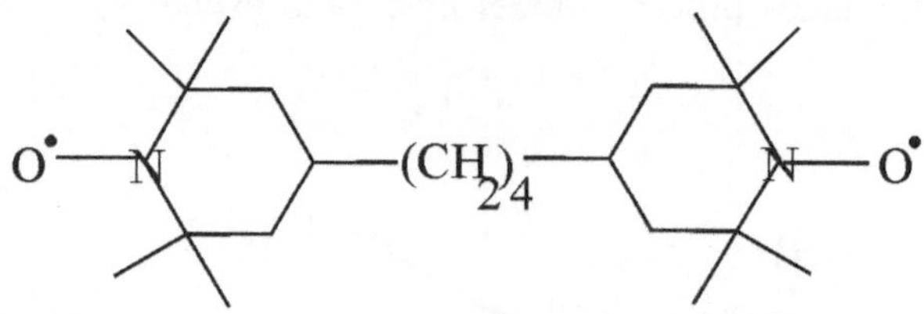

The monoclinic crystals of this biradical have the elementary crystal cell parameters $a = 11.754$, $b = 10.980$, $c = 8.693$ nm with the P_12/b spatial group (Fig. 5.7).

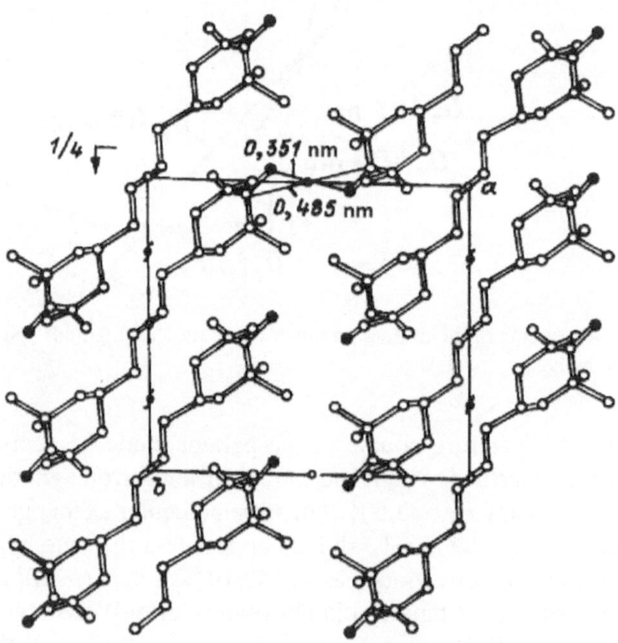

FIGURE 5.7 Projection of the 1, 4-*bis*-2, 2.6, 6-tetramethyl-1-oxyl-4-piperidyl butane biradical structure.

This structure features the existence of two systems of pairs of =NO• radical fragments which are mirror symmetric about the *ab* plane. For the mirror symmetric couples the angle between the lines connecting the centers of the iminoxyl fragments is 50°. Inside each couple, the oxygen atoms are at a distance of 0.351 nm, and the nitrogen atoms are at a distance of 0.485 nm. The short distance between the NO fragments in a couple and the relative location of the C–N–C planes promote direct electronic exchange in these couples (Fig. 5.8) [16, 17].

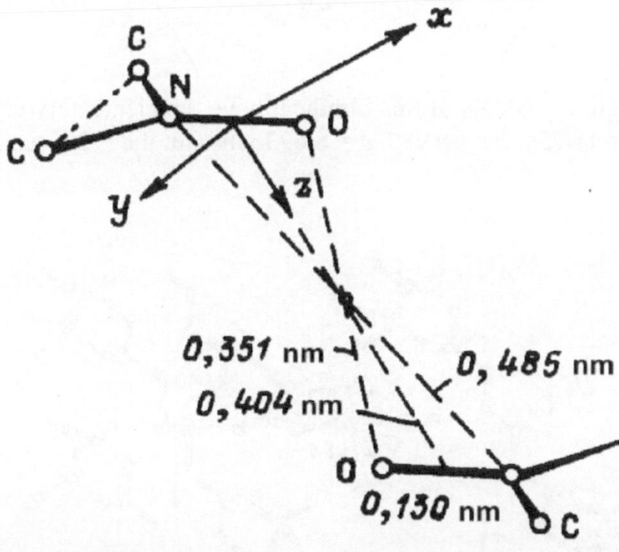

FIGURE 5.8 A scheme of the mutual arrangement of radical fragment =NO• pairs bound with strong exchange.

Really, the temperature course of the paramagnetic susceptibility in the crystals is well described within the model of antiferromagnetic paired exchange with a constant $J = -33.5$ K. The intramolecular exchange interactions J' transferred through the -$(CH_2)_4$-bonds appear less than the hyperfine coupling constant, which corresponds to $J'J' < 2 \cdot 10^{-3}$K, i.e. $J/J' > 10^4$.

The tanolic ester of octanoic acid obeys the Curie–Weiss law with a positive constant $\theta = +1$ K in a temperature range 1.9–300 K. All the magnetic interactions of interest are rather weak and manifest themselves at temperatures below 1 K only. Apparently, a magnetic transition at $T = 0.38 \pm 0.01$ K proceeds in the system due to ferromagnetic ordering. Neutronography has

established that the crystals of this paramagnetic are layered: the neighboring particles inside each layer are bound ferromagnetically with $J_1 = +1.1$ K and $J_2 = 0.07$ K, but the layers are connected among themselves by weaker antiferromagnetic interactions with $J' = -0.015$ K. It is believed [18] that the substance behaves as a metamagnetic with 2D ferromagnetic ordering. For 2,2,6,6-tetramethyl-4-oxo piperidyl-1-oxyl azine (TEMPAD), a maximum at 16.5 K is found on the curve of the temperature course of paramagnetic susceptibility, and its change under the Curie–Weiss law ($\theta = -15$K) is observed within 77–273K. In the case of diluted TEMPAD crystals, two values of the Weiss constant have been found: about −10 K in the high-temperature range and about −1 K in the low-temperature one (Fig. 5.9) [19].

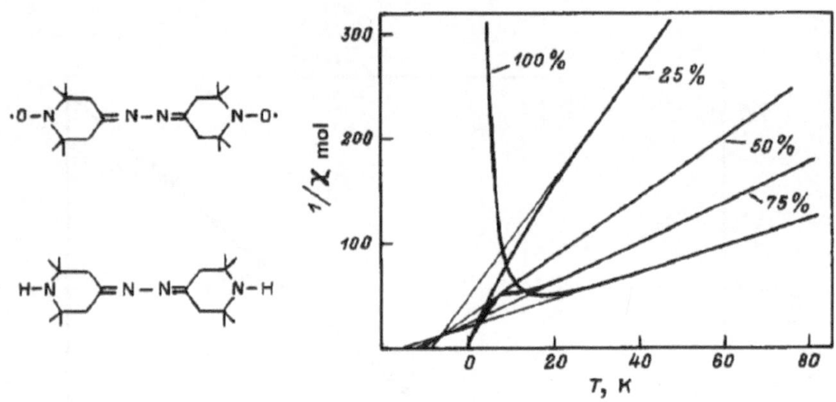

FIGURE 5.9 Temperature dependence of the inverse paramagnetic susceptibility of 2, 2, 6, 6-tetramethyl-4-oxopiperidine-1-oxyl azine crystals in a matrix of triacetonamine azine.

The magnetic behavior of TEMPAD was interpreted within the theory of magnetic triplet paired transitions [46–48]. Nobody can exclude the existence of strong intermolecular exchange interaction along the *a* axis (J_1; $J_1/k = -12.8$ K), weak intramolecular interaction (J_2; $J_2/k \sim 210^{-2}$ K), and interlayer interaction with $J_1/k \sim 1$ K [20].

In 2, 2, 6, 6-tetramethyl-4-oxypiperidyl-1-oxyl phosphite (TEMPOP)

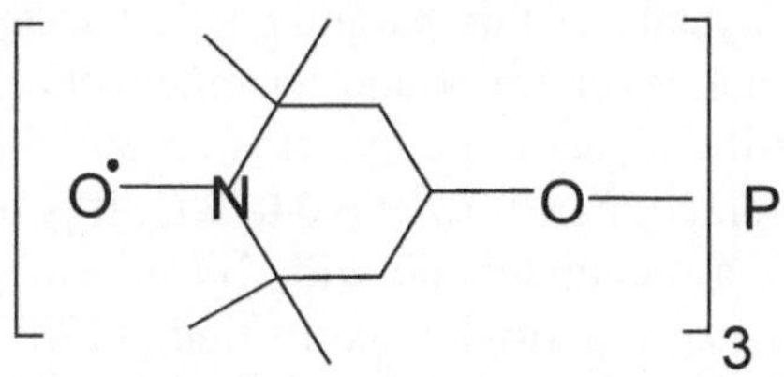

no near order of interactions was found, though at very low temperatures the course of inverse paramagnetic susceptibility deviated from the Curie–Weiss law (Fig. 5.10). Theoretically, the effective magnetic moment of three noninteracting spins with $g = 2.00$ should be equal to 3.00 μ_B; in experiment (at high temperatures), a value of 3.01μ_B was obtained.

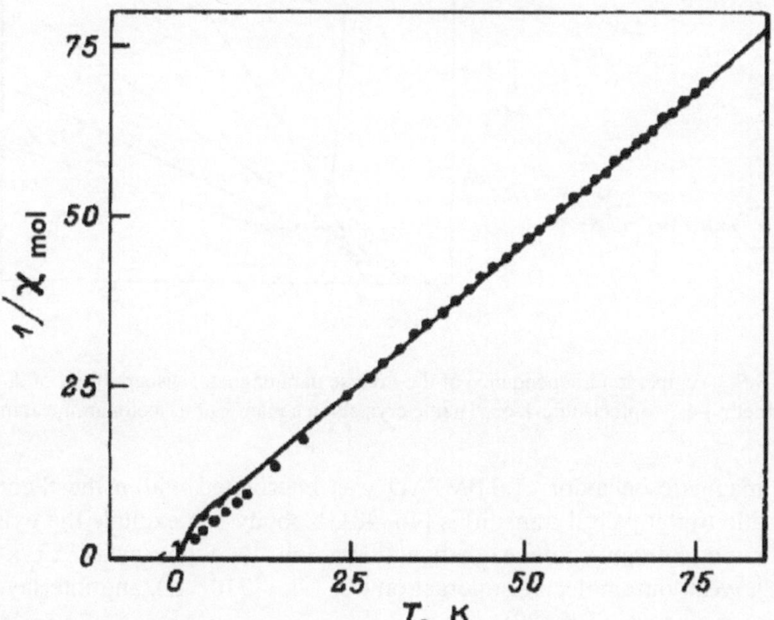

FIGURE 5.10 Temperature dependence of the inverse paramagnetic susceptibility of 2,2,6,6-tetramethyl-4-hydroxypiperidyl-1-oxyl phosphite ($\theta = -3.5$ K). The solid line is calculated for a three-spin cluster with J = −2.7 K.

Interesting studies on nitronyl nitroxyls (or nitroxide nitroxyls) NIT(R) ≡ 2R-4,4,5.5-tetramethyl-4,5-dihydro-1H-imidazolyl-1-oxyl-3-oxides

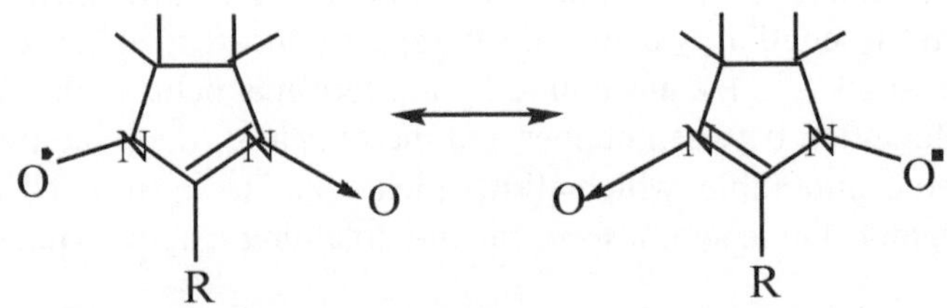

with metal ions are presented in Refs. [21–32]. Owing to the conjugation of a nitroxyl group with a nitroxide one, the exchange interaction between one oxygen atom and a metal ion can be transferred to another oxygen atom without attenuation. Therefore, these radicals with a delocalized uncoupled electron are capable of forming not only mononuclear complexes with metals [21–23] but also magnetic chains of various natures [24–27].

There exist a metal-containing compound Cu(hfac)$_2$(NIT)Me, where **hfac** is a hexaftoracetylacetonate ion [CF$_3$C(0)CHC(0)CF$_3$]$^-$, behaves as a one-dimensional ferromagnetic with an interaction constant of 25.7 cm^{-1}. The effective magnetic moment equal to 2.8 μ_B at 300 K monotonously increases with decreasing temperature and gets 4.9 μ_B at 4 K. This means that the effective spin of the system increases almost up to 3.

Replacement of a copper ion by Ni (II) or Mn (II) in the Ni(hfac)$_2$(NIT)R and Mn(hfac)$_2$(NIT)R complexes with R = Me, Et, i-Pr, n-Pr, Ph leads to stronger antiferromagnetic-type interactions with $J = -424$ cm^{-1} for nickel and $J = -230–330$ cm^{-1} for manganese derivatives. When exploring Mn(hfac)$_2$(NIT) iPr monocrystals, noticeable anisotropy was discovered. As follows from the temperature dependence of susceptibility along the easy magnetization axis (this direction in the crystal coincides with the spin direction orientation), phase transition to a ferromagnetic state occurs at 7.6 K.

In Mn(hfac)$_2$(NIT)Et, Mn(hfac)$_2$(NIT)nPr and Ni(hfac)$_2$(NIT)Me ferromagnetic ordering occurs at 8.1, 8.6, and 5.3 K, respectively. Calculations confirm the dipole-dipole nature of the magnetic interactions in these compounds. Higher temperatures of magnetic phase transition were found for [Mn(F$_5$benz)$_2$]$_2$ NITEt and [Me(F$_5$benz)$_2$]$_2$NITMe (where F$_5$benz is pentafluorobenzoate): 20.5 and 24 K, respectively. But there is no unambiguous confidence which type (ferrimagnetic or weak ferromagnetic) this ordering belongs.

The binuclear complex [CuCl$_2$(NITpPy)$_2$l$_2$], where NITpPy ≡ 2-(4-piridyl)-4,4,5,5-tetramethylimidizolinyl-3-oxide-l-oxyl, was studied. Potentially, the radical NITpFy could be a tridentate ligand. The copper ions are shown to be coordinated with two nitrogen atoms in the pyridine rings and three chlorine atoms, two of which are bridge ones. The NO groups of radicals belonging to

different mononuclear fragments are rather close to each other. On the basis of magnetic susceptibility data and EPR spectra recorded at 4.2 K, it is supposed that six spins $S = 1/2$ are bound by antiferromagnetic exchange interaction. The interaction between copper and the radical through the pyridine cycle's nitrogen is preferable, which allows the authors to consider NITpFy as promising ligands in the synthesis of metal-containing magnetic materials.

5.3 ORGANIC LOW-MOLECULAR-WEIGHT AND HIGH-MOLECULAR-WEIGHT MAGNETS

The interest to low-molecular magnets [33–40] and high-spin compounds [41–57] is associated with the hope of obtaining compounds possessing spontaneous magnetization below their critical temperature. Though the critical temperatures reached are quite low, it is possible to state with confidence that the understanding of the necessary conditions for the design of a high-temperature organic magnetic material has become clearer than several years ago.

Wide-range studies on ion-radical salts like $D^{\cdot+}A^{\cdot-}D^{\cdot+}A^{\cdot-}$, where D is a cation (donor) and A is an anion (acceptor), were carried out by Miller et al. [33–40]. Docamethylferrocene $Fe(II)(C_5\ Me_5)_2$ is often used as a donor and flat 7,7,8,8-tetracyan-*p*-quinodimethane (TCNQ) and tetracyanoethylene (TCNE) serve as acceptors

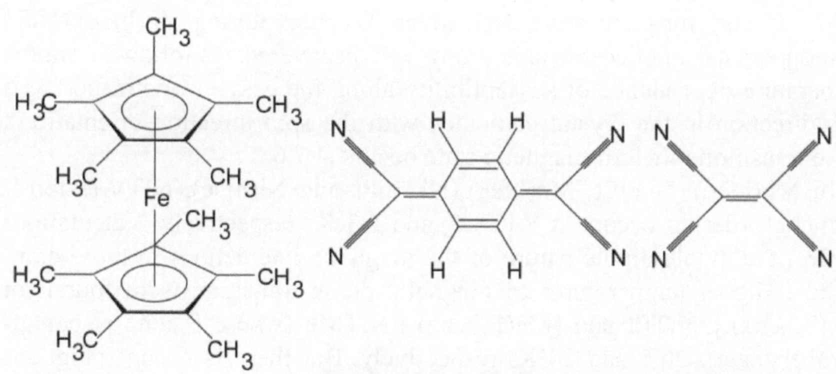

TCNQ TCNE

The complex $[Fe(C_5Me_5)_2]^{\cdot+}[TCNE]^{\cdot+}$ is characterized by a positive Curie–Weiss constant, $\theta = +30$ K, Curie's temperature (T_c) equal to 4.3 K; in a zero magnetic field, spontaneous magnetization is observed for a polycrystalline sample (M~ 2000 emu·Gs/mol) [35]. The saturation magnetization is 16300

emu·Gs/mol in oriented monocrystals. This result is in good agreement with the theoretical magnetic saturation moment at ferromagnetic spin alignment of the donor and acceptor and is by 36% higher than for metal iron (per one gram atom). At 2 K, hysteresis with a coercive force of 1000 Gs is observed, corresponding to the values for magnetically hard materials. Above 16 K, the magnetic properties are described by Heisenberg's one-dimensional model with ferromagnetic interaction ($J = +27.4$ K). At temperatures near T_c, three-dimensional ordering prevails [36].

The compound $[Fe(C_5Me_5)_2]^{+}[TCNQ]^{\cdot-}$ shows metamagnetic signs with a Néel temperature $T_N = 2.55$ K and a critical field of ~ 1600 Gs. As a rule, metamagnetics are substances with strong anisotropy and, in the presence of concurrent interactions therein, first-order transition to a phase with a total magnetic moment can be observed [1]. E.g., for the salt $[TCNQ]^{\cdot-}$ the magnetization in fields with $H < 1600$ Gs is characteristic of a antiferromagnetic, while when $H > 1600$ Gs, a hump-like increase of magnetization occurs up to the saturation value, which is characteristic of the ferromagnetic state [37].

Of the 2,5-disubstituted TCNQ salts of decamethylferripinium $[Fe(C_5Me_5)_2]^{+}[TCNQR_2]^{\cdot+}$ (R = Cl, Br, J, Me, OMe, OPh) [39], $[Fe(C_5Me_5)_2]^{+}[TCNQI_2]^{\cdot-}$ possesses the highest effective moment $_{eff} = 3.96\mu_B$. Above 60 K the magnetic susceptibility obeys the Curie–Weiss law ($\theta = +9.5$ K) and the substance is a one-dimensional ferromagnetic. This feature, in combination with that $[TCNQI_2]^{\cdot-}$ exhibits stronger interchain antiferromagnetic interactions in comparison with $[Fe(C_5Me_5)_2]^{+}[TCNQ]^{\cdot-}$, provides no 3D ferromagnetic ground state at temperatures above 2.5 K.

The complex [40]

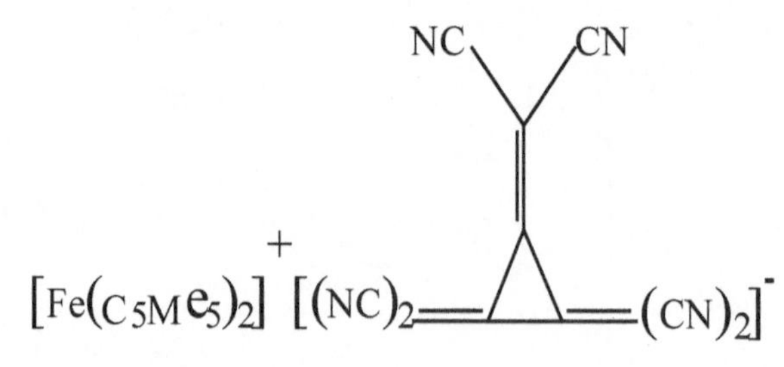

can exist in two polymorphic modifications, namely, monoclinic and triclinic, both obeying the Curie–Weiss law with $\theta = -3.4$ K, $\mu_{eff} = 2.98\ \mu_B$, and $\theta = -3.4$ K, $\mu_{eff} = 3.10\ \mu_B$, respectively. Below 40 K, in a magnetic field of 30 Gs the monoclinic compound shows the Bonner–Fisher type of one-dimensional antiferromagnetic interaction, i.e. has a typical flat maximum about 4 K. This

is attributed to antiferromagnetic interaction along the cation chains with an exchange parameter of $J/k = -2.75$ K. To explain the magnetism of ion radical salts, the model of configuration interaction of the virtual triplet excited state with the ground state offered by McConnell [57] is applied. E.g., in the case of donor-acceptor pair $D^{\cdot+}A^{\cdot-}$ it is supposed that the wave function of the ground state has the maximum "impurity" to the wave function of the lower virtual excited state with charge transfer. This state can arise due to direct virtual charge transfer $(D^{\cdot+} + A^{\cdot-}) \rightarrow (D^{2+} + A^{2-})$, reverse charge transfer $(D^{\cdot+} + A^{\cdot-} \rightarrow D^0 + A^0)$ or disproportionation $(2D^{\cdot+} \rightarrow D^{2+} + D^0)$.

If any of the states with charge transfer (either donor D or acceptor A, but not both) is triplet, the ground ferromagnetic state of the $D^{\cdot+}A^{\cdot-}$ pair will be stabilized. Therefore, for ferromagnetism manifestation, an organic radical should possess a degenerated and partially filled valent orbital. An essential contribution of the lower virtual excited state with charge transfer to the ground state of the system is necessary; and the structure of the radical ion should be high-symmetric, without any structural or electronic dislocations breaking the symmetry and eliminating the degeneration [33, 38].

A whole series of high-spin polycarbenes has been so far synthesized by means of photolysis of the corresponding polydiazo compounds [41–47]. Attempts to get high-spin macromolecules by iodine oxidation of 1, 3, 5-triaminobenzene have failed. Breslow et al. [57–59] have succeeded to synthesize stable organic triplet systems with C_3 and higher symmetry on the basis of hexaminotriphenylene and hexaaminobenzene derivatives.

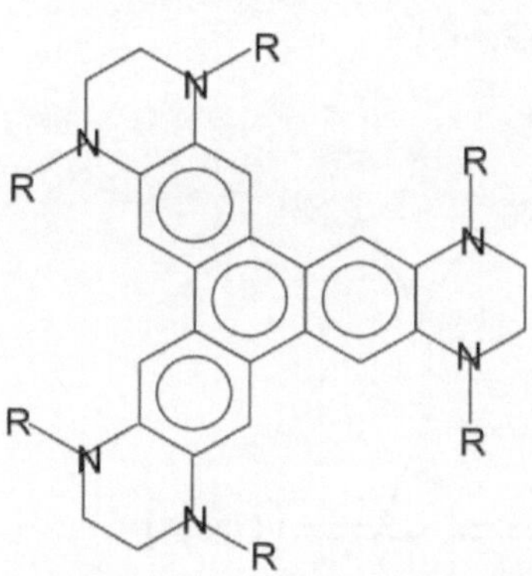

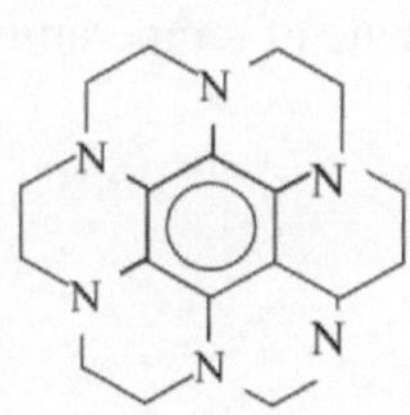

where R = C_2H_5, $C_6H_5CH_2$, CF_3CH_2,

where R = C_2H_5, $C_6H_5CH_2$, CF_3CH_2.

Studies on the material obtained by spontaneous polymerization of diacetylene monomer containing stable iminoxyl fragments of butadiyn-*bis*-2,2,6,6-tetramethyl-1-oxyl-4-oxi-4-piperidyl (BIPO) is a highly mysterious story...

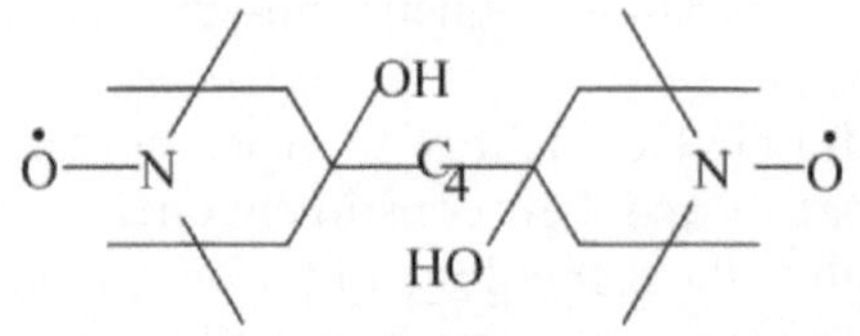

The magnetic susceptibility of BIPO [60] obeys the Curie–Weiss law with $\theta = -1.8$ K. The effective magnetic moment equal to 2.45 μ_B at high temperatures corresponds to two independent spins of $S = 1/2$ per monomer unit. The exchange constant derived from analysis of the EPR line has appeared to be $J \sim 0.165$ K (0.115 cm^{-1}), and an estimation in the approach of molecular field has given a value of $J \sim 0.155$ K (0.108 cm^{-1}).

The thermal or photochemical polymerization of BIPO leads to the formation of black powder whose insignificant fraction (0.1%) shows ferromagnetic properties, its magnetization reaches above 1 $Gs \cdot g^{-1}$. It is noted that ferromagnetism holds up to abnormally high temperatures (up to 200–300 °C) and the paramagnetic centers thus die during polymerization (in some cases no more than 10% of their initial quantity remains).

Contrary to the earlier published analysis of these intriguing data, in a subsequent work [60] it was noted that the products of thermal decomposition of BIPO showed neither signs of 3D ferromagnetism nor magnetic interaction. Detailed static and dynamic magnetic data indicate the existence of weak intradimeric ferromagnetic (triplet) interaction with J ~ 10K only.

Obviously, while solving this problem, the degree of reliability of obtained results will strongly depend on the chemical purity of the materials studied. It is possible to state without exaggeration that natural sciences progress is associated with obtaining and studying chemically pure materials.

Unfortunately, even superficial analysis of the available publications convinces us that the majority of experimental works in this field is associated with studying of structurally disordered "dirty" systems like spin glasses [61] with no coordinated magnetic interactions between chemical particles. The relative simplicity of obtaining "dirty systems" provokes the avalanche-like spreading of "impressive results," various fantastic models and theories, having nothing in common with true science.

It would be thoughtless to consider that chemical purity is sufficient to achieve success in the basic research of high-spin nonmetallic systems. Precision measuring equipment and a methodology including automated X-ray diffraction analysis and modern magnetometry with the usage of superconducting quantum interferometers (squids), without being limited to EPR

equipment and high-temperature magnetic measurements, are, undoubtedly, other necessary conditions.

Only the successful development of the basic research of the magnetic properties of pure systems and their constituent chemical particles can provide real breakthrough in the technology of the design of materials of a new generation suitable for manufacturing competitive organic ferromagnets, antiferromagnets, and ferrimagnets, including metamagnets and speromagnets.

In 1990, Emsley [62] published a paper under an intriguing title where he reported about the synthesis of a stable iminoxyl radical (nitroxide nitroxyl triradical) with its properties of a "molecular" organic magnet. In other words, the discovery of a metal-free "organic magneton" with cooperatively ordered electronic interactions at the level of a discrete chemical particle was claimed:

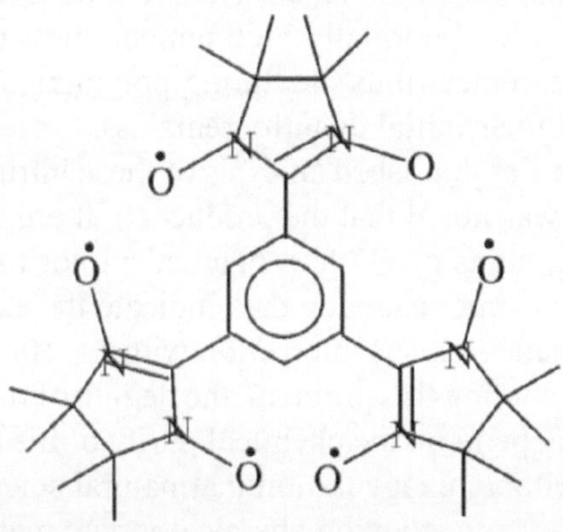

Dulog and Kim [63] have found that some blue powder obtained by them possesses a high value of magnetic susceptibility and can strengthen or weaken the intensity of the applied magnetic field like metal magnets. Provided that the remarkable properties of blue trinitroxyl are not a trivial consequence of metallic pollution, the new material will be able to find applications when designing magnetic registering devices of a new generation, magnetoplanes, and other equipment.

Using the principle of orienting effect of the intraradical electrostatic field of nitroxide groups, it could be possible to design high-molecular-weight magnetic materials with magneto-ordered organic domains (magnetons) like blue nitroxide nitroxyl triradical of Stuttgart's chemists as monomeric links therein.

Stable paramagnets have found practical applications as additives to polarized proton targets in the experimental physics of high energy. The method of reaching ultralow temperatures by ^3He dissolution in ^4He opens new opportunities in the technology of polarized targets. For example, the high polarization obtained by a usual dynamic method in a strong and uniform magnetic field (25kOe) can be kept for a long time after the termination of the dynamic polarization "pumping" if the working substance of the target is cooled rather quickly down to a temperature about 0.1÷0.01 K. Then, the intensity of magnetic field can be lowered down to ~5kOe. This opens new prospects of the use of such targets in physical experiments.

In the existing polarized proton targets operating at temperatures as low as 0.5K, the main working substances are butyl alcohol and ethylene glycol as frozen balls. However, these substances are not technological for their usage in cryostats with ^3He dissolution in ^4He.

A substance, solid at room temperature, rather rich with protons, and containing radicals stable at room temperature as paramagnetic additives, would be most convenient. Therefore, polyethylene used as either a 200 mμ film or powder with ~200 mμ grains was selected as the working substance on the basis of recommendations from Refs. Stable iminoxyl radicals were taken as a paramagnetic additive. To introduce such a radical into polyethylene, the necessary amount of the radical and polyethylene was placed into a tight glass ampoule, heated up to 80°C, and maintained at this temperature for 8–10 h. To study proton polarization in polyethylene, preliminary experiments were conducted at a temperature of 1.3 K in magnetic fields of 13 and 27 kOe. The stable volatile radical 2,2,6,6-tetramethyl-piperidine-1-oxyl appeared most promising.

At an optimum concentration of the radical in a magnetic field of 13kOe in a polyethylene film, a polarization of 5–7% has been reached, which corresponds to an amplification polarization factor of $E = 50 \div 70$. The period T_1 of proton spin-lattice relaxation was 2.5 min. Transfer into a magnetic field of 27kOe led to no increase in the polarization amplification factor, only T_1 increased. However, it was revealed that the use of polyethylene of lower density and higher purity as powder with 200 mμ grains led to an increase in E up to 70÷100. Thus, a 14–20% polarization was achieved in the field of 27 kOe at $T = 1.3$ K.

In experiments at ultralow temperatures, polyethylene powder was exposed to preliminary annealing in vacuum (10^{-5} mm Hg) at 80°C during 5–6 days. Then, the ampoule with this powder was filled with pure gaseous helium and the powder was saturated by the vapors of the radical in a helium atmosphere. Such an annealing procedure does not influence the rate of spin-lattice

relaxation considerably, but increases the final polarization almost twice. The sample prepared in this amount (150 mg) was introduced, at room temperature, into a glass camera of ^3He dissolution in ^4He located in a running-wave microwave cell. The cell was placed into a superconducting solenoid and cooled down to a temperature of 1.3 K. The sample was in direct contact with the solution whose minimum temperature was 0.04–0.05K.

For polarization pumping, a microwave generator of the OV-13 type (λ= 4 mm) with a power about 70 mW was used; bringing 1.5–2.0 mW into the cryostat was enough to increase the solution temperature up to 0.1 K.

Proton polarization up to 50% was reached in polyethylene samples as powder with an optimum concentration of the radical of 10^{-19} spins per polymer gram. The period of polarization pumping was about 3.5 h. After switching the microwave field off, the temperature of the ^3He–^4He system decreased down to 0.05 K. At this temperature in a magnetic field of 27 kOe almost no polarization decay was observed. In a magnetic field lowered down to 5 kOe, the relaxation time was not less than 30 h and only with the magnetic field decreased down to 1.5 kOe it reduced to 1.5 h.

The EPR spectrum of tetramethylpiperidine-1-oxyl introduced into polyethylene in the above quantity at room temperature represents a well-resolved triplet with a distance between the hyperfine coupling components of 15 kOe. At temperatures below 1 K the EPR spectrum is transformed to a line with its half-width of 80 Oe with an ill-defined superfine structure. To simplify the EPR structure, an attempt to replace the radical with ^{14}N by that with ^{15}N in the same concentration was made. However, such replacement gave no increase in the maximum polarization.

The experiments conducted have shown that now there is a real possibility to create a "frozen" polarized proton targets in a magnetic field about 5 kOe.

Now, in geophysics and astronautics, nuclear precession magnetometers possessing a number of essential advantages are widely adopted. Their high sensitivity (to 0.01 gamma) and the accuracy of measurements, the absoluteness of indications, and the independence of temperature, pressure and sensor orientation are advantages of nuclear magnetometers.

Cyclic operation is a feature of precession magnetometers. The process of measurement consists of two consecutive processes, namely: polarization of the working substance of the magnetometer sensor during which nuclear magnetization is established, and measurements of the signal frequency of the nuclear induction determining the absolute value of the field measured.

The use of the phenomenon of dynamic polarization of atomic nuclei allows one to overlap the stages of polarization and measurement, and to essentially increase the speed of the magnetometer. Nuclear generators based on the

phenomenon of dynamic polarization of atomic nuclei, allowing continuous monitoring of changes in the magnetic field were designed.

Earlier, only Fremy's diamagnetic salt dissociating into paramagnetic anions in aqueous solution was used as the working substance of the sensors of nuclear magnetometers based on the phenomenon of dynamic nuclei polarization.

Saturation of any hyperfine coupling line in the EPR spectrum of Fremy's salt solution leads to a significant increase in the nuclear magnetization of the solvent. This is the effect of dynamic polarization. The hydrolytic instability is a demerit of Fremy's salt. Even in distilled water, the radical anion of this salt is hydrolyzed to diamagnetic products during several dozen minutes, the process of degradation having autocatalytic character. A paramagnetic solution stabilized with an additive of potassium carbonate preserves about a month provided that its temperature would not exceed 40 °C. Stable paramagnets of the iminoxyl class have indisputable advantages over Fremy's salt.

Rozantsev and Stepanov [64] proposed 2,2,6,6-tetramethyl-4-oxipiperidine-1-oxyl and 2,2,6,6-tetramethyl-4-oxopiperidine-1-oxyl well soluble in many proton-containing solvents and possessing a resolved hyperfine coupling in their EPR spectra within a wide range of magnetic fields as working substances for nuclear precession magnetometers in 1965.

In weak magnetic fields, the bond between the electronic and nuclear spins of nitrogen is not broken off, and the set of energy levels is characterized by a total spin number S taking on two values (1/2 and 3/2) and a magnetic quantum number ms taking on values $2S + 1$. The set of energy levels of iminoxyl radicals in weak magnetic fields is presented in Fig. 5.11.

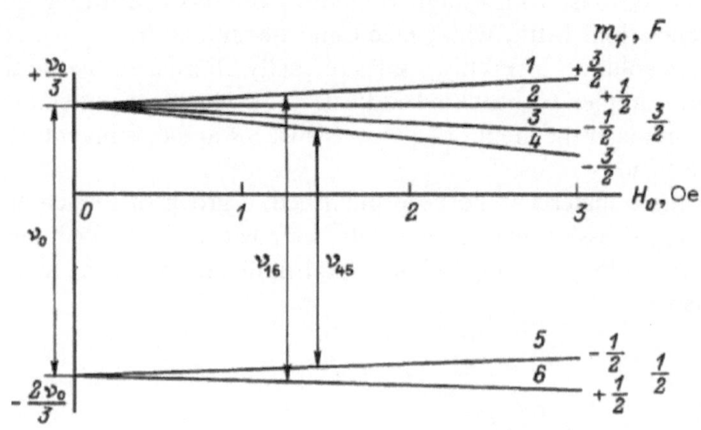

FIGURE 5.11 A scheme of the energy levels of iminoxyl radicals in weak magnetic fields.

The rule of selection $\Delta S = 0\pm1$ and $\Delta m_s = \pm1$ exists for π-electronic transitions. Saturation of one of such transitions by a strong radio-frequency field oriented perpendicular to the constant field significantly changes the electronic magnetization M_z of the solution, which, in turn, results in an increase in the proton magnetization m_z of solvent molecules according to the expression

$$m_z - m_o = fg\,(M_o - M_z),$$

where m_z, m_o, M_o, M_z are the nuclear and electronic magnetizations of the solution in a stationary mode and thermal equilibrium at saturation of the transition in the system of electronic energy levels, respectively, f is a coefficient determining the contribution of electronic–nuclear interaction into the mechanism of nuclear relaxation of the solvent, g a coefficient depending on the nature of the electronic–nuclear interaction.

Studying of iminoxyl radicals has shown that only saturation of the 1–6 and 4–5 transitions (Fig. 5.11) leads to a considerable dynamic polarization of the nuclei of the solvent. By the value of proton dynamic polarization in solutions of iminoxyl paramagnets the latter ones are on a par with Fremy's salt applied earlier. They were more stable in water and organic solvents and did not change their initial characteristics within half a year. In the course of research, solutions of iminoxyl paramagnets were repeatedly heated up to 90°C. It is obvious that organic paramagnets represented essentially new working substances for nuclear precession magnetometers of a new generation. The main advantage of these new working substances consists in the possibility to choose a solvent for them to operate in any climatic conditions, with a high content of protons and a long period of proton relaxation. Little wider electronic transitions in comparison with an aqueous solution of Fremy's salt are easily eliminated by the usage of deuterated samples of iminoxyl radicals [65]. Numerous attempts of the authors to register the mentioned idea as a USSR author's invention certificate were steadily refused.

After these materials had been published, a group of French engineers designed a precession nuclear magnetometer with a deuterated solution of 2,2,6,6-tetramethyl-4-oxopiperidine-1-oxyl in dimethoxyethane as its working substance.

KEYWORDS

- **Free radical**
- **Interaction**
- **Magnetic susceptibility**
- **Magnetism**
- **Magnetometer**
- **Paramagnetic**
- **Properties**

REFERENCES

1. Karlin, R. (1989). Magneto-chemistry Moscow: World, [Russ].
2. McConnell, H. M. (1963). *J. Chem. Phys.,* 39, 1910.
3. Buchachenko, A. L. (1979). Reports of USSR Science Academy 244, 1146. [Russ].
4. Izuoka, A., Murata, S., Sugavara, T., & Iwamura, H. J. (1987). Amer. Chem. Soc., *109,* 2631.
5. Williams, D. (1969). *Vol. Phys. 16.* 1451.
6. Mukai, K., Mishina, T., Ishisu, K., & Deguchi, Y. (1977). Bull. Chem. Soc. Jpn., *50*, 641.
7. Agawa, K., Sumano, T., & Kinoshita, M. (1986). *Chem. Phys. Lett. 128,* 587.
8. Agava, K., Sumano, T., & Kinoshita, M. (1987). *Chem. Phys. Lett. 141*, 540.
9. Azuma, N., Yamauchi, J., Mukai, K., Ohya-Nisciguchi, H., & Deguchi, Y. (1973). Bull. Chem. Soc. Jpn., 46, 2728.
10. Ballester, M., Riera, J., Castaner, J., & Vonso, J. J. (1971). Amer. Chem. Soc., *93*, 2215.
11. Rozantsev, E. G., Goldfein, M. D., & Pulin, V. F. (2000). *The Organic Paramagnets* Saratov State University Russia [Russ].
12. Rozantsev, E. G., Sholle, V. D. (1979). The Organic Chemistry of Free Radicals Moscow Chemistry [Russ].
13. Karimov, U. S., & Rozantsev, E. G. (1966). *Physics of Solid State, 8*, 2787 [Russ].
14. Nakajima, A., Ohya-Nisciguchi, H., & Deguchi, Y. (1985). *Bull. Chem. Soc. Jpn., 107*, 2560.
15. Benelli, C., Gatteschi, D., Carnegie, J., & Carlin, R. (1983). *J. Amer. Chem. Soc. 105*, 2760.
16. Zvarykina, A. V., Stryukov, V. B., Fedutin, D. N., & Shapiro, A. B. (1974). Letters in JET, Ph *19* 3 [Russ].
17. Zvarykina, A. V., Stryukov, V. B., Umansky, S. Ya., Fedutin, D. N., Shibaeva, R. P., & Rozantsev, E. G. (1974). Reports in USSR Academy, 216, 1091 [Russ].
18. Stryukov, V. B., & Fedutin, D. N. (1974). *Physics of Solid State, 16,* 2942 [Russ].
19. Rassat, A. (1990). *Pure and Appl. Chem., 62*, 223
20. Nakajima, A. (1973). *Bull. Chem. Soc. Jpn., 46*, 779.
21. Nakajima, A., Yamauchi, J., Ohea-Nisciguchi, H., & Deguchi Deguchi, Y. (1976). *Bull. Chem. Soc. Jpn. 49,* 886.
22. Caneschi, A., Gatteschi, D., Grand, A., Laugier, J., Rey, P., & Pardi, L. (1988). *In org Chem., 27*, 1031.
23. Caneschi, A., Laugier, J., Rey, P., & Zanchini, C. (1987). *In org Chem., 26*, 938.

24. Caneschi, A., Gatteschi, D., Laugier, J., Pardi, L., Rey, P., & Zanchini, C. (1988). *In org Chem., 27*, 2027.
25. Laugier, J., Rey, P., Bennelli, C., Gatteschi, D., & Zanchini, C. (1986). *J Amer. Chem. Soc., 108*, 5763.
26. Caneschi, A., Gatteschi, D., Laugier, J., & Rey, P. J. (1987). *Amer. Chem. Soc.*, 109, 2191.
27. Caneschi, A., Gatteschi, D., Rey, P., & Sessoli, R. (1988). *In org Chem., 27*, 1756.
28. Caneschi, A., Gatteschi, D., Laugier, J., Rey, P., & Sessoli, R. (1987). *In org. Chem., 27*, 1553.
29. Caneschi, A., Gatteschi, D., Renard, J., & Sessoli, R. (1989). *In org. Chem., 28*, 2940.
30. Caneschi, A., Gatteschi, D., Renard, J., Rey, P., & Sessoli, R. (1989). *In org Chem., 28*, 1976.
31. Caneschi, A., Gatteschi, D., Renard, J., Rey, P., & Sessoli, R. (1989). *In org. Chem., 28*, 3314.
32. Caneschi, A., Gatteschi, D., Renard, J., Rey, P., & Sessoli, R. (1989). *J. Amer. Chem. Soc., 111,* 785.
33. Caneschi, A., Ferraro, F., Gatteschi, D., Rey, P., & Sessoli, R. (1990). *Injrg. Chem 29*, 1756.
34. Miller, J., Epstein, A., & Reiff, W. (1988). *Chem Rev. 88,* 201.
35. Miller, J., Epstein, A., & Reiff, W. (1988). *Acc. Chem. Res., 21*, 114.
36. Miller, J., Calabrese, J., Rommelmann, Y., Chittapeddi, S., Zhang, J., Reiff, W., & Epstein, A. J. (1987). *Amer. Chem. Soc., 109,* 769.
37. Chittapeddi, S., Cromack, K., Miller, J., & Epstein, A. (1987). *Phys. Rev. Lett. 22*, 2695.
38. Candela, G., Swartzendruber, L., Miller, J., & Rice, M. (1979). *J Amer. Chem. Soc.*, 101, 2755.
39. Miller, J., & Epstein, A. J. (1987). *Amer. Chem. Soc., 109*, 3850.
40. Miller, J., Calabrese, J.,Harlow, R., Dixon, D., Zhang, J., Reiff, W., Chittapeddi, S., Selover, M., & Epstein, A. J. (1990). *Amer. Chem. Soc., 112,* 5496.
41. Miller, J., Ward, M., Zhang, J., & Reiff, W. (1990). *In org Chem., 29*, 4063.
42. Sugamara, T., Bandow, S., Kimura, K., Itamura, H., & Itoh, K., (1986). *J. Amer. Chem. Soc. 108*, 368.
43. Sugamara, T., Bandow, S., Kivura, K., H., Itamura, H., & Itoh, K. (1984). *J. Amer. Chem. Soc.* 106, 6449.
44. Itamura, H. (1986). *Pure and Appl. Chem. 58*, 187.
45. Teki, Y., Takui, T., Itoh, K., Iwamura, H., & Kobayaschi, K. (1983). *J. Amer. Chem. Soc.*, 105, 3722.
46. Teki, Y., Takui, T., Yagi, H., Itoh, K., & Imamura, H. J. (1985). *Chem. Phys. 83,* 539.
47. Sugamara, T., Murata, S., Kimura, K., & Itamura, H. J. (1985). *Amer. Chem. Soc., 107,* 5293.
47. Sugamara, T., Tukada, H., & Izuoka, A. J. (1986). Amer. Chem. Soc. *108*. 4272.
48. Magata, N. (1968). Theor. Chim Acta, 10, 372.
49. Itoh, K. (1971). *Bussei 12*, 635.
50. Torrance, J., Oostra, S., & Nazzal, A. (1987). *Synth Met 19*, 709.
51. Torrance, J., & Bugus, P. (1988). *J Appl. Phys. 63*, 2962.
52. Breslow, R. (1982). *Pure and Appl Chem., 54*, 927.
53. Breslow, R., Jaun, B., & Kluttz, R. (1982). *Tetrahedron, 38*, 863.
54. Breslow, R. (1982). *Pure and Appl. Chem. 54,* 927.
55. Breslow, R., & Maslak, P. J. (1984). Amer. Chem Soc., *106*, 6453.
56. Le Pade, T., & Breslow, R. (1987). *J. Amer Chem Soc., 109*, 6412.
57. Breslow, R. (1985). *Mol. Cryst. Liquid cryst., 125*, 261.
58. McConnell, H., & Proc, R. A. (1967). *Welch Found. Chem. Res*. 11, 144.

59. Miller, J., & Glatzhofer, D. et al. (1990). *Chemistry of Materials, 2*, 60.
60. Steyn, D. L., (1989). In a world of science, 9, 26. [Russ].
61. Emsley, J. (1990). New Scient 127(1727), 29.
62. Dulog, L., & Kim, J. S. (1990). *Angev. Chem. Bd. 29*, S. 415.
63. Rozantsev, E. G., & Stepanov, A. P. (1966). The Geophysics apparatus sankt-Peterburg, *Nedra, 29*, 35 [Russ].
64. Rozantsev, E. G. (1970). Freeiminoxylradicals, Moscow: *Chemistry*. [Russ].

CHAPTER 6

PHOTOELECTROCHEMICAL PROPERTIES OF THE FILMS OF EXTRA-COORDINATED TETRAPYRROLE COMPOUNDS AND THEIR RELATIONSHIP WITH THE QUANTUM CHEMICAL PARAMETERS OF THE MOLECULES

V. A. ILATOVSKY, G. V. SINKO, G. A. PTITSYN, and G. G. KOMISSAROV

CONTENTS

ABSTRACT

There were studied photoelectrochemical properties of thin films of 45 tetrapyrrole compounds with extraligands (ETPC), deposited on a platinum substrate by thermal sublimation in vacuum 10^{-6} Torr. The films have a cluster structure, where generally polycrystalline film (a-form) with a thickness of 50 nm is composed of differently oriented clusters (10–20 nm in size) with a monocrystalline structure. Quantum-chemical calculations of the electron density distribution in molecules of ETPC have been carried out, the results of which became the basis for the interpretation of experimental data. It is shown that, as compared with the complexes of bivalent metals, ETPC have significantly higher photoactivity. This is due to a change of the coordination of electron acceptors in completing the coordination sphere of metals and an increase in intermolecular interaction due to the overlapping of the electron clouds of the extraligands that come out of the plane of pigment molecules. The effect decreases on the formation of dimmers coupled by extraligands, and the photo activity falls to zero at diphthalocyanines of lanthanides, which can be regarded as a limiting form of extra coordination with a second phthalocyanine macrocycle as an extraligand. Most photo activity is observed in ETPC with a high degree of iconicity of the bond of a metal to a ligand and one extraligand having large electron affinity.

6.1 INTRODUCTION

In the study of photo electrochemical properties of aza-derivatives of tetrabenzoporphyrin [1] experiments and quantum-chemical calculations have shown a dramatic impact that changes in the electron density distribution in molecules by sequential aza-substitution have on photo potentials, photocurrents, quantum yield on a current of thin films (50 nm) of pigments studied. However, apart from the main trend (increase in the photo activity of the films at a gradual transition from the structure to porphyrin to the structure of phthalocyanine as a result of a fourfold aza-substitution of carbon by nitrogen in the mesaposition) there was a marked difference in the properties of derivatives of tetra-benzoporphyrin of iron (Fe-TBP) and its extracoordinated analog (ClFe-TBP). Probably the introduction of additional ligands in complexes with III-IV valence metals allows to have a strong impact on the electronic structure of molecules [2]. There is a possibility of purposeful changes in catalytic and semiconductor properties of pigment films by varying the intermolecular forces, the focal space of the central atom, and the coordination locations of the primary electron acceptors in photo electrochemical processes.

Previously, there was also repeatedly noted the increased photo activity of extracoordinated pigments in the oxygen reduction reaction in the study of the Becquerel effect and other photovoltaic phenomena in films of tetrapyrrole compounds (TPC) [3–5]. In this regard, we measured the photocurrents and photo potentials of 45 different extracoordinated TPC – porphyrins and phthalocyanines, which allowed to identify some general patterns of extra-ligands' impact on the characteristics of pigment films. Quantum-chemical calculations of the energy spectra and the electron density distribution of the molecules, complementing experiments, allow more arguments to interpret the results.

6.2 EXPERIMENTAL PART

Initial pigments had a concentration of impurities of not more than 10^{-3} wt%. They were further purified by sublimation at 10^{-6} Torr vacuum, which was carried out immediately before deposition of the pigments onto annealed at T = 1000°C platinum substrate with a diameter of 11 mm. Pigmented films (a-form) with a thickness of 50 nm was applied to the substrate by explosive thermal evaporation in vacuum 10^{-6} Torr. The substrates temperature at deposition was 180°C, the deposition rate of pigment – 1000 nm/s, the radiation pattern of the evaporator 2p radians. Detailed description of the method of manufacturing the electrodes is given in Refs. [6–8]. For each type of pigment there were manufactured 20 identical electrodes with variation of parameters not more than 15%.

Measuring of photocurrents (I_{ph}), photo potentials (U_{ph}) and dark values U_d, I_d were performed relative to a saturated Ag/AgCl electrode in a temperature controlled cuvette with electrodes holder, that eliminates electrolyte contact with the substrate. Variation of the pH of the electrolyte (1 M solution of aerated KCl) in the range from 1 to 12 was carried out by stepwise titration of the solutions by HCl or KOH. Periodic change of light regime to darkness with an exposure of at least 5 min provided an output signal to the steady-state values. Stabilized optical system gave a luminous flux of 100 mW/cm^2 in the plane of electrode with a spectrum close to the AMO-1. After statistical processing of the measurement results on the dependences U_{ph}(pH), I_{ph}(pH), U_d(pH), I_d(pH) there were determined maximum values of the electrodes' parameters and the pH at which they were obtained.

6.3 RESULTS AND DISCUSSION

The experimental data (standard deviation s = 3.2%) are shown in Table 6.1.

TABLE 6.1 Maximum Values of Photopotentials (U_{ph}) and Photocurrent (I_{ph}) of Extracoordinated Tetrapyrrole Compounds (ETPC), Phc – phthalocyanines, EP – etioporphyrin, TPhP – tetraphenylporphyrins, pH – pH of the Electrolyte At Which the Maximum Values of the Parameters Are Reached

№	ETPC	U_{ph} mV	pH	I_{ph} µA	pH	I_{ph}/I_d
1	InCl-Phc	560	4	62.0	2	80
2	InBr-Phc	575	4	67.0	2	92
3	InI-Phc	590	5	72.0	3	103
4	GaCl-Phc	450	3	38.0	2	76
5	GaBr-Phc	485	3	48.0	2	81
6	GaI-Phc	530	4	65.0	3	93
7	$GeCl_2$-Phc	210	2	9.0	1	52
8	$GeBr_2$-Phc	245	2	9.8	1	54
9	GeI_2-Phc	254	2	10.2	1	57
10	AlCl-Phc	450	2	40.0	2	70
11	AlBr-Phc	475	2	58.0	2	78
12	AlI-Phc	505	3	76.0	3	89
13	AlOH-Phc	200	3	2.8	1	30
14	Mn-Phc	230	1	6.5	1	28
15	MnCl-Phc	340	2	28.0	1	76
16	$MnCl_2$-Phc	290	2	8.8	1	40
17	$SiCl_2$-Phc	380	2	20.0	1	50
18	$Si(OH)_2$-Phc	160	2	3.2	1	20
19	$SnCl_2$-Phc	350	2	9.0	2	60
20	$Sn(OH)_2$-Phc	210	3	1.8	3	38
21	Fe-Phc	180	0	1.4	0	18
22	FeCl-Phc	250	1	5.4	1	26

TABLE 6.1 *(Continued)*

№	ETPC	U_{ph} mV	pH	I_{ph} μA	pH	I_{ph}/I_d
23	VO-Phc	350	1	28.0	0	75
24	$SnCl_2$-EP	300	2	5.0	1	42
25	VO-EP	400	2	6.0	1	40
26	AlCl-EP	400	3	10.0	2	50
27	AlBr-EP	410	3	12.0	2	54
28	AlOH-EP	180	2	1.5	1	43
29	$SnCl_2$-TPhP	350	1	7.0	1	60
30	$SiCl_2$-TPhP	360	1	7.5	1	65
31	Fe-TPhP	170	1	1.0	1	12
32	FeCl-TPhP	210	1	2.0	1	30
33	VO-TPhP	410	1	14.0	1	46
34	AlCl-TPhP	480	2	26.0	1	65
35	AlBr-TPhP	510	3	34.0	2	73
36	GaCl-TPhP	490	2	32.0	1	68
37	GaBr-TPhP	535	3	45.0	2	78
38	InCl-TPhP	590	3	65.0	2	81
39	InBr-TPhP	620	5	78	3	98
40	$Yb-[Phc]_2$	0		0		0
41	$Lu-[Phc]_2$	0		0		0
42	AlOH-Phc*	140	2	0.8	1	10
43	$Si(OH)_2$-Phc*	35	1	0.4	1	8
44	$Sn(OH)_2$-Phc*	160	2	1.0	2	16
45	AlOH-ЭП*	45	2	0.3	1	11

* Slow deposition of pigments with formation of dimmers.

Since all measurement of parameters of pigmented electrodes were conducted in 1 M KCl aerated electrolyte with no donor-acceptor additives, the main cathodic reaction was oxygen reduction and anodic – Ag oxidation.

When oxygen is adsorbed on the Phc film, its coordination is possible on central atom and on pyrrole rings of the pigment molecules [2, 9]. In case of full or partial overlap of oxygen orbitals with the orbitals of the metal atom coordinated with phthalocyanine (Me-Phc) (e.g., in coordination with Fe-complexes) and transition of an electron to the anti bonding p-orbital of the O_2, binding energy decreases from 5.0 to 3.4 eV, which leads to dissociation: $O_2^- \rightarrow O^- + O$, and formation of a substantial fraction of charged form of adsorption in the dark. Because of this dark currents increase sharply and photo activity of pigment film reduces (Table 6.1). Introduction of an extraligand in these compounds fills the coordination sphere of the central atom and leads to coordination of O_2 on the macro cycle and to a change in the mechanism of oxygen reduction. In this case relatively weak polarization interaction retains a high level of uncharged form of adsorption, which significantly increases the photoactivity of FeCl-Phc and the ratio of the photocurrent to the dark current (I_{ph}/I_d). Lowering of the ionization potential also contributes to that, but instead leads to a decrease in the stability of the pigment and displaces the activity maximum of the pH scale. This effect manifests itself equally well to both porphyrins and phthalocyanines and is apparently independent of the structure of macroligand. Similar conclusions can be made about an increase of photosignal with increasing of ionicity of the bond of central atom with the macroligand in a row of investigated complexes. For many complexes direct observation of influence of an extraligand is not possible, but comparing their U_{ph}, I_{ph} with similar parameters of divalent metal complexes, it may be noted that, jointly, the photo activity of ETPC is much higher.

Quantum chemical calculations of the electron density redistribution for some of the most pronounced cases illustrate well how the intra molecular characteristics is related to the photo electrochemical properties of the ensemble of molecules bound by intermolecular interactions (pigment film).

The following are the results of calculations of geometry and electronic structure of metal complexes of TPC in all spin states using the Gaussian-03 program. Numbering of atoms is shown in Fig. 6.1.

The numbers in the squares mark similar groups of atoms (structures) in accordance with the sequence of their appearance in the tables of charges below.

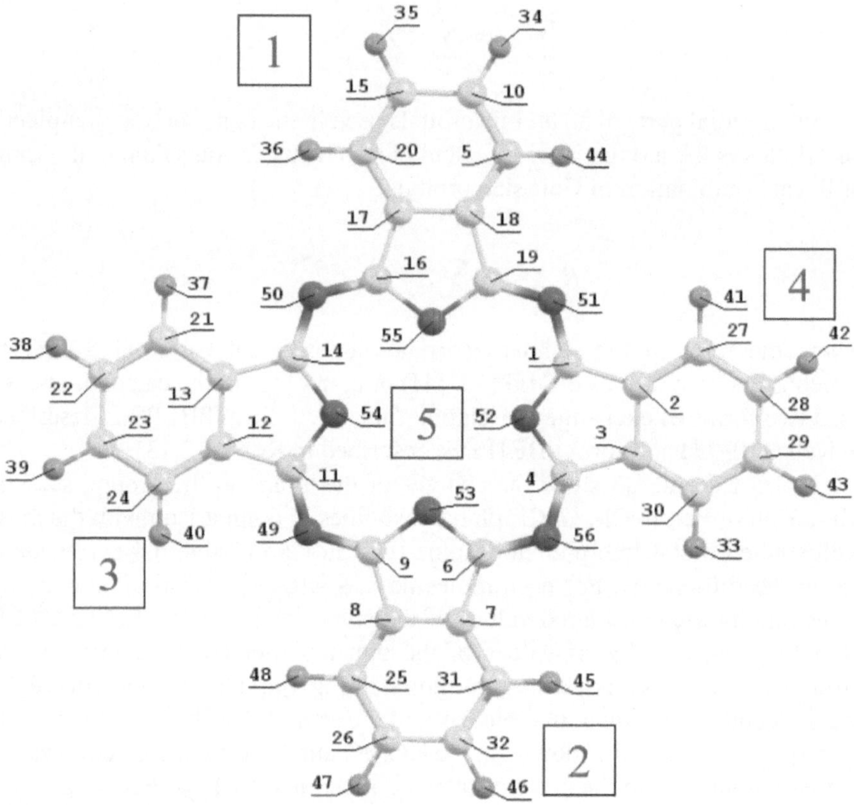

FIGURE 6.1 Numbering of atoms in the molecules considered.

Molecular orbitals were determined by density functional method and Rutan approach, in which molecular orbital are sought in the class of functions of the form

$$\varphi_p(\vec{r}) = \sum_{k=1}^{N_{atom}} \sum_{s=1}^{M(k)} c_{sp}^k \eta_s^k (\vec{r} - \vec{s}_k)$$

Here $\eta_1^k(\vec{r}), \eta_2^k(\vec{r}), ..., \eta_{M(k)}^k(\vec{r})$ – a given set of functions for the k-th atom in the molecule, whose position is determined by the vector $\vec{s}_k$. These functions, called atomic orbitals, are not assumed to be linearly independent or orthogonal, but are choosen to be normalized. c_{sp}^k – varying coefficients, s index is a set of three indices (n, ℓ, m) and the function $\eta_s^k(\vec{r})$ has the form:

$$\eta_s^k(\vec{r}) = R_{n\ell}^k(r) Y_{\ell m}(\vec{\Omega}) \cdot$$

Accordingly, the sum over s is a triple sum:

$$\sum_{s=1}^{M(k)} \equiv \sum_{n=1}^{N(k)} \sum_{\ell=0}^{L(k,n)} \sum_{m=-\ell}^{\ell} .$$

Set of radial parts of all atomic orbitals used in the construction of molecular orbitals is the atomic basis of calculation. The basis was taken in the form of linear combination of Gaussian orbitals:

$$R_{n\ell}(r) \approx \sum_{q=1}^{Q'} d_q^{n\ell} r^{k_q^{n\ell}} e^{-\alpha_q^{n\ell} r^2} ,$$

where the form of the standard correlation-consistent basis set 6–31G**, which is denoted also as 6–31G (1d1pH) or cc-pVDZ. In the calculations, we used two forms of exchange-correlation functional, form PBEPBE, described in Refs. [10, 11] and form PBE1PBE, described in Refs. [12,13].

Tables 6.2 and 6.3 show the charges of the atoms in the ground state of Mn-, MnCl-, Fe-, FeCl-, GaCl- phthalocyanines in comparison with the most well-studied metal-free phthalocyanine (H_2-Phc), and Table 6.4 corresponding to the full-energy. For each molecule it is stressed the minimum energy value among the considered values of the spin. For Fe-, Mn-, FeCl-, MnCl-phthalocyanines series of values of the upper one-electron energies in the ground state of these molecules are shown in Fig. 6.2. Dipole moments (d) of the molecules are shown in Table 6.5 and the spatial distribution of electrons with spin "up" and spin "down" in the HOMO and LUMO one-electron states of these molecules in the ground state, are illustrated by Figs. 6.3–6.12.

TABLE 6.2 Mulliken charges on atoms in Mn-, MnCl-, Fe-, FeCl-, GaCl-, H_2-phthalocyanines in Their Ground States

No	L	Structure No	Phc (singlet)	Phc-Mn (quartet)	Phc-Fe (triplet)	Phc-FeCl (quartet)	Phc-MnCl (triplet)	Phc-GaCl (singlet)
5	C	1	–0.13735	–0.13026	–0.126842	–0.125085	–0.12678	–0.126833
10	C	1	–0.10292	–0.11085	–0.104648	–0.102419	–0.10277	–0.102362
15	C	1	–0.10281	–0.10928	–0.104658	–0.102445	–0.10279	–0.102366
17	C	1	0.058462	0.051653	0.055229	0.055238	0.055424	0.049298
18	C	1	0.058777	0.048002	0.054660	0.054613	0.054808	0.048585
20	C	1	–0.13734	–0.13101	–0.126992	–0.125239	–0.12695	–0.127047
34	H	1	0.101716	0.113452	0.104265	0.107911	0.106553	0.107501
35	H	1	0.101706	0.113872	0.104273	0.119461	0.118064	0.119055

TABLE 6.2 *(Continued)*

No	L	Structure No	Phc (singlet)	Phc-Mn (quartet)	Phc-Fe (triplet)	Phc-FeCl (quartet)	Phc-MnCl (triplet)	Phc-GaCl (singlet)
36	H	1	0.112515	0.126995	0.116188	0.119457	0.11806	0.119053
44	H	1	0.112491	0.127061	0.116193	0.119458	0.118061	0.119056
7	C	2	0.058462	0.051927	0.055228	0.055216	0.055403	0.049266
8	C	2	0.058777	0.047476	0.054661	0.054634	0.054828	0.048619
25	C	2	–0.13735	–0.12983	–0.126841	–0.125086	–0.12678	–0.126838
26	C	2	–0.10292	–0.11106	–0.104648	–0.102422	–0.10277	–0.102369
31	C	2	–0.13734	–0.13157	–0.126992	–0.125234	–0.12694	–0.127038
32	C	2	–0.10281	–0.1092	–0.104658	–0.102443	–0.10279	–0.102362
45	H	2	0.112515	0.126755	0.116188	0.107921	0.106562	0.107505
46	H	2	0.101706	0.113803	0.104273	0.107921	0.106562	0.107505
47	H	2	0.101716	0.113468	0.104265	0.107921	0.106562	0.107503
48	H	2	0.112491	0.12844	0.116193	0.107921	0.106562	0.107503
12	C	3	0.040309	0.065189	0.059236	0.055230	0.055418	0.049283
13	C	3	0.039753	0.045925	0.058662	0.054624	0.054819	0.048612
21	C	3	–0.12672	–0.12048	–0.132286	–0.125083	–0.12678	–0.126835
22	C	3	–0.10295	–0.13533	–0.104152	–0.102420	–0.10277	–0.102366
23	C	3	–0.10296	–0.1185	–0.104172	–0.102443	–0.10279	–0.102362
24	C	3	–0.12684	–0.14524	–0.132434	–0.125236	–0.12695	–0.127042
37	H	3	0.120148	0.129358	0.114580	0.107912	0.106553	0.107502
38	H	3	0.107138	0.110604	0.102897	0.119461	0.118063	0.119054
39	H	3	0.107134	0.126995	0.102906	0.119460	0.118063	0.119054
40	H	3	0.120168	0.118375	0.114577	0.107911	0.106553	0.107502
2	C	4	0.040308	0.048183	0.059234	0.055227	0.055412	0.049282
3	C	4	0.039753	0.051868	0.058660	0.054622	0.054815	0.048591
27	C	4	–0.12684	–0.13037	–0.132431	–0.125238	–0.12695	–0.127043
28	C	4	–0.10296	–0.11218	–0.104172	–0.102445	–0.10279	–0.102365
29	C	4	–0.10295	–0.11017	–0.104152	–0.102420	–0.10277	–0.102365
30	C	4	–0.12672	–0.13181	–0.132283	–0.125088	–0.12679	–0.126835
33	H	4	0.120148	0.125347	0.114580	0.119457	0.118059	0.119055
41	H	4	0.120168	0.125611	0.114577	0.119456	0.118058	0.119053
42	H	4	0.107134	0.112407	0.102906	0.119460	0.118063	0.119053

TABLE 6.2 *(Continued)*

No	L	Structure No	Phc (singlet)	Phc-Mn (quartet)	Phc-Fe (triplet)	Phc-FeCl (quartet)	Phc-MnCl (triplet)	Phc-GaCl (singlet)
43	H	4	0.107138	0.112842	0.102897	0.107912	0.106553	0.107502
1	C	5	0.492541	0.443806	0.455846	0.483253	0.472905	0.476438
4	C	5	0.493137	0.436325	0.456281	0.483738	0.473345	0.476920
6	C	5	0.439002	0.438233	0.458910	0.483259	0.472913	0.476454
9	C	5	0.43906	0.446524	0.459361	0.483730	0.473342	0.476913
11	C	5	0.492541	0.444325	0.455846	0.483251	0.472911	0.476451
14	C	5	0.493137	0.436805	0.456281	0.483732	0.473346	0.476907
16	C	5	0.439002	0.438378	0.458908	0.483246	0.472904	0.476436
19	C	5	0.43906	0.446956	0.459362	0.483739	0.473348	0.476913
49	N	5	–0.53628	–0.53621	–0.529467	–0.530810	–0.52795	–0.534761
50	N	5	–0.53630	–0.53744	–0.529456	–0.530810	–0.52795	–0.534756
51	N	5	–0.53628	–0.54165	–0.529467	–0.530814	–0.52796	–0.534753
52	N	5	–0.61622	–0.73564	–0.733508	–0.720526	–0.71953	–0.638220
53	N	5	–0.63744	–0.73185	–0.749609	–0.720510	–0.7195	–0.638156
54	N	5	–0.61622	–0.73904	–0.733507	–0.720519	–0.71949	–0.638148
55	N	5	–0.63744	–0.7322	–0.749609	–0.720535	–0.71952	–0.638211
56	N	5	–0.5363	–0.53706	–0.529456	–0.530813	–0.52795	–0.534757
57			0.322064	1.191249	1.088319	1.070922	1.070146	0.756375
58			0.322064			–0.371189	–0.26604	–0.247612

Numbering of atoms in accordance with Fig. 6.1.

TABLE 6.3 Relative (To Phc, In %) Deviations of the Mulliken Atomic Charges in the Ground States of Mn-, Mncl-, Fe-, Fecl-, Gacl- Phthalocyanines. Numbering of Atoms in Accordance with Fig.6.1

No	L	Structure No	Phc (singlet)	Phc-Mn (quartet)	Phc-Fe (triplet)	Phc-FeCl (quartet)	Phc-MnCl (triplet)	Phc-GaCl (singlet)
5	C	1	–0.13735	5.16	7.65	8.93	7.70	7.66
10	C	1	–0.10292	–7.71	–1.68	0.49	0.15	0.54
15	C	1	–0.10281	–6.29	–1.80	0.36	0.02	0.43
17	C	1	0.058462	11.65	5.53	5.51	5.20	15.68

TABLE 6.3 *(Contined)*

No	L	Structure No	Phc (singlet)	Phc-Mn (quartet)	Phc-Fe (triplet)	Phc-FeCl (quartet)	Phc-MnCl (triplet)	Phc-GaCl (singlet)
18	C	1	0.058777	18.33	7.00	7.08	6.75	17.34
20	C	1	–0.13734	4.61	7.53	8.81	7.57	7.49
34	H	1	0.101716	–11.54	–2.51	–6.09	–4.76	–5.69
35	H	1	0.101706	–11.96	–2.52	–17.46	–16.08	–17.06
36	H	1	0.112515	–12.87	–3.26	–6.17	–4.93	–5.81
44	H	1	0.112491	–12.95	–3.29	–6.19	–4.95	–5.84
7	C	2	0.058462	11.18	5.53	5.55	5.23	15.73
8	C	2	0.058777	19.23	7.00	7.05	6.72	17.28
25	C	2	–0.13735	5.48	7.65	8.93	7.70	7.65
26	C	2	–0.10292	–7.91	–1.68	0.48	0.15	0.54
31	C	2	–0.13734	4.20	7.53	8.81	7.57	7.50
32	C	2	–0.10281	–6.22	–1.80	0.36	0.02	0.44
45	H	2	0.112515	–12.66	–3.26	4.08	5.29	4.45
46	H	2	0.101706	–11.89	–2.52	–6.11	–4.77	–5.70
47	H	2	0.101716	–11.55	–2.51	–6.10	–4.76	–5.69
48	H	2	0.112491	–14.18	–3.29	4.06	5.27	4.43
12	C	3	0.040309	–61.72	–46.95	–37.02	–37.48	–22.26
13	C	3	0.039753	–15.53	–47.57	–37.41	–37.90	–22.29
21	C	3	–0.12672	4.92	–4.39	1.29	–0.05	–0.09
22	C	3	–0.10295	–31.45	–1.17	0.51	0.17	0.57
23	C	3	–0.10296	–15.09	–1.18	0.50	0.17	0.58
24	C	3	–0.12684	–14.51	–4.41	1.26	–0.09	–0.16
37	H	3	0.120148	–7.67	4.63	10.18	11.32	10.53
38	H	3	0.107138	–3.24	3.96	–11.50	–10.20	–11.12
39	H	3	0.107134	–18.54	3.95	–11.51	–10.20	–11.13
40	H	3	0.120168	1.49	4.65	10.20	11.33	10.54
2	C	4	0.040308	–19.54	–46.95	–37.01	–37.47	–22.26
3	C	4	0.039753	–30.48	–47.56	–37.40	–37.89	–22.23

TABLE 6.3 *(Contined)*

No	L	Structure No	Phc (singlet)	Phc-Mn (quartet)	Phc-Fe (triplet)	Phc-FeCl (quartet)	Phc-MnCl (triplet)	Phc-GaCl (singlet)
27	C	4	–0.12684	–2.78	–4.41	1.26	–0.09	–0.16
28	C	4	–0.10296	–8.95	–1.18	0.50	0.17	0.58
29	C	4	–0.10295	–7.01	–1.17	0.51	0.17	0.57
30	C	4	–0.12672	–4.02	–4.39	1.29	–0.06	–0.09
33	H	4	0.120148	–4.33	4.63	0.58	1.74	0.91
41	H	4	0.120168	–4.53	4.65	0.59	1.76	0.93
42	H	4	0.107134	–4.92	3.95	–11.51	–10.20	–11.13
43	H	4	0.107138	–5.32	3.96	–0.72	0.55	–0.34
1	C	5	0.492541	9.89	7.45	1.89	3.99	3.27
4	C	5	0.493137	11.52	7.47	1.91	4.01	3.29
6	C	5	0.439002	0.18	–4.53	–10.08	–7.72	–8.53
9	C	5	0.43906	–1.70	–4.62	–10.17	–7.81	–8.62
11	C	5	0.492541	9.79	7.45	1.89	3.99	3.27
14	C	5	0.493137	11.42	7.47	1.91	4.01	3.29
16	C	5	0.439002	0.14	–4.53	–10.08	–7.72	–8.53
19	C	5	0.43906	–1.80	–4.62	–10.18	–7.81	–8.62
49	N	5	–0.53628	0.01	1.27	1.02	1.55	0.28
50	N	5	–0.53630	–0.21	1.28	1.02	1.56	0.29
51	N	5	–0.53628	–1.00	1.27	1.02	1.55	0.28
52	N	5	–0.61622	–19.38	–19.03	–16.93	–16.77	–3.57
53	N	5	–0.63744	–14.81	–17.60	–13.03	–12.87	–0.11
54	N	5	–0.61622	–19.93	–19.03	–16.93	–16.76	–3.56
55	N	5	–0.63744	–14.87	–17.60	–13.04	–12.88	–0.12
56	N	5	–0.5363	–0.14	1.28	1.02	1.56	0.29

TABLE 6.4 Calculations and the Corresponding Values of Total Energy

Spin Molecule	Singlet	Doublet	Triplet	Quartet	Quintet	Sextet	Septet	Exchange-correlation functional
Mn–Phc		−2815.97035		−2815.99838		−2815.97552		PBEPBE
Fe–Phc	−2928.69872		−2928.7655		−2928.70917			PBE1PBE
MnC–Phc 1			−3276.01424		−3276.03369		−3275.98854	PBEPBE
FeCl–Phc		−3388.69310		−3388.70722		−3388.68747		PBEPBE
GaCl–Phc	−4047.98795		−4047.94495					PBEPBE

For Each Molecule the Minimum Energy
Among the Considered Spin Values is Underlined

TABLE 6.5 Electric Dipole Moments (d) of the Molecules

Molecule	Fe-Phc	Mn-Phc	FeCl-Phc	MnCl-Phc
d (Debye)	0	0	3.7077	4.5437

(1 Debye = $a_0 e$ = 0.8478–10–^{29} C*m).

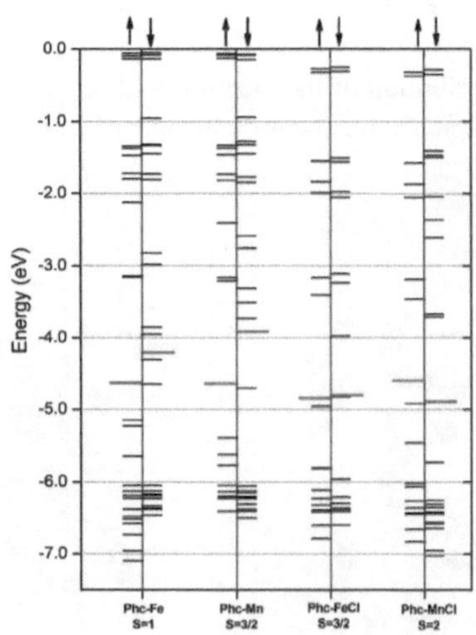

FIGURE 6.2 The upper part of the spectrum of one-electron energies in the ground state of molecules of certain metalloporphyrins (spin ground state S is not equal to 0). The energy of the highest occupied electron state is marked with a long dash of red. The calculations were performed with the exchange-correlation functional PBEPBE [1, 2].

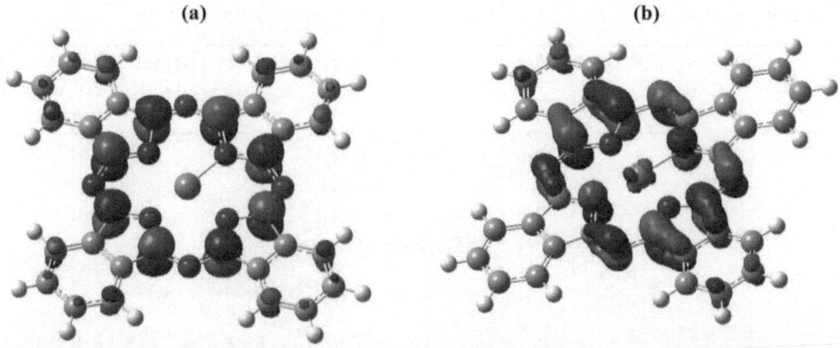

FIGURE 6.3 Spatial distribution of the electron with spin "up" in the single-particle HOMO (a) and LUMO (b) states for the ground state of Fe-phthalocyanine. Ground-state spin S = 1.

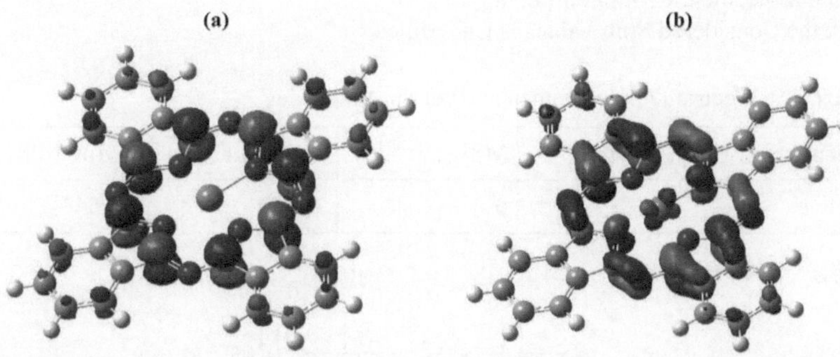

FIGURE 6.4 Spatial distribution of the electron with spin "down" in the single-particle HOMO (a) and LUMO (b) states for the ground state of Fe-phthalocyanine. Ground-state spin S = 1.

FIGURE 6.5 Spatial distribution of the electron with spin "up" in the single-particle HOMO (a) and LUMO (b) states for the ground state of Mn-phthalocyanine. Ground-state spin S = 3/2.

FIGURE 6.6 Spatial distribution of the electron with spin "down" in the single-particle HOMO (a) and LUMO (b) states for the ground state of Mn-phthalocyanine. Ground-state spin S = 3/2.

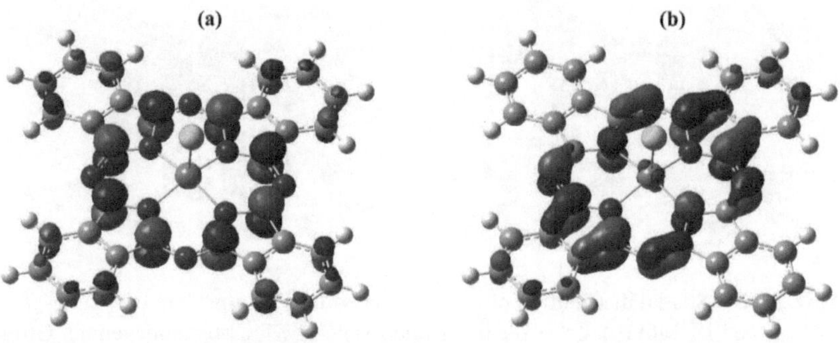

FIGURE 6.7 Spatial distribution of the electron with spin "up" in the single-particle HOMO (a) and LUMO (b) states for the ground state of FeCl-phthalocyanine. Ground-state spin S = 3/2.

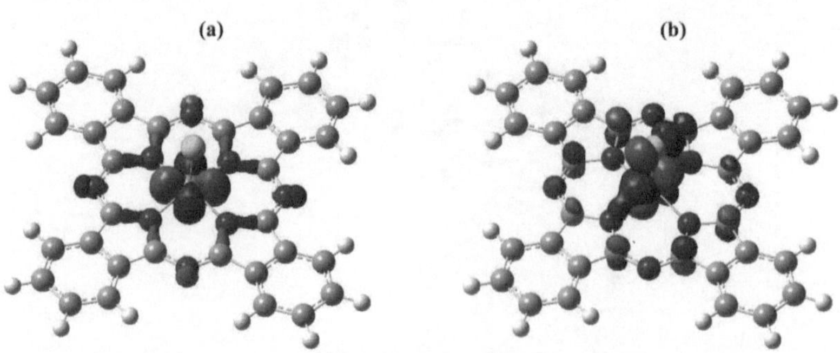

FIGURE 6.8 Spatial distribution of the electron with spin "down" in the single-particle HOMO (a) and LUMO (b) states for the ground state of FeCl-phthalocyanine. Ground-state spin S = 3/2.

FIGURE 6.9 Spatial distribution of the electron with spin "up" in the single-particle HOMO (a) and LUMO (b) states for the ground state of MnCl-phthalocyanine. Ground-state spin S = 2.

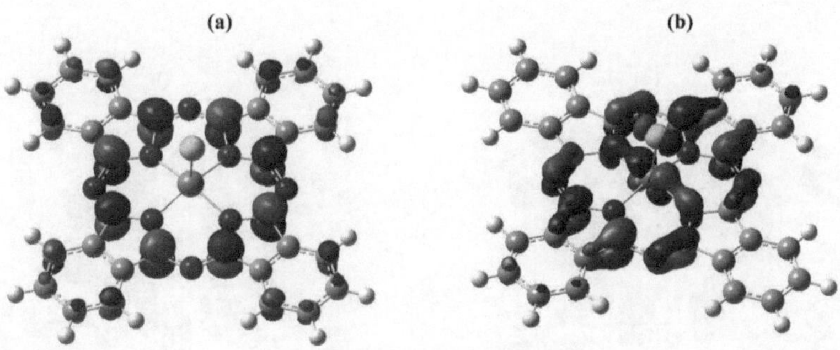

FIGURE 6.10 Spatial distribution of the electron with spin "down" in the single-particle HOMO (a) and LUMO (b) states for the ground state of MnCl-phthalocyanine. Ground-state spin S = 2.

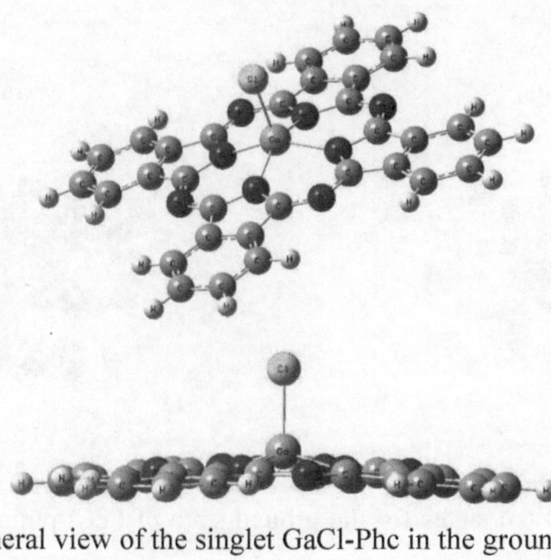

FIGURE 6.11 General view of the singlet GaCl-Phc in the ground state.

FIGURE 6.12 Singlet GaCl-Phc. Cross section of HOMO (a) and LUMO_1 (b) orbitals, isovalue – 0.035.

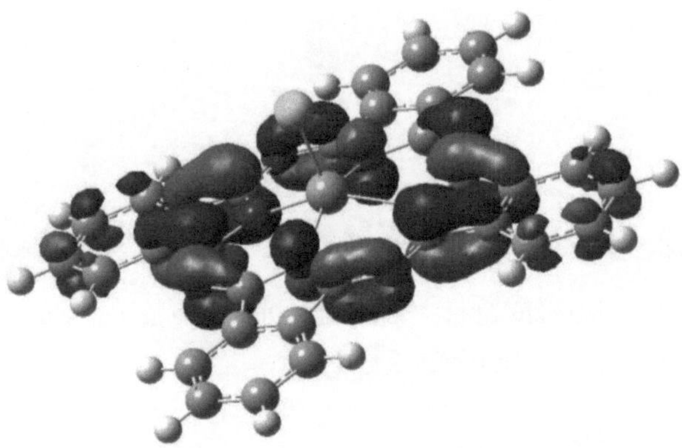

FIGURE 6.13 Singlet GaCl-Phc. Cross section of LUMO_2 orbital, isovalue – 0.035.

Consider now the metal complexes with higher valencies. A typical representative of such metal complexes is phthalocyanine is with GaCl (GaCl-Phc) (ground state in a vacuum – singlet).

Energy spectrum (a.u.) (1a.u.=27.2116 eV):

Degeneration	E(a.u.)
1	−0.067
1	−0.073
2	−0.127 LUMO_1 и LUMO_2
1 E_F–↑↓	−0.178 HOMO
2 –↑↓	−0.224
1 –↑↓	−0.225

Total energy of the isolated molecule $E_{sing} = -4047.98795822$ a.u.

Dipole moment $D_{sing} = 3.7445$ Debye

$E_{LUMO} - E_{HOMO} = 0.051$ a.u.=1.39 eV

FIGURE 6.14 General view of the triplet GaCl-Phc in the ground state.

Energy spectrum (a.u.) (1a.u.=27.2116 eV)

Degeneration	E(a.u.)
1	−0.069
1	−0.071
1	−0.074
1	−0.119
1	−0.121
1	−0.126 LUMO_up
1 –↑	−0.134 HOMO_up
1	−0.169 LUMO_down
1 –↑	−0.184
1 –↑↓	−0.221 HOMO_down
3 –↓	−0.222
1 –↑	−0.223
1 –↑	−0.224
1 –↓	−0.225

Total energy of the isolated molecule $E_{trip} = -4047.94495141$ a.u. E

Dipole moment $D_{trip} = 3.8437$ Debye,

$E_{trip} - E_{sing} = 0.0430$ a.u. $= 1.170$ eV

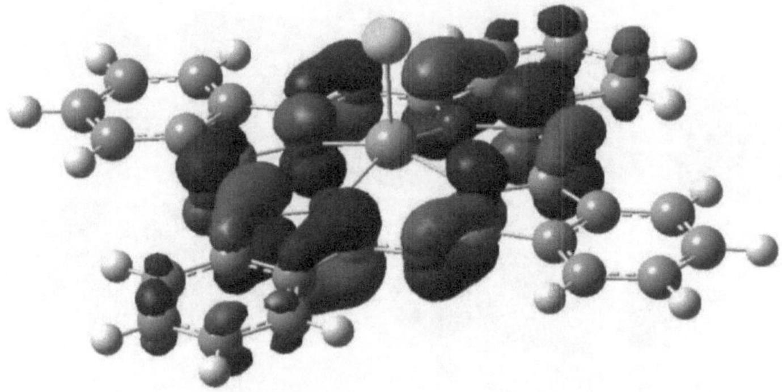

FIGURE 6.15 Triplet GaCl-Phc. Cross section of HOMO_up orbital, isovalue – 0.035.

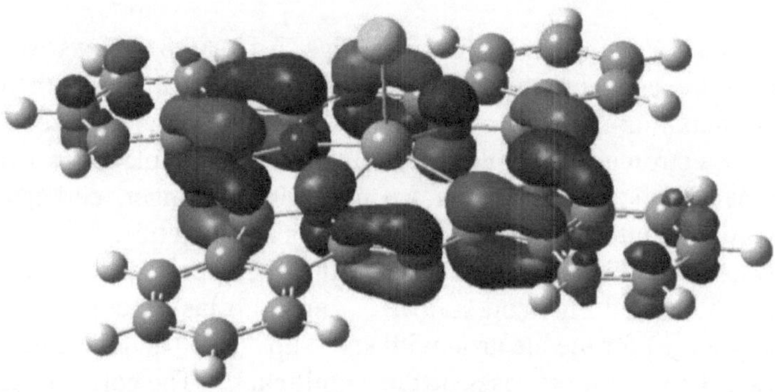

FIGURE 6.16 Triplet GaCl-Phc. Cross section of LUMO_up orbital, isovalue – 0.035.

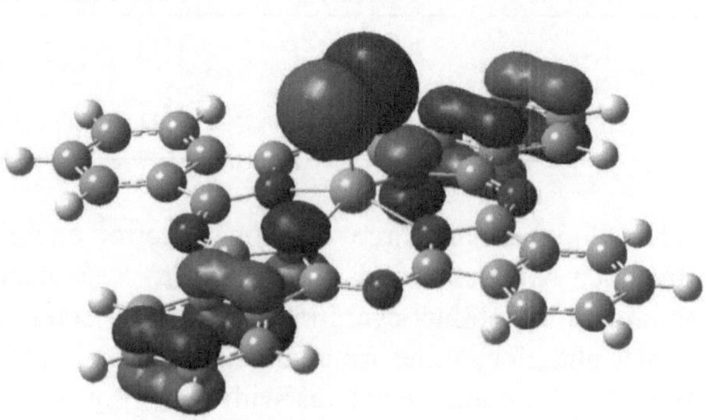

FIGURE 6.17 Triplet GaCl-Phc. Cross section of HOMO_down orbital, isovalue – 0.035.

FIGURE 6.18 Triplet GaCl-Phc. Cross section of LUMO_down orbital, isovalue – 0.035.

Even a cursory view of tables and (Figs. 6.13–6.18) allows to note the main trend – the concentration of charge around the focal area of the metal in the zone of maximum ring currents and the greatest likelihood of coordination of external electron acceptor. For more photoactive complexes (for instance, Mn-Phc, MnCl-Phc) these effects are noticeably stronger, perhaps a wide range of multivalent states of manganese plays a role in this.

Analysis of the Figs. 6.13–6.18 shows that the energy difference between the first unoccupied electron state E_{LUMO} and the last occupied one E_{HOMO} ($\Delta E = E_{LUMO} - E_{HOMO}$) for the electron with spin "up" ($\Delta E^{\uparrow}$) and the electron with spin "down" ($\Delta E^{\downarrow}$) in most cases differs significantly. The corresponding values are given in the following table.

Молекулы	**Fe-Phc**	**Mn-Phc**	**FeCl-Phc**	**MnCl-Phc**
$\Delta E^{\uparrow}$ **(eV)**	1.47	1.45	1.45	1.12
$\Delta E^{\downarrow}$ **(eV)**	0.24	0.16	0.81	1.18

Besides, in Mn-Phc there are three more one-electron energy levels close to $E_{LUMO}^{\downarrow}$, the distance between which is close to $\Delta E^{\downarrow}$. Molecules of FeCl-phthalocyanine and MnCl-phthalocyanine have LUMO state doubly degenerate, while MnCl-phthalocyanine molecule has also a state with a little more energy than E_{LUMO} among free states with spin "down." Such a variety of one-electron spectra means that molecules with nonzero spin will have more complex and significantly different from each other absorption spectra. Distributions of the most weakly bound electron in these molecules are also

differ considerably. Such a diversity of molecular characteristics important for interaction with the radiation in the molecules with nonzero spin, gives reason to look for the most suitable films for use in solar energy converters based on organic semiconductors among the films of such molecules. Due to the presence of magnetic and electric dipole moments in these molecules, the presence of electric and magnetic fields in the formation of these films will contribute to the formation of ordered structures in the films, which may be useful to improve the efficiency of solar energy converters based on such films.

However, the major cause of the increased photo activity of ETPC, probably, is a violation of the planarity of the molecules due to exit of the extra-ligand from the plane of macrocycle. High electron density on these "bridges" between layers of molecules provides a strong overlap of the electron clouds facilitates charge transfer between molecules and drops recombination losses in the film. It is known that ETPC of 3–4 valent metals form lamellar crystals and blocks, unlike needle crystals of divalent metal complexes [9]. Smaller sizes of molecular crystal unit cells correspond to a stronger intermolecular interaction that is seen in recordings of action spectra of these pigments (Fig. 6.19) that have a much wider absorption bands, than TPC of divalent metals, and bathochromic shift to the NIR region.

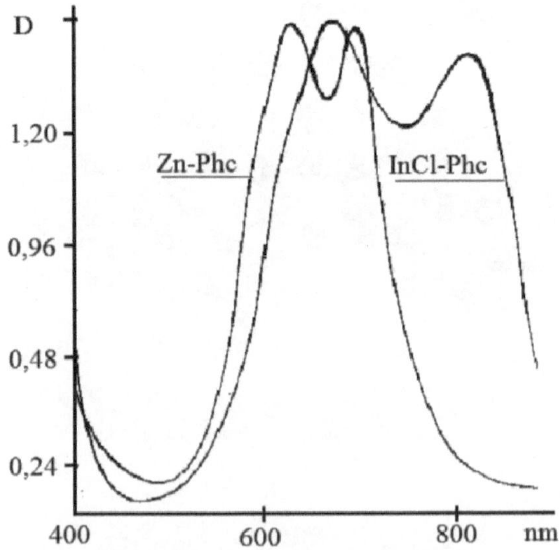

FIGURE 6.19 Absorption spectra of the valence II phthalocyanine Zn-Phc and valence III phthalocyanine InCl-Phc.

At the same time, it should be noted that the axially symmetrical ETPC of tetravalent metals are less photoactive, possibly, because of a strong anisotropy of conductivity similar to complexes with large metal atoms (Pb-Phc). However, calculations show that possible arrangement of extra coordinated molecules in a molecular crystal has a very unique option and demonstrates why the situation is not in favor of compounds with two axially symmetric extraligands. Three variants of formation of phthalocyanine dimers have been considered: (i) a dimmer of a typical complex with divalent metal without extraligand (Zn-Phc); (ii) a complex with trivalent metal and one extraligand (AlCl-Phc, GaCl-Phc); (iii) a complex with tetravalent metal and two axis symmetric extraligands ($GeCl_2$-Phc).

6.3.1 ZINC PHTHALOCYANINE DIMMER (2 ZN-PHC)

(Ground state in a vacuum, singlet) (Figs. 6.20–6.22).

FIGURE 6.20 View of the dimmer 2 Zn-Phc in the ground state.

Energy spectrum (a.u.) (1a.u. =27.2116 eV):
Degeneration E(a.u.)

1	−0.033
1	−0.091
1	−0.093
1	−0.100
1	−0.102 LUMO
1 E_F – ↑↓	−0.182 HOMO
1 – ↑↓	−0.186
2 — ↑↓	−0.250
1 — ↑↓	−0.251
1 — ↑↓	−0.252

Total energy of the isolated dimer $E_{sin\,g} = -6888.46444851$ a.u.
Dipole moment $D_{sin\,g} = 0.0010$ Debye
$E_{LUMO} - E_{HOMO} = 0.080$ a.u. $= 2.18$ eV
Binding energy $E_b = 0.0129$ a.u. $= 0.351$ eV

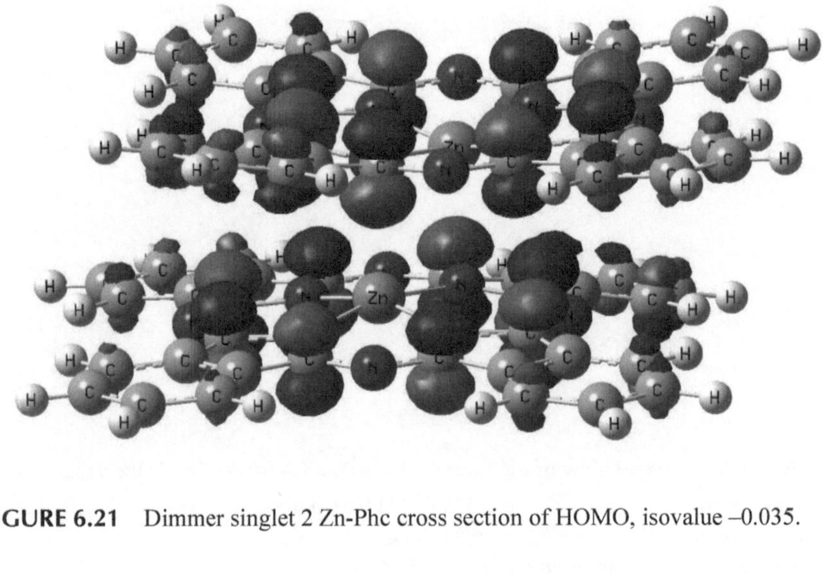

FIGURE 6.21 Dimmer singlet 2 Zn-Phc cross section of HOMO, isovalue –0.035.

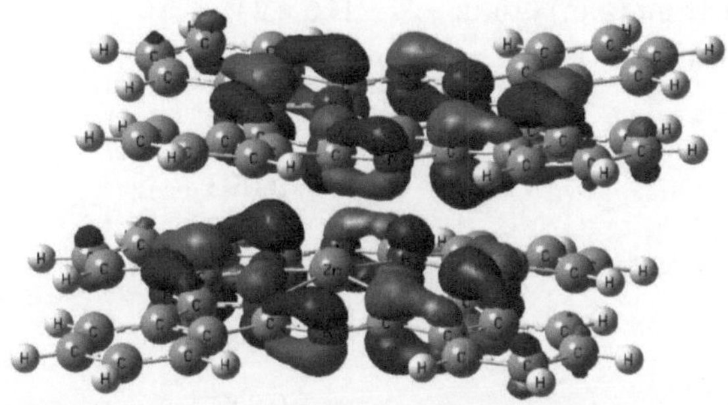

FIGURE 6.22 Dimer singlet 2 Zn-Phc Cross section of LUMO, isovalue 0.035.

6.3.2 DIMMER OF ALCL-PHTHALOCYANINE (2 ALCL-PHC)

(Three ground state conformations considered: (1) both Cl-atoms outwards, (2) both Cl-atoms outwards; (3) one Cl inwards and one outwards.) (Figs. 6.23–6.27).

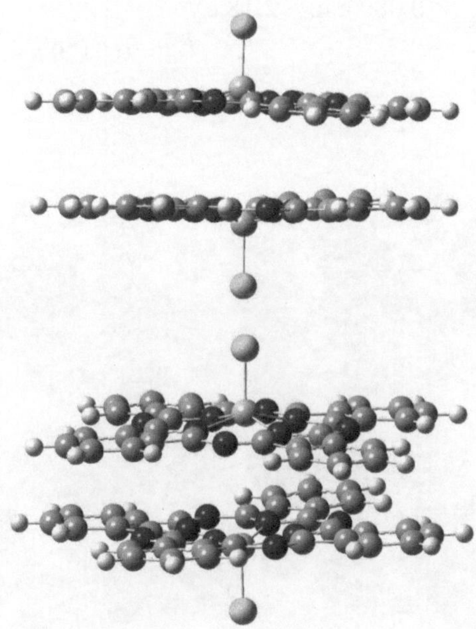

FIGURE 6.23 General view of the dimer 2 AlCl-Phc. Conformation 1, the ground state.

Total energy $E^{(1)} = -4735.73411776$ a.u.
Dipole moment $D_{\sin g} = 0.2349$ Debye
Distance between aluminum atoms – 4.95185 a.u.

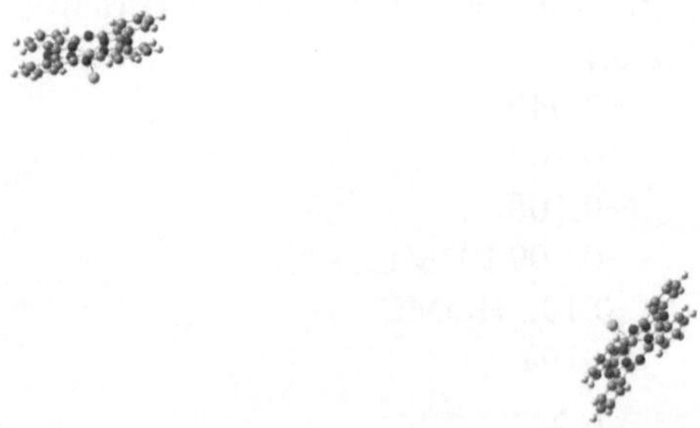

FIGURE 6.24 General view of the dimer 2 AlCl-Phc. Conformation 2, the ground state.

Total energy $E^{(2)} = -4735.72578094$ a.u. = 2E of the monomer
Dipole moment $D_{sing} = 3.3338$ Debye
Distance between aluminum atoms – 51.12 a.u.

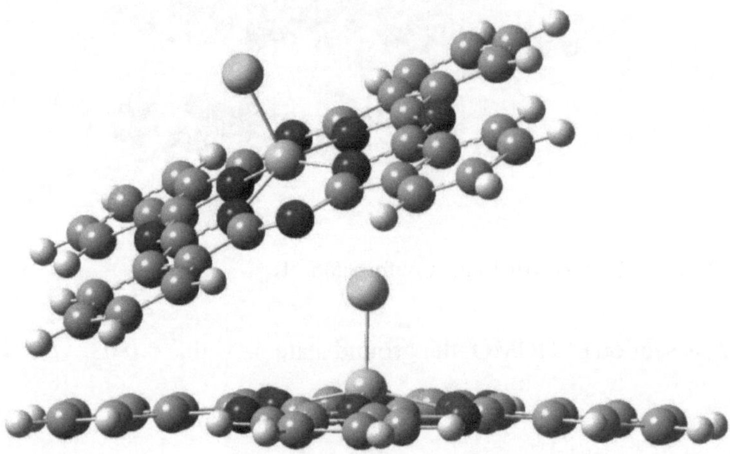

FIGURE 6.25 General view of the dimmer 2 AlCl-Phc. Conformation 3, the ground state.

Total energy $E^{(2)} = -4735.73089211$ a.u. = 2E of the monomer
Dipole moment $D_{sing} = 8.1170$ Debye
Distance between aluminum atoms – 5.46136 a.u.
Conformation 1 has the lowest energy.
Relative energies (1a.u. = 27.2116 eV):

$E^{(2)} - E^{(1)} = 0.00833682$ a.u. = 0.226858 eV

$E^{(3)} - E^{(1)} = 0.00422565$ a.u. = 0.114987 eV

The energy spectrum of Conformation 1 (a.u) (1a.u.=27.2116 eV):

Degeneration	E(a.u.)
1	−0.040
2	−0.107
1	−0.108
1	−0.109 LUMO
1 E_F −↑↓	−0.192 HOMO
1 −↑↓	−0.194
2 —↑↓	−0.262
2 —↑↓	−0.263

$E_{LUMO} - E_{HOMO} = 0.083$ a.u. = 2.26 eV.

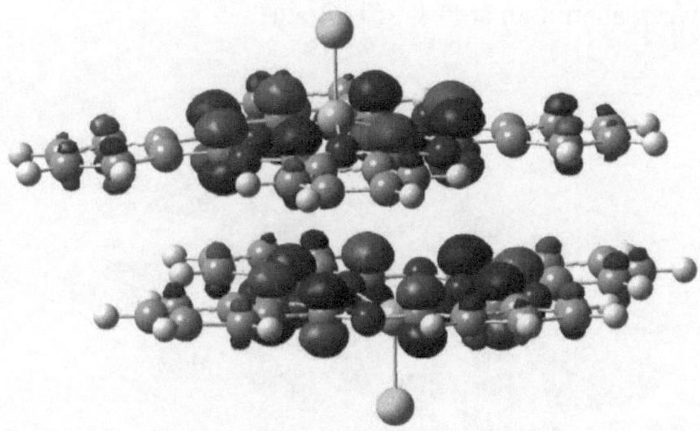

FIGURE 6.26 Dimer 2 AlCl-Phc, Conformation 1.

Cross section of HOMO, the ground state, isovalue – 0.03.

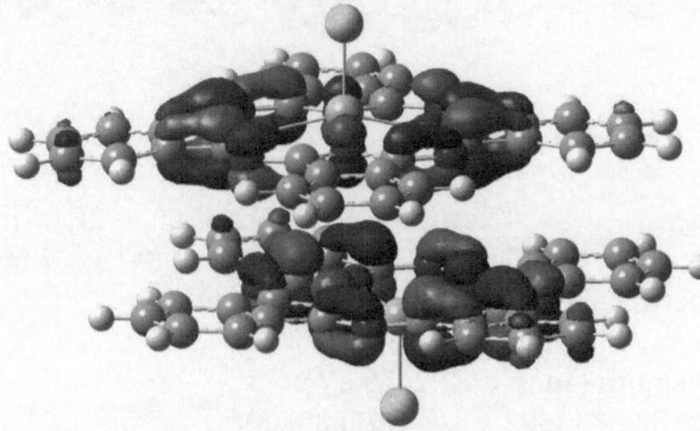

FIGURE 6.27 Dimer 2 AlCl-Phc, Conformation 1.

Cross-section of LUMO, the ground state, isovalue – 0.03.

6.3.3 PHTHALOCYANINE WITH $GECL_2$ ($GECL_2$-PHC)

(ground state in a vacuum – singlet) (Figs. 6.28–6.31).

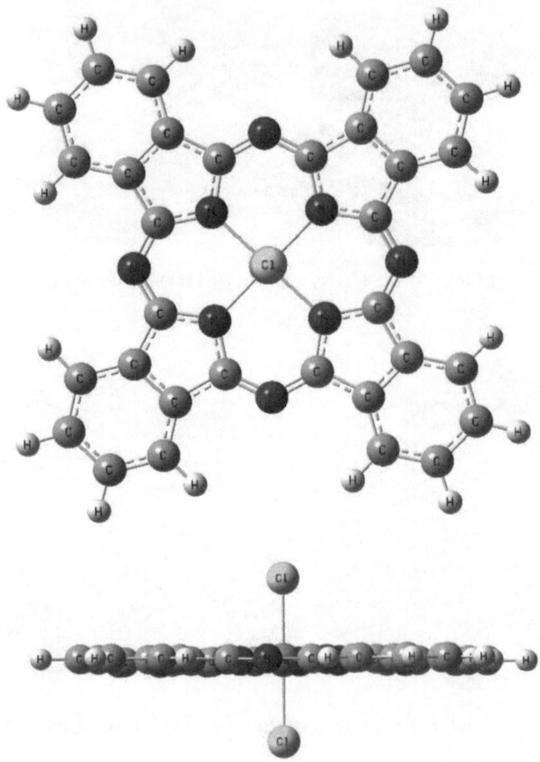

FIGURE 6.28 General view of $GeCl_2$-Phc in the ground state

The energy spectrum (a.u.) (1a.u.=27.2116 eV):

Degeneration	E(a.u.)
1	−0.040
1	−0.046
2	−0.115 LUMO
1 E_F−↑↓	−0.201 HOMO
2 −↑↓	−0.262
1 —↑↓	−0.268

1 —↑↓ −0.272

Total energy of the isolated molecule $E_{sing} = -4660.21111306$ a.u.

Dipole moment $D_{sing} = 0.0030$ Debye

$E_{LUMO} - E_{HOMO} = 0.096$ a.u. = 2.61 eV

FIGURE 6.29 Singlet $GeCl_2$-Phc. Cross section of HOMO, isovalue – 0.03.

FIGURE 6.30 Singlet $GeCl_2$-Phc. Cross section of LUMO, isovalue – 0.03.

FIGURE 6.31 General view of the dimer 2 $GeCl_2$-Phc in the ground state.

The energy spectrum (a.u.) (1a.u.=27.2116 eV):
Degeneration E(a.u.)
2 −0.041
2 −0.043
4 −0.113 LUMO
2 E_F −↑↓ −0.199 HOMO
4 −↑↓ −0.261
2 —↑↓ −0.267
2 —↑↓ −0.271
Total energy of the isolated dimer $E_{sing} = -5749.80351332$ a.u.
Dipole moment $D_{sing} = 0.0012$ Debye
$E_{LUMO} - E_{HOMO} = 0.086$ a.u. = 2.34 eV
Distance between Ge's – 41.94 a.u.

Effect of ETPC structure on the photo-parameters is particularly noticeable when comparing compounds with Cl– and OH– extraligands. Relatively low photo activity of ETPC with OH group can be explained by the formation of bridging bonds on the extraligand: at thermal sublimation such compounds form a certain proportion of dimers of the type Phc-Me-O-Me-Phc, and, in some cases, short chains Phc-Me-O-(Me-Phc)-O-Me-Phc, weakly associated with neighboring molecules. However, due to the incomplete transition to a chemically bonded dimeric form photo activity still appears quite noticeable. Interestingly, at a long-term deposition with a minimum temperature difference between the pigment evaporator and the substrate the photo activity of compounds with OH-extraligands significantly reduces, confirming the potential effect of the chemical dimerization of the pigments. For diphthalocyanines of lanthanides, which can be considered as a limiting form of extra coordination with extraligands in the form of Phc macrocycle, the system acquires completely closed form and U_{ph}, I_{ph} decrease to zero.

6.4 CONCLUSION

Generally, the considered results show that structural sensitization of TPC films by selection of the most photoactive pigments can significantly increase the efficiency of converting light energy into electrical or chemical ones, with the largest photo activity to be in a TPC with the following characteristics:

- high ionicity of the metal atom bond to the ligand;
- reduced electron density on the central atom;
- the possibility of joining one extraligand;

- as extraligands there are suitable elements and compounds that: (i) exclude formation of axial dimmers, (ii) form a compact group with little way out of macrocycle plane (3–4 Å); (iii) have large electron affinity;
- high density of π-electron charge on the periphery of the molecules and pyrrole rings;
- Molecules of tetrapyrrole compounds with spin different from zero have magnetic and electric dipole moments that allow you to manage the process of films formation on conductive substrates using magnetic and electric fields. This is the fundamental solution to the problem of growing ordered structures in a vacuum with high solar energy conversion efficiency. It only remains to determine exactly what should be such a structure; the rest is a matter of technique.

KEYWORDS

- **Chemical parameters**
- **Diphthalocyanines**
- **Extra coordination**
- **Extraligands**
- **Molecules**
- **Photo activity**
- **Photo electrochemical properties**
- **Quantum chemistry**

REFERENCES

1. Ilatovsky, V. A., Sinko, G. V., Ptitsyn, G. A., Komissarov, G. G. (April 25, 2014). Predication of Photo electrochemical Properties of Selected Molecules by Their Structure, in Research Methodology in Physics and Chemistry of Surfaces and Interfaces ISBN: 9781771880114, Editor: Nekane Guarrotxena, Editorial Advisory Board: Pourhashemi, A., Haghi, A. K. & Gennady E. Zaikov, Chapter 2, 255pp.
2. The porphyrins Ed. Dolphin, D. N., Y.: Acad. Press, 1978.
3. Ilatovsky, V. A., Ptitsyn, G. A., & Komissarov, G. G. (2008). Influence of the molecular structure of the films of tetrapyrrole compounds on their photo electrochemical characteristics for various types of sensitization. *Russian Journal of Physical Chemistry B Focus on Physics*, *27(12)*, 66–70 (in Russian)
4. Ilatovsky, V. A., Dmitriev, I. B., Kokorin, A. I., Ptitsyn, G. A., Komissarov, G. G. (2009). Influence of the nature of coordinated metal on photoelectrochemical activity of thin films of tetrapyrrole compounds. *Russian Journal of Physical Chemistry B Focus on Physics, 28(1)*, 89–96 (in Russian)

5. Lobanov, A. V., Nevrova, O. V., Ilatovsky, V. A., Sin'ko, G. V., & Komissarov, G. G. (2011). Coordination and photocatalytic properties of metal porphyrins in hydrogen peroxide decomposition *Macroheterocycles 4(2)*, 132–134.
6. Ilatovsky, V. A., Ovcharov, L. F., Shlyakhovoy, V. V., & Komissarov, G. G. (1975). Dependence of the parameters and reproducibility of pigmented electrodes on the conditions of pigment film deposition. *Russian Journal of Physical Chemistry A*, 49(5), 1351. VINITI № 3323–74 from 30.12. 74, p.1–18 (in Russian).
7. Ilatovsky, V. A., Apresian, E. S., & Komissarov, G. G. (1988). Increase in photo activity of phthalocyanines at structural modification of thin-film electrodes. *Russian Journal of Physical Chemistry A, 62(6)*, 1612–1617 (in Russian)
8. Ilatovsky, V. A., Sinko, G. V., Ptitsyn, G. A., & Komissarov, G. G. (2012). Structural sensitization of pigment films in the formation of nanoscale single crystal clusters. / Collection Dynamics of chemical and biological processes, XXI century. ICP RAS, Moscow, 173 (in Russian).
9. Moser, F. H., & Thomas, A. L. (1963). *Phthalocyanines A.C.S Monograph. (157)*, N.Y. Reinhold Pbbl. Corp.
10. Ilatovsky, V. A., Apresian, E. S., & Komissarov, G. G. (1988). *Russian Journal of Physical Chemistry A, 62(6)*, 1612 (in Russian)
11. Apresian, E. S., Ilatovsky, V. A., & Komissarov, G. G. (1989). *Russian Journal of Physical Chemistry A*, 63(8), 2239 (in Russian)
12. Rudakov, V. M., Ilatovsky, V. A., & Komissarov, G. G. (1987). *Russian Journal of Physical Chemistry B Focus on Physics*, 6(4), 552 (in Russian)
13. Perdew, J. P., Burke, K., & Ernzerhof, M. (1996). *Phys. Rev. Lett. 77*, 3865.
14. Perdew, J. P., Burke, K., & Ernzerhof, M. (1997). *Phys. Rev. Lett. 78*, 1396.
15. Ernzerhof, M., Perdew, J. P., & Burke, K. (1997). *Int. J. Quantum Chem., 64*, 285.
16. Ernzerhof, M., & Scuseria, G. E. (1999). *J. Chem. Phys., 110*, 5029.

CHAPTER 7

BIO-STRUCTURAL ENERGY CRITERIA OF FUNCTIONAL STATES IN NORMAL AND PATHOLOGICAL CONDITIONS

G. A. KORABLEV and G. E. ZAIKOV

CONTENTS

ABSTRACT

With the help of spatial-energy notions it is demonstrated that molecular electronegativity and energy characteristics of functional states of bio-systems are defined basically by P-parameter values of atom first valence electron. The principles of stationary biosystem formation are similar to the conditions of wave processes in the phase.

7.1 SPATIAL-ENERGY P-PARAMETER

The analysis of kinetics of various physical and chemical processes demonstrates that in many cases the reverse values of velocities, kinetic or energy characteristics of corresponding interactions are added.

A few examples: ambipolar diffusion, summary rate of topochemical reaction, change in the light velocity when moving from vacuum to a given medium, effective penetration of biomembranes.

In particular, such assumption is confirmed by the probability formula of electron transport (W_∞) due to the overlapping of wave functions 1 and 2 (in stationary state) at electron-conformation interactions:

$$W_\infty = \frac{1}{2}\frac{W_1 W_2}{W_1 + W_2} \tag{1}$$

Equation (1) is used [1] when assessing the characteristics of diffusion processes accompanied with electron nonradiating transport in proteins.

The modified Lagrangian equation is also illustrative. For the relative movement of isolated system of two interacting material points with the masses m_1 and m_2 in coordinate x it looks as follows:

$$m_{eq}x'' = -\frac{\partial U}{\partial x} \qquad \frac{1}{m_{eq}} = \frac{1}{m_1} + \frac{1}{m_2}$$

where U – mutual potential energy of points; m_{eq} – equivalent mass. Here $x'' = a$ (characteristic of system acceleration). For interaction elementary areas Δx can be taken as follows:

$$\frac{\partial U}{\partial x} \approx \frac{\Delta U}{\Delta x} \text{ That is: } m_{np}a\Delta x = -\Delta U \text{. Then:}$$

$$\frac{1}{1/(a\Delta x)}\frac{1}{(1/m_1+1/m_2)}\approx-\Delta U \;;\; \frac{1}{1/(m_1 a\Delta x)+1/(m_2 a\Delta x)}\approx-\Delta U$$

Or:

$$\frac{1}{\Delta U}\approx\frac{1}{\Delta U1}+\frac{1}{\Delta U2} \tag{2}$$

where ΔU_1 and ΔU_2 – potential energies of material points on the interaction elementary area, ΔU – resultant (mutual) potential energy of these interactions.

"Electron with mass m moving near the proton with mass M is equivalent to the particle with mass $m_{eq}=\frac{mM}{m+M}$." [2].

In this system the energy characteristics of subsystems are: electron orbital energy (W_i) and effective nucleus energy that takes screening effects into account (by Clementi).

Therefore, assuming that the resultant interaction energy of the system orbital-nucleus (responsible for interatomic interactions) can be calculated following the principle of adding the reverse values of some initial energy components, we substantiate the introduction of P-parameter [3] as averaged energy characteristics of valence orbitals according to the following equations:

$$\frac{1}{q^2/r_i}+\frac{1}{W_i n_i}=\frac{1}{P_E} \tag{3}$$

$$P_E=\frac{P_0}{r_i} \tag{4}$$

$$\frac{1}{P_0}=\frac{1}{q^2}+\frac{1}{(Wrn)_i} \tag{5}$$

$$q=\frac{Z^*}{n^*} \tag{6}$$

where W_i – electron bond energy [4]; or atom ionization energy (E_i) – [5]; r_i – orbital radius of i–orbital [6]; n_i – number of electrons of the given orbital, Z^* and n^* – nucleus effective charge and effective main quantum number [7, 8].

P_0 is called the spatial-energy parameter, and P_E – the effective P-parameter. The effective P_E-parameter has a physical sense of some averaged energy of valence electrons in the atom and is measured in energy units, e.g. electron-volts (eV).

The values of P-parameters are calculated based on the Eqs. (3–6), some results are given in Table 7.1. At the same time, the values of parameters during the orbital hybridization are used (marked with index "H") [9]. The calculations are carried out taking into account the possibility of fold bond formation (single, double, triple). For carbon atom the covalent radius of triple bond (0.60 Å) nearly equals its orbital radius of 2P-orbital (0.59 Å), therefore, the possibility of carbon compound formation becomes more effective.

TABLE 7.1 P-Parameters of Atoms Calculated Via the Electron Bond Energy

Atom	Valence electrons	W (eV)	r_i (Å)	q^{2-}_0 (eVÅ)	P_0 (eVÅ)	R (Å)	P_0/R (eV)
H	$1S^1$	13.595	0.5292	14.394	4.7969	0.5292	9.0644
						0.375	12.792
						0.28	17.132
C	$2P^1$	11.792	0.596	35.395	5.8680	0.77	7.6208
	$2P^2$	11.792	0.596	35.395	10.061	0.67	8.7582
	$2P^3_r$	19.201	0.620	37.240	13.213	0.60	9.780
	$2S^1$				9.0209	0.77	13.066
	$2S^2$				14.524	0.67	15.016
	$2S^1+2P^3_r$				22.234	0.60	16.769
	$2S^1+2P^1_r$				13.425	0.77	17.160
	$2S^2+2P^2$				24.585	0.77	11.715
					24.585	0.77	18.862
						0.77	28.875
						0.77	17.435
						0.77	31.929
						0.67	36.694
						0.60	40.975

TABLE 7.1 *(Continued)*

Atom	Valence electrons	W (eV)	r_i (Å)	q^{2-}_0 (eVÅ)	P_0 (eVÅ)	R (Å)	P_0/R (eV)
	$2P^1$	15.445	0.4875	52.912	6.5916	0.70	9.4166
	$2P^2$	25.724	0.521	53.283	11.723	0.55	11.985
	$2P^3$				15.830	0.70	16.747
	$2S^2$				17.833	0.63	18.608
N	$2S^2+2P^3$				33.663	0.70	22.614
						0.55	28.782
						0.70	25.476
						0.70	48.090
	$2P^1$	17.195	0.4135	71.383	6.4663	0.66	9.7979
	$2P^1$	17.195	0.4135	71.383	11.858	0.55	11.757
	$2P^2$	17.195	0.4135	71.383	20.338	0.66	17.967
	$2P^4$	33.859	0.450	72.620	21.466	0.59	20.048
O	$2S^2$				41.804	0.66	30.815
	$2S^2+2P^4$					0.59	34.471
						0.66	32.524
						0.66	63.339
						0.59	70.854

And now we will briefly discuss the reliability of such approach. According to the calculations [3] the values of P_E-parameters numerically equal (in the limits of 2%) the total energy of valence electrons (U) by the atom statistic model. Using a well-known correlation between the electron density (b) and interatomic potential by atom statistic model, we can obtain the direct dependence of P_E-parameter upon the electron density at the distance r_i from the nucleus:

$$\beta_i^{2/3} = \frac{AP_0}{r_i} = AP_E$$

where A – constant.

The rationality of this equation can be confirmed when calculating the electron density using Clementi's wave functions [10] and comparing it with the value of electron density calculated via the values of P_E-parameter.

The modules of maximum values of radial part of Ψ-function are correlated with the values of P_0-parameter and the linear dependence between these values is established. Using some properties of wave function as applied to P-parameter, we obtain the wave equation of P-parameter that is formally analogous with Ψ-function equation.

Based on the calculations and comparisons, two principles of adding spatial-energy criteria depending on wave properties of P-parameter and system character of interactions and particle charges are substantiated:

1. Interaction of oppositely charged (heterogeneous) systems containing I, II, III, … atom grades is satisfactorily described by the principle of adding their reverse energy values based on equations (5–8), thus corresponding to the minimum of weakening the oscillations in antiphase;

2. When similarly charged (homogeneous) subsystems are interacted, the principle of algebraic addition of their P-parameters is followed based on the equations:

$$\sum_{i=1}^{m} P_0 = P_0' + P_0'' + \ldots + P_0^{m} \quad (7)$$

$$\Sigma P_E = \frac{\Sigma P_0}{R} \quad (8)$$

where R – dimensional characteristic of an atom (or chemical bond) – thus corresponding to the maximum of enhancing the oscillations in phase.

7.2 P-PARAMETER AS AN OBJECTIVE ELECTRONEGATIVITY ASSESSMENT

The notion of electronegativity (EN) was introduced by Poling in 1932, as a quantitative characteristic of atoms' ability to attract electrons in a molecule. Currently there are a lot of methods to calculate the electronegativity (thermal-chemical, geometrical, spectroscopic, etc.) producing comparable results. The notion of electronegativity is widely applied in chemical and crystal-chemical investigations.

Taking into account Sandersen's idea that EN changes symbately with atom electron density and following the physical sense of P-parameter as a direct characteristic of atom electron density at the distance r_i from the nucleus,

we assume that electro negativity for the lowest stable oxidation degree of the element equals the averaged energy of one valence electron:

$$X = \frac{\sum Po}{3nR} \tag{9}$$

or – for the first valence electron:

$$X = \frac{Po}{3R} \tag{9a}$$

where: $\sum P_0$ – total of P_0-parameters for n-valence electrons, R – atom radius (depending on the bond type – metal, crystal or covalent), value of P_0–parameter is calculated via the electron bond energy [4]. Digit 3 in the denominator (9, 9a) demonstrates that probable interatomic interaction is considered only along the bond line, that is, in one of the three spatial directions. The calculation of molecular electro negativity for all elements by these equations is given in Table 7.2, there is no table for metal one.

For some elements (characterized by the availability of both metal and covalent bonds) the electro negativity is calculated in two variants – with the use of the values of atomic and covalent radii.

The deviations in calculations of electro negativity values by Batsanov [11, 12] and Allred-Rokhov in most cases do not exceed 2–5% from those generally accepted.

Thus, simple correlations (9 and 9a) quite satisfactorily assess the electro negativity value in the limits of its values based on reference data.

The advantage of this approach – greater possibilities of P-parameter to determine the electronegativity of groups and compounds, as P-parameter quite simply (based on the initial rules) can be calculated both for simple and complex compounds.

At the same time, the individual features of structures can be considered, and consequently, important physical-chemical properties of these compounds can be not only characterized but also predicted (isomorphism, mutual solubility, eutectic temperature, etc.). For instance, with the help of the notion of P-parameter we can evaluate the electronegativity of not only metal but also crystal structures.

From the data obtained we can conclude that molecular electronegativity for the majority of elements numerically equals the P_E-parameter of the first valence electron divided by three.

It should be noted that there is a significant difference between the notions of electronegativity and P_E-parameter: EN – stable characteristic of an atom (or radical), and the value of P_E-parameter depends not only on quantum number of valence orbital, but also the bond length and bond type. Thus P_E-parameter is an objective and most differentiated energy characteristic of atomic structure.

7.3 SPATIAL-ENERGY CRITERIA OF FUNCTIONAL STATES OF BIOSYSTEMS

With the help of P-parameter methodology the spatial-energy conditions of isomorphic replacement are found based on experimental data of about a thousand different systems [3]:

1. Complete (100%) isomorphic replacement and structural interaction take place at approximate equality of P-parameters of valence orbitals of exchangeable atoms: $P_A' \approx P_A''$
2. P-parameter of the least value determines the orbital, which is mainly responsible for the isomorphism and structural interactions.

Modifying the rules of adding the reverse values of the energy magnitudes of subsystems as applicable to complex structures, we can obtain the equations to calculate a P_C-parameter of a complex structure [13]:

$$\frac{1}{P_C} = \left(\frac{1}{NP_E}\right)_1 + \left(\frac{1}{NP_E}\right)_2 + \ldots \tag{10}$$

where N_1 and N_2 – number of homogeneous atoms in subsystems.

The calculation results in this equation for some atoms and radicals of biosystems are given in Table 7.3.

When forming the solution and other structural interactions, the same electron density should be set in the contact spots of atom-components. This process is accompanied with the redistribution of electron density between the valence areas of both particles and transition of a part of electrons from some external spheres into the neighboring ones.

Apparently, with the approximation of electron densities in free atom-components, the transition processes between boundary atoms of particles are minimal, thus facilitating the formation of a new structure. So the task of evaluating the degree of structural interactions mostly comes to the compara-

tive assessment of electron density of valence electrons in free atoms (on averaged orbitals) participating in the process based on the following equation:

$$\alpha = \frac{P'_E - P''_E}{P'_E + P''_E} 200\% \quad (11)$$

The degree of structural interactions (ρ) in many (over a thousand) simple and complex systems is evaluated following this technique. The nomogram of ρ dependence on the coefficient of structural interaction (α) is created Fig. 7.1.

Isomorphism as a phenomenon is usually considered as applicable to crystal structures. But, obviously, similar processes can also flow between molecular compounds, where the bond energy can be evaluated via the relative difference of electron densities of valence orbitals of interacting atoms. Therefore the molecular electronegativity is quite easily calculated via the values of the corresponding P-parameters.

Since P-parameter possesses wave properties (similar to Y'-function), mainly the regularities in the interference of the corresponding waves should be fulfilled.

The *interference minimum*, oscillation weakening (in antiphase) takes place if the difference in wave move (Δ) equals the odd number of semiwaves:

$$\Delta = (2n+1)\frac{\lambda}{2} = l(n + \frac{1}{2}), \text{ where } n = 0, 1, 2, 3, \ldots \quad (12)$$

As applicable to P-parameters, this rule means that the minimum of interactions take place if P-parameters of interacting structures are also "in antiphase" – either oppositely charged systems or heterogeneous atoms are interacting (e.g., during the formation of valence-active radicals CH, CH_2, CH_3, NO_2 …, etc.).

In this case, P-parameters are summed up based on the principle of adding the reverse values of P-parameters – equations (3–5).

The difference in wave move (Δ) for P-parameters can be evaluated via their relative value ($g=\frac{P_2}{P_1}$) or via relative difference of P-parameters (coefficient a), which give an odd number at minimum of interactions:

$$g=\frac{P_2}{P_1} = \left(n+\frac{1}{2}\right) = \frac{3}{2}; \frac{5}{2} \ldots\ldots \text{ At n=0 (basic state) } \frac{P_2}{P_1} = \frac{1}{2} \quad (12a)$$

Let us point out that for stationary levels of one-dimensional harmonic oscillator the energy of theses levels $e=hn(n+\frac{1}{2})$, therefore, in quantum oscil-

lator, as opposed to the conventional one, the least possible energy value does not equal zero.

In this model the interaction minimum does not provide the energy equaled to zero corresponding to the principle of adding the reverse values of P-parameters.

The *interference maximum*, oscillation enhancing (in phase) takes place if the difference in wave move equals an even number of semiwaves:

$$\Delta = 2n\frac{\lambda}{2} = \lambda n \quad \text{or} \quad \Delta = \lambda(n+1) \tag{13}$$

As applicable to P-parameters, the maximum enhance of interaction in the phase corresponds to the interactions of similarly charged systems or systems homogeneous by their properties and functions (e.g., between the fragments or blocks of complex organic structures, such as CH_2 and NNO_2 in octogene).

Then:

$$g = \frac{P_2}{P_1} = (n+1) \tag{13a}$$

In the same way, for "singular" systems (with similar values of functions) of two-dimensional harmonic oscillator the energy of stationary states is found as follows: $e = hn(n+1)$

By this model the maximum of interactions corresponds to the principle of algebraic addition of P-parameters – equations (7–8). When n=0 (basic state), we have $P_2 = P_1$, or: the maximum of structure interaction occurs if their P-parameters are equal. This postulate and equation (13a) are used as basic conditions for the formation of stable structures [13].

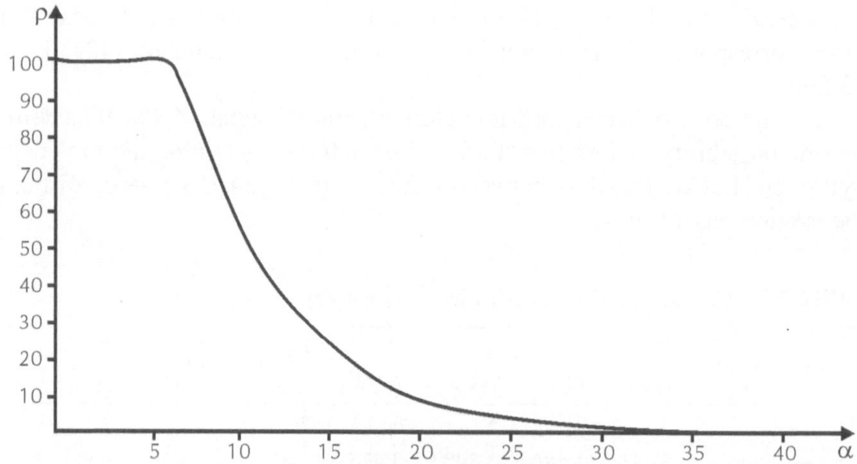

FIGURE 7.1 Dependence of the degree of structural interactions on coefficient α

Hydrogen atom, element No 1 with orbital $1S^1$ defines the main energy criteria of structural interactions (their "ancestor").

Table 7.1 shows its three P_E-parameters corresponding to three different characteristics of the atom.

$R_1 = 0.5292$ A^0 – orbital radius – quantum-mechanical characteristic gives the initial main value of P_E-parameter equaled to 9.0644 eV;

$R_2 = 0.375$ A^0 – distance equaled to the half of the bond energy in H_2 molecule. But if hydrogen atom is bound with other atoms, its covalent radius is $\approx$ 0.28 A^0. Let us explain the reason:

In accordance with equation (13a) $P_2 = P_1$ (n+1), therefore $P_1 \approx 9.0644$ eV, $P_2 \approx 18.129$ eV.

These are the values of possible energy criteria of stable (stationary) structures. The dimensional characteristic 0.375 A^0 does not satisfy them, therefore, there is a transition onto to the covalence radius $\approx$ 0.28 A^0, which provides the value of P-parameter approximately equaled to P_2.

From a big number of different combinations of interactions we can obtain series with approximately equal values of P-parameters of atoms (or radicals). Such series, by initial values of hydrogen atom, are given in Table 7.4 (at α < 7.5%).

First series for P_E= 9.0644 eV – the main, initial, where H, C, O, N atoms have P_E-parameters only of the first electron and interactions proceed in the phase.

Second series for P'_E = 12.792 eV is the nonrational, pathological as it more corresponds to the interactions in antiphase: by equation (12a) P''_E = 13.596 eV.

Coefficient α between the parameters P'_E and P''_E equals 6.1%, thus defining the possibility of forming "false" biostructures containing the molecular hydrogen H_2. Coefficient *α* between series I and II is 34.1%, thus confirming the irrationality of series II.

TABLE 7.2 Calculation of Molecular Electro Negativity

Atom	Orbital	W (eV)	r_i (Å)	q^2 (eVÅ)	P_0 (eVÅ)	R_K (Å)	$P_0/3Rn$ (eV)	X (Batsanov)
1	2	3	4	5	6	7	8	9
Li	$2S^1$	5.3416	1.586	5.8902	3.475	1.33	0.87	0.98
Be	$2S^1$	8.4157	1.040	13.159	5.256	1.13 (M)	1.55	1.52
B	$2P^1$	8.3415	0.770	21.105	4.965	0.81	2.04	2.03
C	$2P^1$	11.792	0.596	35.395	5.868	0.77	2.54	2.56
N	$2P^1$	15.445	0.4875	52.912	6.5903	0.74	2.97	3.05
O	$2P^1$	17.195	0.4135	71.383	6.4660	0.66	3.27	3.42
F	$2S^22P^5$				50.809	0.64	3.78	3.88
Na	$3S^1$	4.9552	1.713	10.058	4.6034	1.54	1.00	0.98
Mg	$3S^1$	6.8859	1.279	17.501	5.8588	1.6 (M)	1.22	1.28
Al	$3P^1$	5.7130	1.312	26.443	5.8401	1.26	1.55	1.57
Si	$3P^1$	8.0848	1.068	29.377	6.6732	1.17	1.90	1.89
P	$3P^1$	10.659	0.9175	38.199	7.7862	1.10	2.36	2.19
S	$3P^1$	11.901	0.808	48.108	8.432	1.04	2.57	2.56
Cl	$3P^1$	13.780	0.7235	59.844	8.546	1.00	2.85	2.89
K	$4S^1$	4.0130	2.162	10.993	4.8490	1.96	0.83	0.85
Ca	$4S^1$	5.3212	1.690	17.406	5.9290	1.74	1.14	1.08
Sc	$4S^1$	5.7174	1.570	19.311	6.1280	1.44	1.42	1.39
Ti	$4S^1$	6.0082	1.477	20.879	6.227	1.32	1.57	1.62
V	$4S^1$	6.2755	1.401	22.328	6.3077	1.22 1.34	1.72 1.57	(1.7) 1.54
Cr	$4S^23d^1$				17.168	1.19	1.60	1.63
Mn	$4S^1$	6.4180	1.278	25.118	6.4180	1.18	1.81	1.73

TABLE 7.2 *(Continued)*

Fe	$4S^1$	7.0256	1.227	26.572	6.5089	1.17	1.85	1.82
						1.26	1.72	1.74
Co	$4S^1$	7.2770	1.181	27.983	6.5749	1.16	1.89	1.88
Ni	$4S^1$	7.5176	1.139	29.348	6.6226	1.15	1.92	1.98
Cu	$4S^1$	7.7485	1.191	30.117	7.0964	1.31	1.81	(1.8)
	$4S^13d^1$				13.242	1.31	1.68	1.64
Zn	$4S^1$	7.9594	1.065	32.021	6.7026	1.31	1.71	1.72
Ga	$4P^1$	5.6736	1.254	34.833	5.9081	1.25	1.58	(1.7)
	$4S^24P^1$				20.760	1.25	1.82	1.87
Ge	$4P^1$	7.819	1.090	41.372	7.0669	1.22	1.93	2.08
	$4S^24P^2$				30.370	1.22	2.07	
As	$4P^1$	10.054	0.9915	49.936	8.275	1.21	2.28	2.23
Se	$4P^1$	10.963	0.909	61.803	8.5811	1.17	2.44	2.48
Br	$4P^1$	12.438	0.8425	73.346	9.1690	1.11	2.75	2.78
Rb	$5S^1$	3.7511	2.287	14.309	5.3630	2.22	0.81	0.82
Sr	$5S^1$	4.8559	1.836	21.224	6.2790	2.00	1.05	1.01
Y	$5S^1$	6.3376	1.693	22.540	6.4505	1.69	1.27	1.26
Zr	$5S^1$	5.6414	1.593	23.926	6.5330	1.45	1.50	1.44
Nb	$5S^1$	5.8947	1.589	20.191	6.3984	1.34	1.52	1.56
Mo	$5S^1$	6.1140	1.520	21.472	6.4860	1.29	1.68	1.73
Tc	$5S^1$	6.2942	1.391	30.076	6.7810	1.27	1.78	1.76
Ru	$5S^1$	6.5294	1.410	24.217	6.6700	1.24	1.79	1.85

Ru (II)	$5S^1 4d^1$				15.670	1.24	2.11	(2.1)
Rh	$5S^1$	6.7240	1.364	33.643	7.2068	1.25	1.92	1.96
Rh ($5S^1 4d^8$)	$5S^1$	6.7240	1.364	25.388	6.7380	1.25	1.87	(1.82)
Pd	$5S^2 4d^2$				30.399	1.28	1.98	1.95
Pd	$5S^1$	6.9026	1.325	35.377	7.2670	1.28	1.89	1.95; 1.85
Ag* ($5S^2 4d^9$)	$5S^1$	7.0655	1.286	37.122	6.9898	1.25	1.86	1.90
Ag ($5S^1 4d^{10}$)	$5S^1$	7.0655	1.286	26.283	6.7520	1.34	1.68	(1.66)
Cd	$5S^1$	7.2070	1.184	38.649	6.9898	1.38	1.69	(1.68); 1.62
Jn	$5S^2 5P^1$				21.841	1.42	1.71	1.76; 1.68
Sn	$5P^1$	7.2124	1.240	47.714	7.5313	1.40	1.79	(1.80); 1.88

Third series for $P'_E = 17.132$ eV – stationary as the interactions are in the phase: by equation (13a) $P''_E = 18.129$ eV ($\alpha = 5.5\%$).

With specific local energy actions (electromagnetic fields, radiation, etc.), the structural formation processes can grow along the pathological series II. Maybe this is the reason of oncological diseases? If this is so, the practical recommendations can be given. Some of them are simple and common but are now being substantiated. They propose to transform the molecular hydrogen H_2 into the atomic one. In former times the ashen alkaline water was used in Russian sauna, i.e. OH^- hydroxyl groups. A so-called "live" alkaline water fraction successfully used for the treatment of a number of diseases has the same value. The water containing fluorine and iodine ions is similar.

During the transplantation and use of stem cells the condition of approximate equality of P-parameters of the corresponding structures should be observed (not by the series II).

From Table 7.4, it is seen that the majority of atoms and radicals, depending on the bond types and bond lengths, have P_E-parameters of different series. When introducing the stem cells, it is important for the molecular hydrogen not to be present in their structures. Otherwise atoms and radicals can transfer into the series II and disturb the vital functions of the main first system.

Tables 7.2–7.4 contain only those atoms and radicals that have the main value in the formation of the molecules of DNA, RNA and nitrogenous bases of nucleic acids (Ц-Г, А-Г). For these pairs the calculations give α equaled to 0.3%. The average values of P-parameters of contacting hydrocarbon rings Ц-$Г_1$ and А-$Т_1$ are also nearly the same. The calculations demonstrate that the systems with P-parameters approximately two times less than in system II are also stable, as in the pair of structural formations they produce the nominal value of the parameter close to the initial one (9.0644 eV).

TABLE 7.3 Structural P_C-Parameters Calculated Via the Bond Energy of Electrons

Radicals, molecule fragments	P_i' (eV)	P_i'' (eV)	P_C (eV)	Orbitals
	9.7979	9.0644	4.7084	O ($2P^1$)
OH	30.815	17.132	11.011	O ($2P^4$)
	17.967	17.132	8.7710	O ($2P^2$)
H_2O	2·9.0644	17.967	9.0237	O ($2P^2$)
	2·17.132	17.967	11.786	O ($2P^2$)
	17.160	2·9.0644	8.8156	C ($2S^12P^3_r$)
CH_2	31.929	2·17.132	16.528	C ($2S^22P^2$)
	36.694	2·9.0644	12.134	C ($2S^12P^3_r$)
	31.929	3·17.132	19.694	C ($2S^22P^2$)
CH_3	15.016	3·9.0644	9.6740	C ($2P^2$)
	40.975	3·9.0644	16.345	C ($2S^22P^2$)
	36.694	17.132	11.679	C ($2S^22P^2$)
CH	31.929	12.792	9.1330	C ($2S^22P^2$)
	40.975	17.132	12.081	C ($2S^22P^2$)

TABLE 7.3 *(Continued)*

Radicals, molecule fragments	$P_i^{'}$ (eV)	$P_i^{''}$ (eV)	P_C (eV)	Orbitals
	16.747	17.132	8.4687	N($2P^2$)
NH	19.538	17.132	9.1281	N($2P^2$)
	48.090	17.132	12.632	N($2S^22P^3$)
	19.538	2·9.0644	9.4036	N($2P^2$)
NH_2	16.747	2·17.132	12.631	N($2P^2$)
	28.782	2·17.132	18.450	N($2P^3$)
C_2H_5	2·31.929	5·17.132	36.585	C ($2S^22P^2$)
NO	19.538	17.967	9.3598	N($2P^2$)
	28.782	20.048	11.817	N($2P^3$)
CH_2	31.929	2·9.0644	11.563	C ($2S^22P^2$)
CH_3	16.769	3·17.132	12.640	C ($2P^2$)
CH_3	17.160	3·17.132	12.865	C ($2P^3_r$)
CO–OH	8.4405	8.7710	4.3013	C ($2P^2$)
CO	31.929	20.048	12.315	C ($2S^22P^2$)
C=O	15.016	20.048	8.4405	C ($2P^2$)
C=O	31.929	34.471	16.576	O ($2P^4$)
CO=O	36.694	34.471	17.775	O ($2P^4$)
C–CH_3	31.929	19.694	12.181	C ($2S^22P^2$)
C–CH_3	17.435	19.694	9.2479	–
C–NH_2	31.929	18.450	11.693	C ($2S^22P^2$)
C–NH_2	17.435	18.450	8.8844	–

TABLE 7.4 Biostructural Spatial-Energy Parameters (eV)

Series number	H	C	N	O	CH	CO	NH	C–NH$_2$	C–CH$_3$	$<P_E>$	α
I	9.0644 ($1S^1$)	8.7582 ($2P^1$) 9.780 ($2P^1$)	9.4166 ($2P^1$)	9.7979 ($2P^1$)	9.1330 ($2S^22P^2$-$1S^1$)	8.4405 ($2P^2$-$2P^2$)	8.4687 ($2P^2$-$1S^1$) 9.1281 ($2P^2$-$1S^1$)	8.8844 $2S^12P^1_r$– ($2P^3$-$1S^1$)	9.2479 $2S^12P^1_r$– ($2S^2$-$2P^2$-$1S^1$)	9.1018	0.34–7.54
II	12.792 ($1S^1$)	13.066 ($2P^2$) 11.715 ($1S^1$)	11.985 ($2P^1$)	11.757 ($2P^1$)	11.679 ($2S^22P^2$-$1S^1$) 12.081 ($2S^22P^2$-$1S^1$)	12.315 ($2S^22P^2$-$2P^2$)	12.632 ($2S^22P^3$-$1S^1$)	11.693 $2S^22P^2$– ($2P^3$-$1S^1$)	12.181 $2S^22P^2$– ($2S^22P^2$-$1S^1$)	12.173	0.07–7.08
III	17.132 ($1S^1$)	16.769 ($2P^2$) 17.435 ($2S^12P^1$)	16.747 ($2P^2$)	17.967 ($2P^2$)	C and H blocks	16.576 ($2S^22P^2$-$2P^4$)	N and H blocks	C and NH$_2$ blocks	C and NH$_2$ blocks	17.104	0.16–4.92

Note: The designations of interacting orbitals are given in brackets.

7.4 CONCLUSIONS

Based on spatial-energy notions it is shown that:

1. Molecular electronegativity of the majority of elements numerically equals the P-parameter of the first valence electron (divided by three).
2. P-parameters of the first valence electron of atoms define the energy characteristics of stationary states (in normal state) under the condition of the maximum of wave processes.
3. Under the condition of the minimum of such interactions, the pathological (but not stationary) biostructures containing the molecular hydrogen can be formed.

KEYWORDS

- **Biosystems**
- **Electronegativity**
- **Spatial-energy parameter**
- **Stationary and pathological states**

REFERENCES

1. Rubin, A. B. (1987). Biophysics, *Theoretical Biophysics* Vysshaya shkola, 319
2. Aring, G., Walter, D., & Kimbal, D. (1948). Quantum Chemistry, M: I. L., 528.
3. Korablev, G. A. (2005). Spatial-Energy Principles of Complex Structures Formation Netherlands, Brill Acad. Publ. and VSP, 426
4. Fischer, C. F., (1972). *Atomic Data, 4,* 301
5. Allen, K. U., (1977). *Astrophysical magnitudes* M: Mir, 446.
6. Waber, J. T., & Cromer, D. T., (1965). *J. Chem. Phys., 12*, 4116.
7. Clementi, E., & Raimondi, D. L. (1963). *J. Chem. Phys., 1963, 11*, 2686.
8. Clementi, E., & Raimondi, D. L. (1967). *J. Chem. Phys., 14*, 1300.
9. Korablev, G. A., & Zaikov, G. E. (2006). *J. Appl. Polym. Sci., 3*, 2101
10. Clementi, E. (1965). J.B.M. S. Res. Develop Suppl., 2, 76.
11. Batsanov, S. S., & Zvyagina, R. A. (1966). Overlap integrals and problem of effective charges. Novosibirsk: Nauka, 386.
12. Batsanov, S. S. (1976). Structural Refractometry, M: Vysshaya shkola, 304.
13. Korablev, G. A. (2010). Exchange Spatial-energy Interactions. P. H.: Udmurt State University, 530.

CHAPTER 8

THE TEMPORAL DEPENDENCE OF ADHESION JOINING STRENGTH: THE DIFFUSIVE MODEL

KH. SH. YAKH'YAEVA, G. V. KOZLOV, G. M. MAGOMEDOV, R. A. PETHRICK, and G. E. ZAIKOV

CONTENTS

ABSTRACT

The quantitative model of the dependence of adhesion contact shear strength on its formation duration was proposed, which uses fractal analysis and strange (anomalous) diffusion notions. This model gives a clear physical picture of contact formation and allows precise enough estimations of adhesion joining strength.

8.1 INTRODUCTION

At present the general law governed nature is well-known: adhesional contact formation duration increasing at other equal conditions results in its strength enhancement [1, 2]. As a rule, this effect is explained by macromolecular coils diffusion in boundary layer of samples, forming adhesional contact [3]. However, such explanation has usually a qualitative character. Therefore the present work purpose is the development of the quantitative model adhesional contact strength temporal dependence. Within the frameworks of fractal analysis and anomalous (strange) diffusion conception on the example of amorphous polymer polystyrene (PS)[1].

8.2 EXPERIMENTAL PART

Amorphous PS (M_w=23×10^4, M_w/M_n=2.84), obtained from Dow Chemical (USA) was used. The glass transition temperature T_g was measured using Perkin-Elmer DSC-4 differential scanning calorimeter, at a heating rate of 20 K/min (T_g=376 K for PS) [2]. For adhesional joining formation two samples with width of 5 mm were bonded in a lap-shear joint geometry with the area of 5′5 mm^2, using a Carver laboratory press at constant temperature and pressure of 0.8 MPa. Division boundaries PS-PS were healed during 10 min and 24 h within the range of temperatures of 335–373 K – in all cases below glass transition temperature. Mechanical tests of the formed contacts were conducted at temperature 293 K on an Instron tensile tester, Model 1130 at tension rate of 3×10–^{2} m/s with shear strength determination in the contact zone (or on division boundary) [1, 2].

8.3 RESULTS AND DISCUSSION

As it is known [4], within the frameworks of fractal analysis shear strength τ_c of adhesional joining is described by the following general equation:

$$\ln \tau_c = A \ln N_c - B, \tag{1}$$

where A and B are constants, which can be changed depending on polymer nature, temperature, testing specific conditions and so on, N_c is macromolecular coils contacts number, which is determined as follows [5]:

$$N_c \sim R_g^{D_{f1}+D_{f2}-d}, \tag{2}$$

where R_g is a macromolecular coil gyration radius, D_{f1} and D_{f2} are fractal dimensions of the coils structure, forming autohesional bonding, d is dimension of Euclidean space, in which a fractal is considered (it is obvious, that in our case d=3).

For the autohesion case $D_{f1}=D_{f2}=D_f$ and d=3 the equation (2) is simplified up to:

$$N_c \sim R_g^{2D_f-3} \tag{3}$$

Let us consider estimation methods of the parameters, included in the equation (3), i.e. D_f and R_g. For the dimension D_f estimation the following approximated technique will be used, which consists in the following. As it is known [7], between D_f and structure dimension d_f of linear polymers in the solid state the following intercommunication exists:

$$D_f = \frac{d_f}{1.5} \tag{4}$$

where d_f estimation can be conducted according to the formula [6]:

$$d_f = 3 - 6\left(\frac{\phi_{cl}}{SC_\infty}\right)^{1/2}, \tag{5}$$

where φ_{cl} is a relative fraction of local order domains (clusters), S is a macromolecule cross-sectional area, $C_{¥}$ is characteristic ratio, which is an indicator of polymer chain statistical flexibility [8].

φ_{cl} value was estimated according to the following percolation relationship [6]:

$$\phi_{cl} = 0.03\left(T_g - T\right)^{0.55}, \tag{6}$$

where T_g and T are temperatures of glass transition and autohesional contact formation, respectively.

For PS $C_¥$=9.8 [9], S=54.8 Å² [10]. Further the macromolecular coil gyration radius R_g was calculated as follows [11]:

$$R_g = l_0\left(\frac{C_\infty M_w}{6m_0}\right)^{1/2}, \tag{7}$$

where l_0 is the length of the main chain skeletal bond, which is equal to 0.154 nm for PS [9], m_0 is the molar mass per backbone bond, which is equal to 52 for PS [11].

Let us note the important methodological aspect. At R_g value calculation according to the Eq. (7) value $C_¥$ was accepted as a variable one and calculated according to the following equation [6]:

$$C_\infty = \frac{2d_f}{d(d-1)(d-d_f)} + \frac{4}{3} \tag{8}$$

The constants A and B in the Eq. (1) for PS are equal to 2.15 and 6.0, respectively [12].

As it is known [11], at M_w>12 M_e (where M_e is molecular weight of a chain part between macromolecular entanglements) the boundary layer thickness α_i in the autohesion case can be determined according to the formula [11]:

$$\alpha_i = l_0\left(\frac{12C_\infty M_e}{6m_0}\right)^{1/2} \tag{9}$$

In PS case M_e=18000 [11], that is, the indicated above condition M_w>12 M_e is fulfilled and the equation (9) can be used for α_i value estimation. In Fig. 8.1, the comparison of macromolecular coil gyration radius R_g, calculated according to the Eq. (7), and the boundary layer thickness α_i, calculated according to the Eq. (9), is adduced. As one can see, at contact formation duration t=10 min α_i value reaches R_g value in PS case. Let us also note, that the Eq.

(9) does not assume α_i dependence on contact formation duration t, although the dependence $\tau_c(t)$ exists [1]. This circumstance provides boundary layer structure change for autohesional joining at t>10 min.

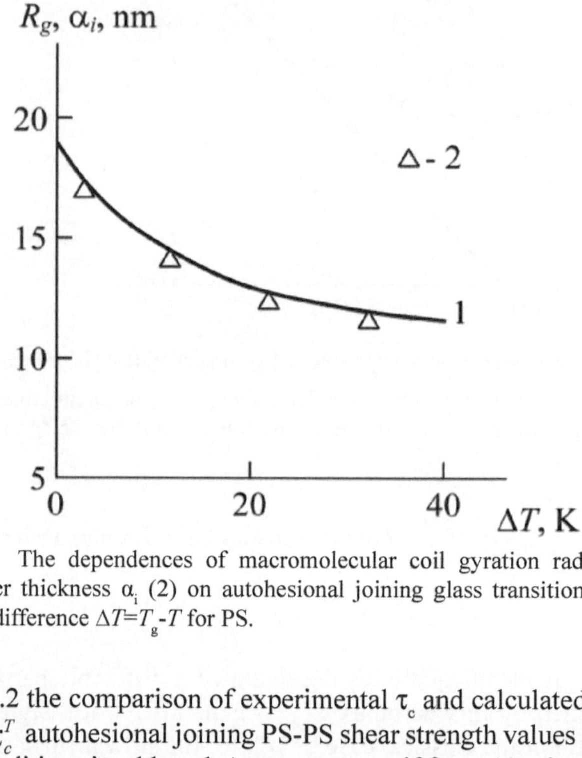

FIGURE 8.1 The dependences of macromolecular coil gyration radius R_g (1) and boundary layer thickness α_i (2) on autohesional joining glass transition and formation temperatures difference ΔT=T_g-T for PS.

In Fig. 8.2 the comparison of experimental τ_c and calculated according to the Eq. (1) τ_c^T autohesional joining PS-PS shear strength values at the indicated above conditions is adduced. As one can see, if for t=10 min the theory and experiment good correspondence is observed, then for t=24 h such correspondence is absent – the values τ_c^T are essentially smaller than the experimentally obtained ones. As it was noted above, this effect was due to boundary layer structure change in virtue of proceeding in it macromolecular coils diffusion processes. Let us consider the indicated processes within the frameworks of anomalous (strange) diffusion conception [13].

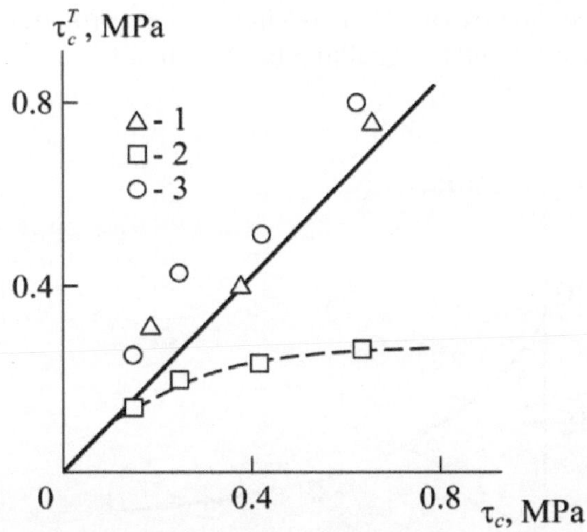

FIGURE 8.2 The comparison of experimental τ_c and calculated according to the equation (1) τ_c^T shear strength values of autohesional joining for PS. The joining formation duration t=10 min (1) and 24 h (2, 3). The calculation of dimension d_f according to the Eq. (5) (1, 2) and Eq. (13) (3).

The basic equation of this conception can be written as follows [14]:

$$\langle r^2(t)\rangle^{1/2} = D_{gen} t^{\beta}, \tag{10}$$

where $\langle r^2(t)\rangle^{1/2}$ is mean-square displacement of a diffusible particle (the size of region, visited by this particle), D_{gen} is generalized diffusivity, b is diffusion exponent. For the classical case b=1/2, for the slow diffusion – b<1/2 and for the fast one – b>1/2. The condition $\beta \neq 1/2$ is the definition of anomalous (strange) diffusion.

The border value for slow and fast diffusion processes is the condition d_f=2.5 [13]. Since for the considered in the present work PS samples $d_f \geq 2.88$, then all proceeding in them diffusion processes are slow ones. The exponent b value is connected with the main parameter in a fractional derivatives theory (fractional exponent a) by the following relationship [13]:

$$\beta = \frac{1-\alpha}{2} \tag{11}$$

In its turn, the fractional exponent a is determined according to the equation [13]:

$$\alpha = d_f - (d-1)\cdot \tag{12}$$

The Eqs. (11) and (12) combination allows to obtain direct interconnection between rate (intensity) of diffusive processes in the autohesional joining border layer, characterized by exponent b, and the polymer structure, characterized by dimension d_f:

$$\beta = \frac{d - d_f}{2} \tag{13}$$

It is obvious, that in the autohesion case the value $\langle r^2(t)\rangle^{1/2}$ will be equal to α_i and then generalized diffusivity D_{gen} at t=10 min in relative units according to the Eq. (10) can be determined. Further the exponent b for t=24 hours can be determined according to the same equation and the dimension d_f of boundary layer structure for the same conditions from the Eq. (13) can be calculated. In Fig. 8.3, the dependence of D_{gen} on autohesional contact formation temperature T for PS is adduced. As it was to be expected from the most common considerations [13, 14], D_{gen} increasing at T growth is observed.

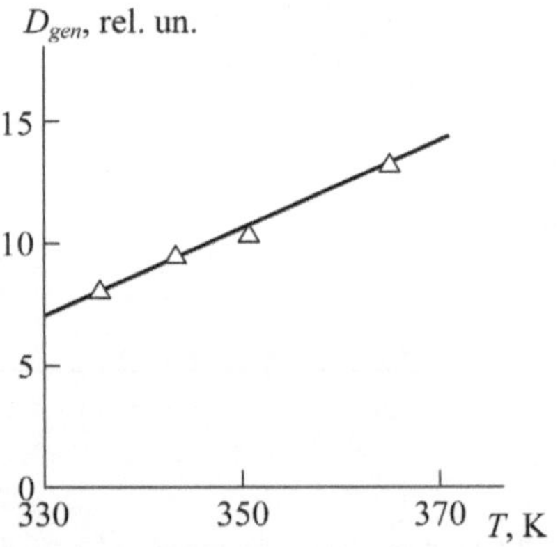

FIGURE 8.3 The dependence of generalized diffusivity D_{gen} on autohesional joining formation temperature T for PS.

Let us note, that the obtained according to the Eq. (13) d_f values for t=24 h proved to be higher than a the ones calculated according to the Eq. (5) (d_f=2.931–2.950 and d_f=2.878–2.914, respectively), that was expected. Then with the described above technique usage (the Eqs. (1), (3), (4) and (13)) the values of autohesional contact shear strength τ_c^T were calculated, the comparison of which with the corresponding experimental values τ_c is adduced in

Fig. 8.2. As one can see, the consideration of autohesional joining boundary layer structure change, which is due to the proceeding in it macromolecular coils anomalous (strange) diffusion processes allows to improve essentially the theory and experiment correspondence. But the main merit of the proposed model is a clear physical picture of autohesional joining shear strength temporal dependence causes.

8.4 CONCLUSIONS

In the present paper the quantitative model of the autohesional joining shear strength dependence on its formation duration was proposed. This model uses notions of fractal analysis and anomalous (strange) diffusion conception. The indicated model application gives a clear physical picture of the observed effect: the obtained experimentally shear strength enhancement is due to boundary layer structure change owing to macromolecular coils diffusion. This model allows also precise enough estimations of autohesional joining strength.

KEYWORDS

- **Adhesion**
- **Anomalous diffusion**
- **Fractal analysis**
- **Polystyrene**
- **Strength**

REFERENCES

1. Boiko, Yu. M. (1997). Prud'homme R. E. *Macromolecules, 30(12),* 3708–3710.
2. Boiko, Yu. M. (2000). *Mechanics of Composite Materials, 36(1)*, 127–134.
3. Boiko, Yu. M. (1998). Prud'homme R. E. *J. Polymer Sci.: Part B: Polymer Phys., 36(4),* 567–572.
4. Yakh'yaeva, Kh. Sh., Kozlov, G. V., Magomedov, G. M., & Zaikov, G. E. (2012). The temperature dependence of auto hesional bonding strength for miscible polymer pairs. *Polymer Research J., 6(2)*, 97–100.
5. Vilgis, T. A. (1988). Physica A, *153(2),* 341–354.
6. Kozlov, G. V., & Zaikov, G. E. (2004). Structure of the Polymer Amorphous State. Utrecht, Boston: Brill Academic Publishers, 465 p.
7. Kozlov, G. V., Temiraev, K. B., Shustov, G. B., & Mashukov, N. I. (2002). *J. Appl. Polymer Sci., 85(6)*, 1137–1140.

8. Kozlov, G. V., Dolbin, I. V., & Zaikov, G. E. (2014). The Fractal Physical Chemistry of Polymer Solutions and Melts. Toronto, New Jersey: Apple Academic Press, Inc., 334 p.
9. Aharoni, S. M. (1983). *Macromolecules, 16(9)*, 1722–1728.
10. Aharoni, S. M. (1985). *Macromolecules, 18(12)*, 2624–2630.
11. Schnell, R., Stamm, M., & Creton, C. (1998). *Macromolecules, 31(7)*, 2284–2292.
12. Yakh'yaeva, Kh. Sh., Kozlov, G. V., Magomedov, G. M., & Zaikov, G. E. (2014). Polymer processing and formation: part 1. *Nanopolymers and Modern Materials. Preparation, Properties and Applications*. Ed. Stoyanov, O. V., Haghi, A. K., Zaikov, G. E. Toronto, New Jersey: Apple Academic Press, 24–28.
13. Kozlov, G. V., Zaikov, G. E., & Mikitaev, A. K. (2009). The Fractal Analysis of Gas Transport in Polymers. *The Theory and Practical Applications*. New York: Nova Science Publishers, Inc., 238 p.
14. Zelenyi, L. M. & Milovanov, A. V. (2004). *Achievements of Physical Sciences, 174(8)*, 809–852.

CHAPTER 9

WAYS OF REGULATION OF RELEASE OF MEDICINAL SUBSTANCES FROM THE CHITOSAN FILMS

E. I. KULISH, A. S. SHURSHINA, and ELI M. PEARCE

CONTENTS

ABSTRACT

The chitosan films obtained from solutions in acetic acid and containing antibiotics both of cephalosporin series and of amino glycoside have been investigated and the principal possibility of regulating their transport properties as regards the medicinal preparations release has been demonstrated. This regulation can be carried out by films thermal modification consisting in heating of the formed films at the temperature of about 120°C and treatment with sodium dodecylsulfate solution. It has been shown that the antibiotics release from a film will be determined by the amount of antibiotics connected with chitosan by hydrogen bonds, on the one hand, and by the state of the polymer matrix, on the other.

9.1 INTRODUCTION

Systems with controlled transport of medicines are the extremely demanded. The main task of antibacterial chemical therapy is the selective suppression of microorganisms without damaging the organism as a whole. The decrease of the antibiotics therapy efficiency observed recently is mainly caused by the possibility of origination of bacteria strain tolerant (resistant) to these antibiotics. Polymer derivatives of antibiotics can help to solve this task. We've made an attempt to use chitosan as a carrier of antibacterial preparations. In this situation the choice of chitosan (ChT) as a polymer carrier of a medicinal preparation is not accidental because this polymer possesses a whole spectrum of unique properties making it indispensable in polymer medicine [1]. In the present study we've considered some approaches to creating antibacterial ChT-based coatings of prolonged action suitable for treating surgical, burning and slowly healing wounds of different etiology.

9.2 EXPERIMENTAL PART

The objects of investigation were a ChT specimen produced by the company "Bioprogress" (Russia) and obtained by acetic deacetylation of crab chitin and antibiotics both of cephalosporin series cephazolin sodium salt (CPhZ), cephotoxim sodium salt (CPhT), and of aminoglycoside series amikacin sulfate (AMS), gentamicin sulfate (GMS). The investigation of the interaction of medicinal preparations with ChT was carried out according to the techniques described in Refs. [2, 3].

ChT films were obtained by means of casting of the polymer solution in acetic acid onto the glass surface with the formation of chitosan acetate

(ChTA). The polymer mass concentration in the initial solution was 2 g/dL. The acetic acid concentration in the solution was 1, 10 and 70 g/dL. Aqueous antibiotic solution was added to the ChT solution immediately before films formation. The content of the medicinal preparation in the films was 0.1 mol/mol ChT. The film thickness in all the experiments was maintained constant and equal to 0.1 mm. The kinetics of antibiotics release from ChT film specimens into aqueous medium was studied spectrophotometrically at the wave length corresponding to the maximum absorption of the medicinal preparation.

In order to regulate the ChT ability to be dissolved in water the anion nature was varied during obtaining ChT salt forms. So, a ChT-CPhZ film is completely soluble in water. The addition of aqueous sodium sulfate solution in the amount of 0.2 mol/mol ChT to the ChT–CPhZ solution makes it possible to obtain an insoluble ChT-CPhZ-Na_2SO_4 film. On the contrary, a ChT-AMS film being formed at the components ratio used in the process of work isn't soluble in water. Obtaining a water-soluble film is possible if amikacin sulfate is transformed into amikacin chloride (AMCh). In this case the obtained ChT-AMCh film will be completely soluble in water. Thus, the following film specimens have been analyzed in the investigation: ChT-CPhZ and ChT-CPhT (soluble forms); ChT-CPhZ-Na_2SO_4 (insoluble-in-water form); ChT-AMCh (soluble form); ChT-AMS and ChT-GMS (insoluble-in-water forms).

Also To prevent solubility of the film in water chitosan film was subjected to thermal modification consisting in heating of the formed film at the temperature of about 120 °C for 15–200 min and it was treated with the solution of the surface-active substance – sodium dodecyl sulfate (SDS) with concentration equal to 1–15 g/dL.

With the aim of determining the amount of medicinal preparation held by the polymer matrix there was carried out the synthesis of adducts of the ChT-antibiotic interaction in the mole ratio 1:1 in acetic acid solution. The synthesized adducts were isolated by double reprecipitation of the reaction solution in NaOH solution with the following washing of precipitated complex residue with isopropyl alcohol. Then the residue was dried in vacuum up to constant mass. The amount of preparation strongly held by chitosan matrix was determined according to the data of the element analysis on the analyzer EUKOEA – 3000.

IR spectra of specimens were recorded on the spectrometer "Specord-M 80" and "Shimadzu" (KBr tablets, films) in the area of 700–3600 sm^{-1}. UV spectra of all specimens were taken in quarts dishes with thickness of 1 sm relative to water on the spectrophotometer "Specord M-40" in the area of 220–350 nm.

9.3 RESULTS AND DISCUSSION

On the basis of the chemical structure of the studied medicinal compounds [4] one can suggest that they are able to combine with ChT forming polymer adducts of two types – ChT-antibiotics complexes and polymer salts produced due to exchange interaction. As a result some quantity of medicinal substance will be held in the polymer chain. The interaction taking place between the studied medicinal compounds and ChT was demonstrated by UV- and IR-spectroscopy data.

The interaction of antibiotics with ChT is evidenced by UV spectroscopic data. The maximum absorption of CPhZ and CPhT at its concentration of 10^{-5} M in 1% acetic acid is observed at 273 and 261 nm, respectively. Upon addition of an equivalent amount of ChT to the solution, the intensity of the absorption peak of the medicinal preparation noticeably grows and its position is bathochromically shifted by approximately 5–10 nm. The UV spectrum of AMS and GMS at a concentration of 10^{-2} M in 1% acetic acid shows an absorption peak at 267 and 286 nm, respectively. Addition of a ChT solution to the solution of medicinal substances results in that a precipitate is formed; however, analysis of the supernatant fluid shows that the peak of the corresponding absorption band in its spectrum is shifted by 5–7 nm. The observed changes unambiguously indicate that ChT affects the electron system of MS and adducts are formed. The binding energies in the complexes, evaluated by the shift of the absorption peaks in the UV spectra, are about 10 kJ mol^{-1}. This suggests that the complexation occurs via hydrogen bonds.

The interaction of the antibiotics under study with ChT is also confirmed by IR spectroscopic data. For example, the IR spectrum of ChT shows absorption bands at 1640 and 1560 cm^{-1}, associated with deformation vibrations of the acetamide and amino groups, and at 1458 and 1210 cm^{-1}, associated with planar deformation vibrations of hydroxy groups. Analysis of the IR spectra of adducts formed by interaction of ChT with antibiotics suggests that some changes occur in the IR spectra. For example, absorption bands associated with stretching vibrations of the C=O and C=N groups of the medicinal compound appear at 1750 and 1710 cm^{-1} in the IR spectra of ChT–CPhZ polymeric adducts, and a strong absorption band related to the SO_4^{2-} group appears at 619 cm^{-1} in the IR spectrum of the CTS–AMS reaction product. In addition, the IR spectra of all the compounds being analyzed show that the absorption band at 3000–3500 cm^{-1}, associated with stretching vibrations of OH and NH groups, is broadened as compared with the corresponding bands of the antibiotics and ChT, which, taken together, suggests that ChT–antibiotic complex compounds are formed via hydrogen bonds.

Table 9.1 gives the data on the amount of antibiotics determined in polymer adducts obtained from acetic acid solution.

TABLE 9.1 The Amount of Antibiotics Determined in Reaction Adducts

CH_3COOH, g/dL in the initial solution	The antibiotics used	The amount of antibiotics in reaction adduct, % mass.
1	CPhZ	10.1
	CPhT	15.9
	AMS	61.5
	GMS	59.4
10	CPhZ	5.88
	CPhT	57.5
	AMS	55.8
	GMS	31.3
70	CPhZ	3.03
	CPhT	3.7
	AMS	41.3
	GMS	40.1

Attention should be paid to the fact that the amount of medicinal preparation in the adduct of the ChT-medicinal preparation reaction is considerably higher in the case of antibiotics of aminoglycoside series than in the case of antibiotics of cephalosporin series. This can be connected with the fact that CPhZ and CPhT anions interact with ChT polycation forming salts readily soluble in water. In the case of using AMS and GMS because of two-base character of sulfuric acid one may anticipate the formation of water-insoluble "double" salts ChT-AM or ChT-GM sulfates due to which additional quantity of antibiotics is held on the polymer chain.

Table 9.2 gives the data on the value of the rate of AMS and GMS release from film specimens formed from acetic acid solutions of different concentrations. The rate was evaluated only for water-insoluble films because at using soluble films the antibiotic release was determined not by medicinal preparation diffusion from swollen matrix but by film dissolving.

TABLE 9.2 Transport Properties of Chitosan Films in Relation to Medicinal Preparation Release

Acetic acid concentration g/dL	The antibiotics used	Release, % mass./h for chitosan specimens
1	AMS	0.5
	GMS	0.4
10	AMS	0.8
	GMS	0.5
70	AMS	1.5
	GMS	1.3

Attention must be given to the fact of interaction between the rate of antibiotics release from chitosan films and their amount, which is strongly held in ChT chain. For example, at increasing the concentration of acetic acid used as a solvent the amount of medicinal preparation connected with the polymer chain decreases in all the cases considered by us. Correspondingly, the rate of antibiotics release from films insoluble in water, increases.

The influence of the amount of medicinal preparation strongly held in ChT matrix, on the rate of medicinal substance release from the film must be most pronounced at comparing the rates of release of antibiotics of aminoglycoside series and cephalosporin one. However, ChT–CPhZ and ChT–CPhT films are soluble in water while ChT–AMS and ChT–GMS ones do not dissolve in water and it isn't correct to compare them. At ChT transition into insoluble form (by adding sodium sulfate) the rate of release of antibiotics of cephalosporin series decreases considerably (Fig. 9.1, curve 1) as compared with a soluble form but still it is higher than that in the case of antibiotics of aminoglycoside series (Fig. 9.1, curve 2). It should be also noted that the rate of antibiotics release from soluble ChT-CPhZ film (Fig. 9.1, curve 3) is also higher than in the case of ChT-AMCh film (Fig. 9.1, curve 4). Thus, considerable difference between the rate of release of aminoglycoside series antibiotics and that of cephalosporin series antibiotics is evidently explained by the difference in the amount of ChT-antibiotics adduct.

Thus, at forming film coatings one should proceed from the fact that a medicinal preparation can be distributed in the polymer matrix in two ways. One part of it connected with polymer chain, for example, by complex formation is rather strongly held in that polymer chain. The rest of it is concentrated in polymer free volume (in polymer pores). The rate of release of antibiotics from the film will be determined by the amount of antibiotics connected with

ChT by hydrogen bonds, on the one hand, and by the state of the polymer matrix including its ability to dissolve in water, on the other hand.

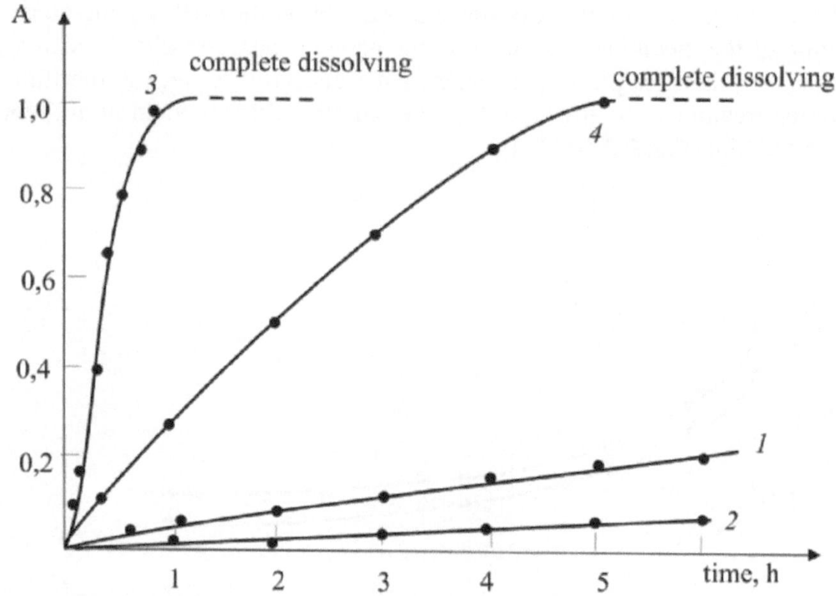

FIGURE 9.1 The kinetic curve of the release of CPhZ (1, 3) and AM (2, 4) from insoluble (1, 2) and soluble (3,4) films.

In essential decrease of affinity of a film to water results the heat treatment of films, which is accompanied by removing acetic acid bound by amino groups and can in principle result in the amidation process with chitosan units formation. The fact of formation of amide groups in heated films is confirmed by considerable strengthening of amide I band in IR-spectra in the area 1630–1550 sm^{-1}. Besides, one can observe the decrease in intensity of absorption of stretching vibrations of hydroxyl groups and the appearance of the shoulder characteristic of stretching vibrations of aminogrops, in the area 3500–3000 sm^{-1}. These data testify to the fact that at heating there really occurs the increase both in the depth of the amidation reaction and in the splitting out of acetic acid.

Since the heat treatment of films results in the loss of solubility in water the increased time of thermal treatment hinders the liberation of antibiotics from polymer matrix and, consequently, decreases the rate of medicinal substances release from films. By varying the thermal treatment conditions it is possible to vary the transport properties of films as well.

Thermally modified films behave in a different way there takes place considerable slowing down of the release of medicinal substance from the film. During the first 10–20 h there occurs the release of medicinal preparation at constant rate, then the process slows down and on the7–10 day the concentration of the medicinal substance in the aqueous phase reaches its constant value. As this takes place, the longer is the period of subjecting the film to thermal treatment, the lower are the rate and the ultimate yield of antibiotic from the film (Fig. 9.2, Table I).

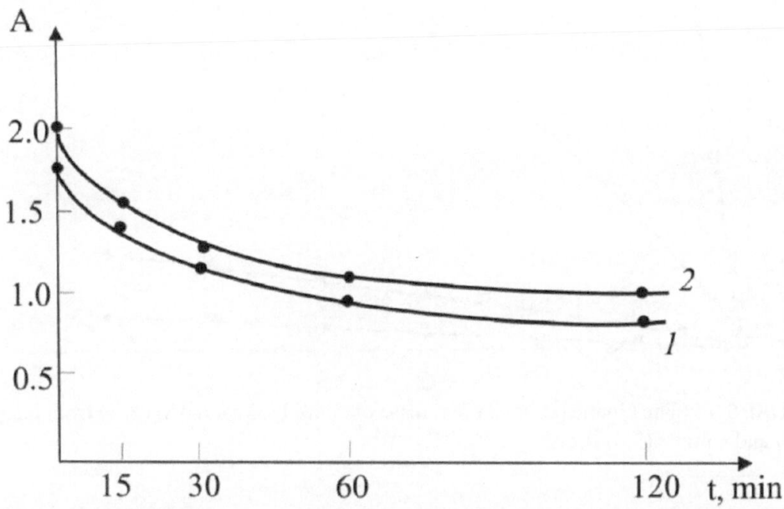

FIGURE 9.2 The dependence of the value of solution optical density corresponding to the ultimate yield of CPhZ (1) and CPhT (2) from the films formed in 1% acetic acid, on the time of the film heat treatment.

Regulate the rate and degree of the medicinal substance release from polymer matrix can also through a modification of the obtained films by treatment of its surface-active substance sodium dodecylsulfate. As well as in case of the hate treatment films, the rate of medicinal substances from the films modified SAS occurs considerable more slowly. In this case the longer the film was being treated with SAS and the higher was its concentration, the lower is the release rate and the ultimate antibiotic yield from the film (Table 9.3). It is evident that the use of micellar SDS solution for films modification results in the formation of a strong water-insoluble SAS polyelectrolytic complex. The surface of the modified film in this case represents a semipenetrable membrane, which allows water to penetrate into the film. The increase of the inner layer volume leads to the surface membrane extension and the release of the

TABLE 9.3 The Influence of Conditions of Chitosan Films Obtaining and Modification on Transport Properties of These Films (Initial Rate and Ultimate Yield of Antibiotics from a Film)

Antibiotic	Way of modification	Treatment time, min	Initial rate, Q^*/h	Ultimate yield, Q*
CPhZ	Heat treatment	15	2.80	80
	30	76		
	2.55	65		
	60	57		
	2.20			
	120			
	2.0			
	Treatment with the solution of SAS 5% mass	15	2.70	82
	10% mass 30	2.55	70	
	15% mass 60	2.25	58	
	30	2.40	56	
	30	2.30	52	
CPhT	Heat treatment	15	2.85	81
	30	77		
	2.60	64		
	60	56		
	2.20			
	120			
	2.0			
	Treatment with the so- 5% mass	15	2.60	80
	lution of SAS 30	2.50	68	
	60	2.30	58	
	10% mass	30	2.45	54
	15% mass	30	2.25	50

Q* – %mass CPhZ of its initial amount introduced into the film.

medicinal substance from the film is facilitated. In this case, the higher is the concentration of SDS used for modification and the longer is the time of the film holding in SAS solution, the thicker is the semipenetrable membrane and the less is the medicinal preparation yield from the film (Figs. 9.3, and 9.4).

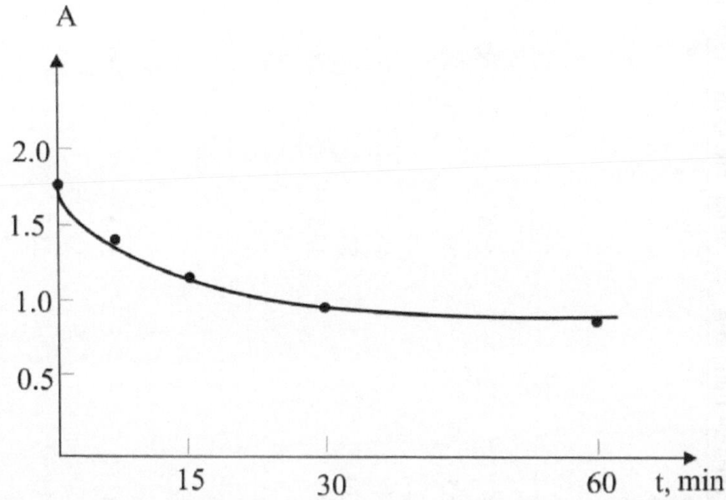

FIGURE 9.3 The dependence of the solution optical density value corresponding to the ultimate CPhZ yield from the films formed in 1% acetic acid, on the time of films treating in 5% SDS solution.

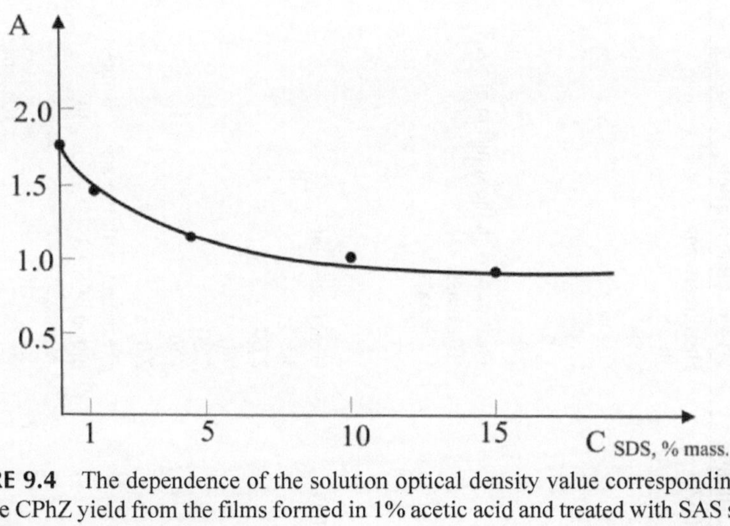

FIGURE 9.4 The dependence of the solution optical density value corresponding to the ultimate CPhZ yield from the films formed in 1% acetic acid and treated with SAS solution for 30 min, on SDS concentration.

Thus, this ways allowing to regulate the degree and rate of medicinal substance release from polymer matrix.

9.4 CONCLUSIONS

Thus, ChT transfer in a insoluble form allows to get film antibacterial coverings with the prolonged exit of medicine, which are deprived of the problems arising at use of individual antibiotics.

Essentially important is represented that fact that eventually in all studied cases of change in a range of absorption are observed not with a length of the wave corresponding to absorption of an individual antibiotic, and with a length of the wave corresponding to absorption of a polymeric complex. It means, what even in that case when the formed film formally insoluble in water, after all passes ChT quantity into a water phase where forms a complex, with an antibiotic allocated from a film. Thus, it is possible to believe that anyway on a wound surface formation of in situ of "polymeric medicine" ChT-medicinal substance is possible.

KEYWORDS

- **Chitosan**
- **Medicinal preparation**
- **Modification**
- **The state of polymer matrix**

REFERENCES

1. Skryabin, K. G., Vikhoreva, G. A., & Varlamov, V. P. (2002). Chitin and chitosan obtaining, properties and application M: Nauka, 365p.
2. Mudarisova, R. Kh., Kulish, E. I., Kolesov, S. V., & Monakov, Yu, B. (2009). Investigation of chitosan interaction with cephazolin. JACh *82(5),* 347–349.
3. Mudarisova, R. Kh., Kulish, E. I., Ershova, N. R., Kolesov, S. V., & Monakov, Yu, B. (2010). The study of complex formation of chitosan with antibiotics amicacin and hentamicin. JACh. *83(6)*, 1006–1008.
4. Mashkovsky, M. D. (1997). Medicinal Preparations, Kharkov: Torsing, *2*, 278p.

CHAPTER 10

A RESEARCH NOTE ON ENZYMATIC HYDROLYSIS OF CHITOSAN IN ACETIC ACID SOLUTION IN THE PRESENCE OF AMIKACIN SULFATE

E. I. KULISH, I. F. TUKTAROVA, V. V. CHERNOVA, M. I. ARTSIS, and R. A. PETHRICK

CONTENTS

10.1 INTRODUCTION

Nowadays, a polysaccharide of natural origin chitosan attracts much attention of researchers. One of the promising applications of chitosan is to obtain of chitosan solutions of bioactive and capable to biodegradation film materials for medical applications, including the treatment of wounds [1, 2]. Inclusion drug into chitosan film, for example, are widely used in the treatment of burn injuries antibiotic amikacin sulfate, to reduce the probability of suppuration and contributes to the suppression of infection. However, the addition of amikacin to chitosan solution can fundamentally affect the rate of its enzymatic hydrolysis, which dictates the need for specific kinetic studies. The need for this work due to the fact that, in contrast to the relatively well-developed learning tasks enzymatic hydrolysis of chitosan under the action of a specific enzymes chitinase and chitosanase [3], the process of the enzymatic hydrolysis of chitosan under the action of a nonspecific enzymes (such as hyaluronidase present at the wound surface), not quantitatively studied.

10.2 EXPERIMENTAL PART

The object of investigation was a Cht specimen produced by the company "Bio-progress" (Russia) and obtained by acetic deacetylation of crab chitin with a molecular weight with M_{sd}=113,000. As the enzyme preparation was used hyaluronidase enzyme preparation production "Microgen" (Moscow, Russia). The concentration of the enzyme preparation was 0.1, 0.2 and 0.3 g/L. Acetic acid of 1 g/dL concentration was used as a solvent. Cht concentration in solution for the enzymatic hydrolysis process C_{eh} ranged from 0.1 to 5 g/dL. Antibiotic amikacin sulfate (AMS), dissolved in a small amount of water was added to the Cht solution in a molar ratio Cht: AMS equal to 1:0.1, 1:0.05 and 1:0.01.

The degree of enzymatic destruction was evaluated by the degree of intrinsic [η] viscosity decrease according to the viscometer data. Intrinsic viscosity in solution of acetic acid was determined at 25 °C, using the method of Irzhak and Baranov [4]. To determine the initial values of the intrinsic viscosity $[\eta]_0$ of Cht a solution of concentration c=0.15 g/dL was used. To determine the values [η] of intrinsic viscosity during the enzymatic hydrolysis, CHT solution in acetic acid to which was added a solution of an enzyme preparation maintained for a certain time. Thereafter, the process of enzymatic hydrolysis was quenched by boiling the starting solution for 30 min in a water bath. Further, the initial concentration of the solution prepared by diluting a solution of C_{eh} to determine the intrinsic viscosity, the concentration c = 0.15 g/dL. The

process of enzymatic hydrolysis was carried out at a temperature of 36 °C. The initial speed of the enzymatic hydrolysis of chitosan V_0 was evaluated in the linear region for the fall of its intrinsic viscosity [η] and calculated by the formula [5]:

$$V_0 = \frac{C_{eh}K^{1/a}\left([\eta]^{-1a} - [\eta]_0^{-1/a}\right)}{t} \tag{1}$$

where is C_{eh} – concentration of chitosan in solution subjected enzymatic hydrolysis, g/dL; t – hydrolysis time, min., K and α – constants in the equation Mark-Houwink-Kuhn $[\eta]= KM^{\alpha}$, M – molecular weight of chitosan.

To determine the constants in equation Mark-Houwink-Kuhn it is necessary to calculate the values of the initial velocity of the enzymatic hydrolysis according to equation (1), the sample was fractionated on 10 fractions ranging in molecular weight from 20000 to 150000 Daltons. The absolute value of molecular weight of Cht fractions was determined by a combination methods sedimentation velocity and viscometry. The molecular weight of the fractions was determined by the formula:

$$M_{s\eta} = \left[\frac{S_0\eta_0[\eta]^{1/3}N_A}{A_{hi}(1-\bar{v}\rho_0)}\right]^{3/2} \tag{2}$$

where is S_0-sedimentation constant; η_0-dynamic viscosity of the solvent, which is equal to 1.2269×10^{-2} PP; [η] – intrinsic viscosity, dl/g; N_A – Avogadro's number, equal to 6.023×10^{23} mol^{-1}; $(1-v\rho_0)$ – Archimedes factor or buoyancy factor, v – the partial specific volume, cm^3/g, ρ_0 – density of the solvent g/cm^3; A_{hi} – hydrodynamic invariant equal to 2.71×10^6.

10.3 RESULTS AND DISCUSSION

Figure 10.1 shows the dependence of the intrinsic viscosity of the solution Cht preparation CHT solution on the time of standing with enzyme.

As can be seen, with increasing exposure time to the enzyme solution CHT, the viscosity decreases regularly, indicating a decrease in the molecular weight of Cht. The most significant drop in viscosity occurs in the initial period. Further enzymatic solution CHT exposure, affects the degree of viscosity drop significantly lesser extent, that is, the reaction rate decreases with time. Such a course of kinetic curve is typical for most enzymatic reactions

[6]. Increasing the concentration of the enzyme preparation leads to a natural increase in the rate of incidence of the intrinsic viscosity.

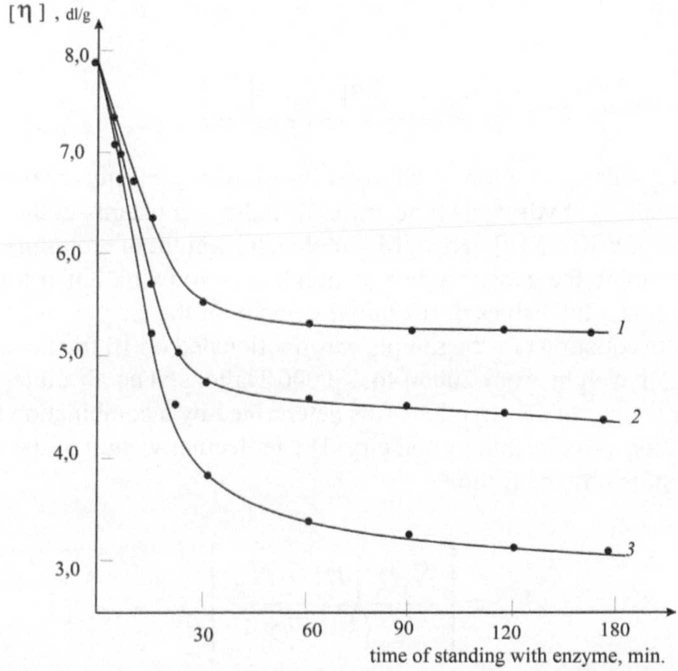

FIGURE 10.1 Dependence of the intrinsic viscosity of the solution Cht preparations Cht solutions on the time of standing with enzyme preparation concentration 0.1 *(1)*, 0.2 *(2)* and 0.3 *(3)* g/L.

As it was established in the course of this work, for all the studied solutions for small times of hydrolysis, the curves of reducing of the intrinsic viscosity are linear. It is on this plot was determined value of the initial velocity of the enzymatic hydrolysis V_0, serving as a measure of enzymatic activity of the enzyme towards Cht. To determine the rate of enzymatic hydrolysis by the Eq. (1), it was necessary to determine the value of the constants in the equation of Mark-Houwink-Kuhn. The constants defined for Cht solution in 1% acetic acid are shown in Table 10.1.

TABLE 10.1 Values of the Constants in the Mark-Houwink-Kuhn Equation for Chitosan Solution in 1% Acetic Acid Temperature 25 °C

Molar ratio of components Cht-AMS	α	K×10⁴
1:0	1.02	0.56
1:0.01	0.89	1.73
1:0.05	0.86	2.28
1:0.10	0.83	3.19

As it was shown by the study, the observed dependence of the initial rate of enzymatic hydrolysis of the substrate concentration can be described within the scheme of the Michaelis–Menten. Figure 10.2 shows the initial rate of concentration Cht solution.

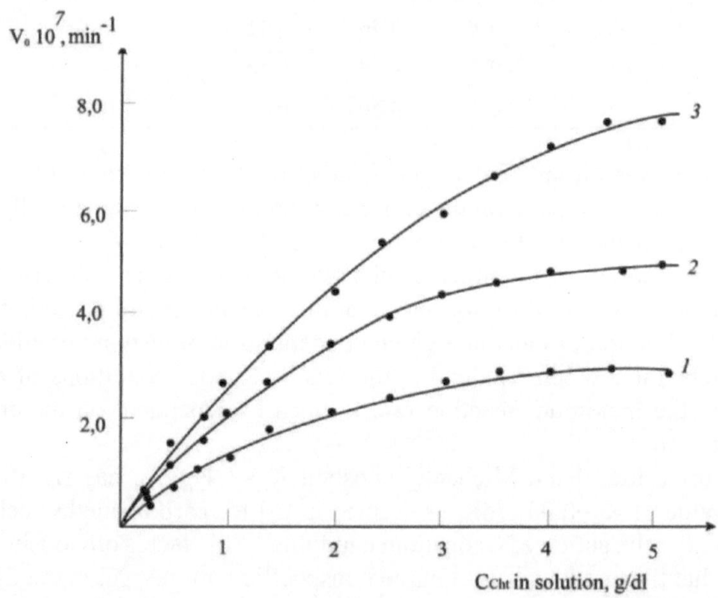

FIGURE 10.2 The dependence of the initial rate of the enzymatic hydrolysis of Cht *(1)* with an enzyme preparation at a concentration 0.1 *(1)*, 0.2 *(2)* and 0.3 *(3)* g/L of Cht concentration in solution.

From this Fig.10.2 it is seen, that it has a classic look and is a part of the rectangular hyperbola. Submission to double check the coordinates (graphical method of Lineweaver-Burk) can accurately determine the value of the Michaelis constant K_m (Table 10.2).

TABLE 10.2 Values of the Constants of Enzymatic Hydrolysis in the Michaelis-Menten Equation for Chitosan Solutions in 1% Acetic Acid

Molar ratio of components Cht-AMS	C_e, г/дл	$К_m$, г/дл	V_{max}, 10^6, мин$^{-1}$	V_{max}/ $К_m$, (г/дл*мин)
	0.1	3.37	0.50	0.15
1:0	0.2	3.47	0.90	0.26
	0.3	3.42	1.50	0.44
	0.1	4.09	0.43	0.10
1:0.01	0.2	4.01	0.80	0.19
	0.3	4.03	1.20	0.30
	0.1	4.16	0.37	0.09
1:0.05	0.2	4.13	0.60	0.14
	0.3	4.18	1.12	0.27
	0.1	4.46	0.31	0.07
1:0.1	0.2	4.48	0.58	0.13
	0.3	4.50	0.88	0.19

As can be seen from the data, the value of the Michaelis constant is a constant that does not depend on the concentration of enzyme and actually characterizes the affinity of the enzyme to the substrate. Value $V_{max}=k_2C_e$, where is k_2 – the decay constant enzyme-substrate complex, gives a description of the catalytic activity of the enzyme, i.e. defines the maximum possible formation of the reaction product at a given concentration of enzyme provided the excess substrate. When conducting the reaction under conditions of excess substrate, the maximum reaction rate is linearly dependent on the enzyme concentration.

Certain value of the Michaelis constant $К_m \approx 3.4$ g/dL, has significantly higher value of $К_m=0.03$ g/dL, as defined in [5] for carboxymethyl cellulase system under the action of Geotrilium candidum. This fact is obviously due to the fact that the enzyme is used in the nonspecific with respect to the Cht and the fact that the conditions of hydrolysis did not correspond to the temperature and the pH optimum of the enzyme hyaluronidase.

Adding AMS to the Cht solution doesn't affect the appearance of the decline curve of the Cht intrinsic viscosity from the time of exposure to the enzyme – it is similar to curves shown in Fig. 10.1. The curve of the initial rate of the enzymatic hydrolysis of the concentration of Cht in the solution also does not change the form. However, comparison of the values V_{max} and $К_m$ obtained for the system Cht-AMS with values V_{max} and $К_m$ for individual

Cht (Table 10.2) shows that the addition of the AMS affects both the value V_{max} and on the value of K_m. Moreover, the values of V_{max}/K_m, actually has the meaning of pseudo first order rate constants for the reaction Cht + enzyme ® product + enzyme also change regularly with increasing content of AMS in the mixture.

From a formal point of view the action AMS can be explained in the framework of a model of mixed inhibition [7, 8]. However, from our point of view, the reason of reducing the rate of Cht enzymatic hydrolysis associated with a change of the conformational state of the substrate, rather than a possible inhibition of the enzyme in the presence of the antibiotic.

This is evidenced by comparison of the values of the constants Mark-Houwink-Kuhn defined for Cht in the presence of AMS (Table 10.1). As it seen from Table 10.1), adding AMS to Cht solution is accompanied by some compression of the macromolecular coil. Such action of AMS is understandable because the antibiotic is a low molecular weight electrolyte, whose presence in the Cht solution inevitably causes an increase of ionic strength, the suppression effect of polyelectrolyte swelling and, consequently, a reduction in the size of the coil.

Reducing the size of macromolecular coil, in turn, leads to a certain increase in its density, and hence the availability of Cht units to interact with enzyme preparation decreases. As a result there is a reduction of the rate of Cht enzymatic hydrolysis.

Thus, for the first time identified by the kinetic characteristics of the activity of the hyaluronidase enzyme for the enzymatic hydrolysis reaction CHT solution of 1% acetic acid in the presence and absence of antibiotic amikacin sulfate. The fact of reducing V_{max} and increasing K_m and regular decreasing of V_{max}/K_m, which takes place by adding AMS to Cht solution, indicates the possibility of delayed catalytic conversion of the substrate in a ternary complex enzyme-Cht-AMS.

KEYWORDS

- **Acetic acid**
- **Amikacin sulfate**
- **Chitosan**
- **Enzymatic reactions**
- **Hydrolysis**
- **Kinetics**
- **Mechanism**

REFERENCES

1. Tapan Kumar Giri, Amrita Thakur, Amit Alexander, Ajazuddin, Hemant Badwaik, Dulal Krishna Tripathi (2012). Modified Chitosan Hydrogels as Drug Delivery and Tissue Engineering Systems: Present Status and Applications. *Acta Pharmaceutical Sinica B 2(5),* 439–449.
2. Hiroshi Ueno, Takashi Mori, & Toru Fujinaga (2011). Topical Formulations and Wound Healing Applications of Chitosan. *Advanced Drug Delivery Reviews 52(2)*, 105–115.
3. I'lina, A. V., & Varlamov, V. P. (2002). Enzymology Synthesis and Degradation of Chitin and Chitosan. Chitin and chitosan. *Preparation, properties and applications*. Moscow Nauka, 79–90.
4. Baranov, V. G., Brestkin, Yu. V., Agranova, S. A., & Pinkevich, V. N. (1986). Behavior of Macromolecules in Polystyrene "*Thickened*" *Good Solvent Polymer Science*. *28B(10),* 841–843.
5. Rabinovich, M. L., Klesov, A. A., & Berezin, I. V. (1977). Kinetics of Action of Cellulitics Enzyme Action Geotrilium Candidum. Viscometric Analysis of the Kinetics of Hydrolysis of Carboxyl Methyl Cellulose. *Bioorganic Chemistry 3(3)*, 405–414.
6. Severin, E. S., ed. (2004). *Biochemistry* Moscow. Geotar Med. 779.
7. Cornish-Bowden, E. (1979). *Principles of Enzyme Kinetics*. Mir, 280
8. Dixon, M., & Webb, E. C. (1982). *Enzymes. 2*, Moscow: Mir, 569.

CHAPTER 11

THE STRUCTURE OF THE INTERFACIAL LAYER AND OZONE-PROTECTIVE ACTION OF ETHYLENE–PROPYLENE–DIENE ELASTOMERS IN COVULCANIZATES WITH BUTADIENE–NITRILE RUBBERS

N. M. LIVANOVA, A. A. POPOV, V. A. SHERSHNEV, M. I. ARTSIS, and G. E. ZAIKOV

CONTENTS

ABSTRACT

The relation of ozone resistance to the volume and structure of the interfacial layer and the amounts of cross links in the interlayer was studied for co vulcanizates of butadiene–acrylonitrile rubbers of various polarities with ethylene–propylene–diene (EPDM) elastomers that differed in the co monomer composition and stereo regularity of propylene units. It was shown that the ozone resistance is determined by the compatibility of the components, phase structure, the interlayer volume and density, the amount of cross links in the interlayer, and the strength of the EPDM network.

11.1 INTRODUCTION

To protect unsaturated rubbers against ozone-induced degradation, they are blended with saturated thermoplastics and rubbers. There are different ideas on the mechanism of antiozonant protection: the surface protection of an ozone-unstable rubber though enrichment of the vulcanizate surface with an ozone-resistant component [1] and the bulk protection mechanism associated with the phase structure and the structure of the interphase layers [2–7]. The bulk protection mechanism was studied using systems based on butadiene–acrylonitrile rubbers (BNRs) with poly (vinyl chloride) (PVC) as an example and ethylene–propylene-diene elastomers (EPDM) [2–7]. This mechanism is supported by the data obtained by Zateev [8], who found, in an electron microscopy study, that an ozone-resistant component does not prevent the formation, on the vulcanizates surface, of primary ozone-induced micro cracks by which ozone penetrates into the sample.

In order to obtain insight into the mechanism of the antiozonant action of a low unsaturation component (the ethylene–propylene–diene elastomer) in cross-linked blends with butadiene–nitrile rubbers. We studied the dependence of the rate of ozone-induced degradation of the diene matrix on the phase structure of blends, volume and the structure of the interphase layer, the amount of cross links formed in it, and the strength of the ethylene–propylene–diene elastomer network. It was of interest to relate the specific features of the structure of the interphase interaction region and its volume to the efficiency of inhibition of ozone degradation of the diene elastomer in cured heterophasic blends of NBR copolymers of different polarities with ethylene–

propylene–diene elastomers manufactured by Uniroyal, DSM and domestic EPDM which differ in the co monomer ratio and in the stereo regularity of propylene sequences [9–11].

11.2 EXPERIMENTAL PART

A heterophase cross linked blend of BNR and EPDM with a ratio of 85:15 (by weight) having a dispersed EPDM structure and 70:30, which represented a system of interpenetrating cross linked networks, was studied. Commercial nitrile–butadiene rubbers (trademarks BNKS-18, BNKS-28, and BNKS-40) were used. The AN-unit contents were 18, 28, and 40 wt%, respectively, and the values of the Mooney viscosity (at 100°C) were 40–50, 45–65, and 45–70 rel. units, respectively. The content of *trans*1, 4-, 1, 2-, and *cis*-1, 4-units of butadiene was estimated via IR spectroscopy (bands at 967, 911, and 730 cm^{-1}) [12] with the use of extinction coefficients from [13] (Table 11.1).

EPDM of the Royalen brand (Uniroyal, USA), the DSM 778, 714, and 712 brands (DSM N.V., Netherlands) and domestic EPDMs having different relative amounts of ethylene, propylene, and ethylidene norbornene (ENB) units and different degrees of microtacticities of the propylene sequences respectively, were used. The composition, the molecular-mass characteristics, the Mooney viscosity, the isotacticity of EPDM propylene units according to IR data [12, 14, 15] are given in Table 11.2. For domestic EPDMs the data on the Mooney viscosity; the content of ethylene, propylene, and ethylidenenorbornene units according to the manufacturer's data.

TABLE 11.1 Isomeric Composition of Butadiene Units in Different Butadiene–AN Copolymers

Copolymer	Content of units, %		
	*trans*1,4	1,2-	*cis*-1,4-
BNKS −18	82.0	8.2	9.8
BNKS −28	76.4	14.4	9.2
BNKS −40	93.0	4.4	2.6

TABLE 11.2 Composition and Basic Characteristics of Ethylene–Propylene–Diene Elastomers

EPDM brand	Ethylene: propylene, wt %	Isotacticity, %	ENB content, wt %	Mooney viscosity
R 512	68/32	20	4	57
R 505	57/43	24	8	55
R 521	52/48	22	5	29
778	65/35	13	4.5	63
714	50/50	12	8	63
712	52/48	11	4,5	63
EPDM-40	70/30	29	4	36–45
Elastokam 6305	74/26	9,5	5,4	67
EPDM-60(I)	60/40	13	4	60
EPDM-60(II)	60/40	13	6,7	62
Elastokam 7505	60/40	9,5	5,1	83

A vulcanizing system for NBRs had the following composition, phr: stearic acid, 2.0; Sulfenamide Ts (*N*-cyclohexylbenzothiazole-2-sulfenamide), 1.5; zinc oxide, 5.0; and sulfur, 0.75. EPDM of the Royalen brand, the DSM and domestic EPDM was vulcanized with supported Peroximon F-40 taken in an amount of 5.5 phr. Each rubber was mixed with its vulcanizing system by roll milling at 40–60 °C for 15 min. Then, a rubber blend was prepared under the same conditions. The blends were vulcanized at 170 °C within 15 min.

The ozone resistance of the blends was studied via the method of stress relaxation at an ozone concentration of 8.5×10^{-5} mol/L, 30 °C and at a tensile strain of 30–150% on an IKhF-2 relaxometer [2–7]. Relaxation rate v_r in the ozone-containing medium in the steady-state region next to the region of the fast physical relaxation reflects the kinetics of accumulation of chain ruptures in the diene matrix. The efficacy of retardation of degradation during the introduction of EPDM, v_r^{rel}, corresponds to the intensity of the decrease in the rate in the presence of the ozone-resistant component relative to that in the presence of the BNR vulcanizate.

The region of a low relaxation rate with quasi equilibrium stress σ* reflects the strength of the EPDM network because it represents the superposition of stresses in the network and in the matrix connected through interfacial layers [5–7]. The absence of the continuous structure of EPDM or its breakdown during tensile drawing is accompanied by a stress drop to zero due to propagation and merging of ozone micro cracks.

The method of swelling in a selective solvent, *n*-heptane, (Zappa's method [6, 16, 17]) was used for studying the formation of the interfacial layer in the cross linked heterophase blends of BNR with different contents of polar

acrylonitrile units and EPDM which are characterized by different co monomer compositions and stereo regularities of propylene units.

The density of the interfacial layer and the number of cross links in this layer were characterized by a difference –α between the equilibrium degree of swelling Q_{eq} of covulcanizate and the additive value Q_{ad} (deviation of the equilibrium degree of swelling from the additive value of each of homo vulcanizates) ш a nonpolar solvent *n*-heptane [16, 17]. When the above parameters were higher, this tendency suggested the occurrence of a weak interfacial interaction between thermodynamically incompatible polymer components, one of which contained polar units. In such systems, only local mutual solubility of segments of nonpolar chain fragments is possible [18, 19]. The friability of the interfacial layer is associated with a reduced number of crosslinks.

Difference –α can be estimated through the following equation [17]:

$$-\alpha = [(Q_{ad} - Q_{eq})/(Q_{ad} - Q_2)]\ 100\%,\ \alpha = \text{alpha}$$

where Q_2 is the degree of swelling of the second elastomer (BNR).

The Flory–Huggins interaction parameter χ for polybutadienes and EPDM with *n*-heptane and solubility parameters δ for *cis*-PB, EPDM, and BNKS were reported in Refs. [21, 22].

11.3 RESULTS AND DISCUSSION

Formation of a strong interfacial layer is the key factor of the mechanism describing retardation of ozone degradation of a diene rubber by elastomer additives with a low degree of unsaturation.

The formation of the interfacial layer in the crosslinked heterophase blends of BNR with different contents of polar acrylonitrile units (Table 11.1) and EPDM, which are characterized by different comonomer compositions and stereo regularities of propylene units (Table 11.2) was studying. The density of the interfacial layers and the content of the formed cross-link for EPDM influence on the kinetics of ozone degradation of the diene matrix and the efficacy of protective action of EPDM.

The objects of research in this study were heterophase cross-linked BNR–EPDM (70:30) blends. At this content of the nonpolar component, a system of interpenetrating cross-linked networks appears. Both isolated EPDM particles and their conglomerates are distributed in the BNR matrix. The size of EPDM particles in the blend depends on compatibility of the components and increases with the polarity of BNR.

The value of $-\alpha$ depends on the amount of cross-links in the interfacial layer and its volume. Let us assume that the major fraction of polar units of BNR is uninvolved in its formation and the value of $-\alpha$ was recalculated to the 100% content of butadiene units $-\alpha^{100\%}$ (Fig. 11.1) and used to characterize the structure of the interfacial interaction zone and the amount of cross links contained in it. In such a manner, the effect of the interfacial layer volume could be minimized. As will be shown below, this situation may not be attained in all cases (Figs. 11.1. and 11.2).

It is seen that a linear decrease in the value of $-\alpha^{100\%}$ with an increase in the total content of 1,4-*cis* and 1,2 units is observed for the cross linked blends of BNR with all DSM EPDM samples and EPDM-60(I) characterized by a low isotacticity of propylene (Table 11.2). The fact that the value of $-\alpha^{100\%}$ decreases in proportion to the amount of 1,4-*cis* and 1,2 isomers of butadiene units for EPDM-based blends provides evidence that the density of the transition layer increases. This circumstance implies that the mutual solubility of EPDM co monomers and butadiene units that occur for the most part in the 1,4-*trans* configuration in the neighborhood of these isomers is improved.

For DSM and EPDM-60(I) the region of interfacial interaction is bounded by nonpolar BNR units. The compatibility of chain portions of these EPDM samples with the polar acrylonitrile groups is ruled out. As the proportion of propylene units in EPDM is increased, the compatibility of the components, the density of the interfacial layer, and the amount of cross links in it drop sharply (the absolute value of $-\alpha^{100\%}$ increases). Thus, the higher the content of atactic propylene units in EPDM, the lower the adhesion interaction of the components and the worse its compatibility with BNR 1,4-*Trans* units (Table 11.1), which show the tendency toward ordering at the low content of acrylonitrile and other isomers [20, 23, 24]. Ordered structures worsen compatibility of polymers even to a higher extent [18].

For BNR covulcanizates with EPDM R 512 containing a large proportion of ethylene units and distinguished by the presence of stereo regular propylene sequences, the value of $-\alpha^{100\%}$ linearly decreases with the content of butadiene units by a factor of 2.6 (Fig. 11.2). This fact leads us to infer that, firstly, ethylene units adjoining predominantly short isotactic propylene sequences are well compatible with all isomers of butadiene units. Secondly, an increase in the density of the interfacial layer and in the amount of crosslinks in it (a decrease in $-\alpha^{100\%}$) with a rise in the proportion of nonpolar units in BNR implies that EPDM molecular fragments may penetrate into BNR regions that apparently contain single polar acrylonitrile groups.

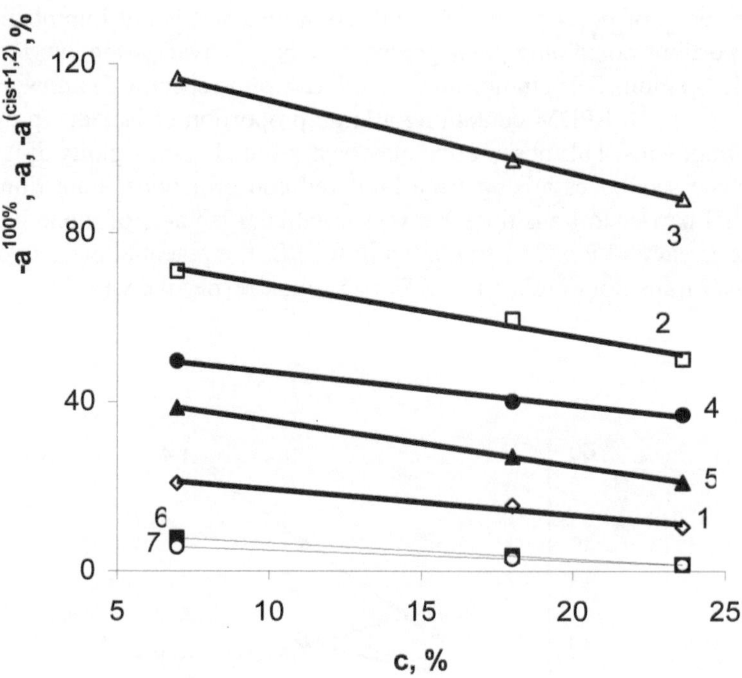

FIGURE 11.1 The plots the value of $-\alpha^{100\%}$ for covulcanizates EPDM with DSM (*1*) 778, (*2*) 714, (*3*) 712, (4) EPDM-60(I); $-\alpha$ (*5*) for R 521; $-\alpha^{cis+1,2}$ for (*6*) R 505 and (*7*) Elastokam 6305 as a function of the total content of 1,4-*cis*- and 1,2-butadiene isomers in BNR.

EPDM-40 with a higher degree of isotacticity of propylene units (as compared with EPDM R 512) is characterized by a far more intensive penetration of propylene chain fragments in their rigid isotactic configuration into BNR regions containing single polar groups. An increased volume of the interfacial layer and its local density reduction are confirmed by the fact that the value of ($-\alpha$) for blends with EPDM-40 is higher than that for the EPDM R 512-based blends.

For the covulcanizates of R 521 EDPM, which contains a large amount of isotactic propylene sequences [11, 12], the proportional decrease in the value of $-\alpha$ by a factor of 1.85 is observed in the $-\alpha$–Σ(1,4-*cis* and 1,2 units) coordinates (Fig. 11.1). Consequently, at a high content of isotactic propylene units, EPDM shows better compatibility with BNR if the butadiene co monomer is enriched with 1,4-*cis* and 1,2 units (a reduction in the value of $-\alpha$). The values of $-\alpha$ do not take into consideration restrictions related to the mutual interpenetration of segments of dissimilar chains associated with

the presence of polar groups in BNR. As a result, it is not improbable that BNR portions containing polar groups may be involved in the interfacial interaction region. This circumstance is related to the specific features of BNR interaction with EPDM containing a large proportion of isotactic propylene chain fragments. This observation may be attributed to the rigidity of isotactic propylene sequences arising from hindered conformational transitions. The potential barrier to transitions between rotational isomers of monomer units for the isotactic PP is 21 kJ/mol, while for PE, the potential barriers of *T–G* and *G–G* transitions are 2.5 and 8.8−10 kJ/mol, respectively [25].

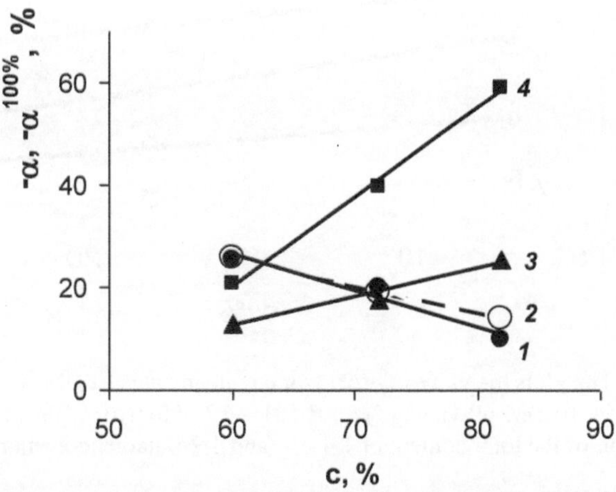

FIGURE 11.2 (*2*) ($-\alpha$) and (*1, 3, 4*) ($-\alpha^{100\%}$) vs. overall content of butadiene units *c* for (*1*) EPDM R 512, (*2*) EPDM-40, (*3*) EPDM-60 (II), (*4*) Elastokam 7505, in their covulcanizates with BNR.

For blends with EPDM R 505, which contains a large amount of ENB, the linear dependence is attained when the values of $-\alpha^{100\%}$ are recalculated to the 1% of the sum of 1,4-*cis* and 1,2 units in BNR ($-\alpha^{cis + 1,2}$) (Fig. 11.1) [4]. An analysis of the $-\alpha^{cis + 1,2}$ versus Σ(1,4-*cis* and 1,2 units) curves demonstrates that $-\alpha^{cis + 1.2}$ decreases by a factor of 5.3 with an increase in the total amount of these butadiene unit isomers. The data presented above suggest that the chain fragments of this EPDM are well compatible only with those portions of butadiene chains that contain 1,4-*cis* and 1,2 isomers, while in the case of 1,4-*trans* units, compatibility is much worse. It appears that the bulky diene group, as in EPDM 714, significantly hinders the incorporation of EPDM chain portions into butadiene regions of BNR.

For Elastokam 6305 with a high content of ethylene units (74%), with a higher amount of ENB (5,4%) and a very low degree of isotacticity of propylene sequences (9.5%) (Table 11.2), as follows from Fig. 11.1, the dependence on the overall content of *cis*-1,4- and 1,2-butadiene units in BNR also is linear if $-\alpha^{cis\text{-}1,4+1,2}$ This copolymer is compatible only with butadiene chain fragments of BNR containing preferably *cis*-1,4- and 1,2-isomers. Hence, when propylene units are characterized by a very low microtacticity (9.5%), compatibility between the components decreases. As was shown in Ref. [4], the high content of ethylene units, whatever the configuration of propylene sequences, provides a better compatibility with butadiene copolymer BNKS. However, as atactic configuration in the blend dominates, the depth of penetration of EPDM segments is limited by the butadiene part of the copolymer. Because, in EPDM-60 (I), half of all propylene units exist in the isotactic configuration and the content of diene groups is lower, its segments are able to diffuse into nonpolar SKN regions to a greater depth than segments of Elastokam 6305.

EPDM-60 (II) with a higher content of ENB (6.7%) is compatibilized only with butadiene units (Fig. 11.2) and ($-\alpha^{100\%}$) increases by a factor of 1.8 as their content grows. This tendency suggests that this interfacial layer is characterized by an increased friability and the number of cross-links in this layer is low. The higher the volume of this layer (the lower the polarity of BNR), the higher the ($-\alpha^{100\%}$) values.

The ratio between ethylene and propylene units in Elastokam 7505 is similar to that in EPDM-60 (II), but the degree of isotacticity of propylene units is lower (9.5%) and the Mooney viscosity is very high (Table 11.2). Elastokam 7505 is characterized by the minimum compatibility with all BNR samples. As follows from Fig. 11.2, the friability of the interfacial layer of this EPDM is maximum; as a result, ($-\alpha^{100\%}$) markedly increases with a decrease in the polarity of SKN (by a factor of ~3).

Thus, the structure of interfacial interaction zone depends on the comonomer composition of EPDM, the stereo regularity of propylene units, and the isomerism of butadiene units. The amount of interphase cross links grows with an increase in the total content of 1,4-*cis* and 1,2-butadiene units in BNR except EPDM with a high content of ethylene units and a low content of predominantly isotactic propylene units. A high compatibility of BNR with this EPDM may be provided by interaction with butadiene units regardless of their isomeric composition.

The ability of EPDM that contains a large amount of diene units and stereo regular propylene units to compatibilize with butadiene units of BNR is close to that of EPDM with a low degree of isotacticity of the propylene co monomer and is determined by steric hindrances related to the presence of the bulky

diene. However, a higher content of diene groups ensures better cross linking of EPDM with the matrix. As a result, the total amount of cross links between phases is higher than that for blends with EPDM having the same ratio of ethylene and propylene units but a smaller amount of diene (cf. EPDM R 505 and R 521, EPDM 714 and EPDM 712).

The largest amount of cross links in the interfacial layer forms when EPDM with a high content of ethylene units is used despite the moderate amount of diene contained in it.

Data on the efficiency of retardation of the ozone-induced degradation of diene rubber in the absence (85:15 blend) and presence of the EPDM network (70:30 blend) at different strains are given in Table 3 (EPDM of the Royalen brand and the DSM). In the case of the continuous EPDM network, the degradation rate for the BNKS-18 or BKNS-28 matrix reaches a minimum (value v_d^{rel} is maximum) [5]. The efficiency of retardation of the ozone degradation of BNKS-18 in the presence of the Royalen EPDMs, which except R 505 have a high stereoregularity of propylene units, is much higher than that in the case of DSM EPDMs of the same composition with a low degree of isotacticity of propylene sequences. The degradation inhibition efficiency v_d^{rel} in the BNKS-18 blend with EPDM 778 is five times below that in the blend with EPDM R512.

TABLE 11.3 Efficiency of retardation of ozone degradation v_d^{rel} of acrylonitrile–butadiene rubber by ethylene–propylene–diene elastomers at component ratios of 85: 15 and 70: 30 and values for $-\alpha$

NBR covulcanizates with EPDM	v_d^{rel} (85:15) at ε=30%	v_d^{rel} (70:30) at different strains, %				$-\alpha$, %
		30	50	70	90	
BNKS −18						
778	1.5	5.1	4.2	3.8	4.2	12.6
714	1.8	5.1	3.1	3.1	2.9	48.9
712	1.7	6.8	5.7	3.6	5.6	79.8
R 512	1.7	26.0	7.9	4.0	3.4	8.1
R 505	0.9	4.5	2.6	2.9	2.0	52.5
R 521	1.3	20.4	7.5	4.9	4.5	27.1

TABLE 11.3 *(Continued)*

BNKS −28						
778	1.6	2.3	1.7	2.2	1.9	7.4
714	1.5	2.5	2.0	1.9	-	36.0
712	2.3	3.2	2.9	3.0	1.8	63.4
R 512	1.2	3.8	2.5	2.2	1.6	14.5
R 505	1.0	1.5	1.4	1.5	1.5	25.3
R 521	3.0	4.4	3.6	2.3	1.7	20.9
BNKS −40						
778	1.9	1.9	0.6	-	-	12.4
714	2.2	2.1	1.7	-	-	42.6
712	2.1	2.1	1.8	-	-	69.9
R 512	1.8	3.1	2.4	1.3	0.7	15.4
R 505	1.5	1.3	1.1	1.1	1.7	32.6
R 521	1.9	2.7	1.6	1.5	1.0	38.6

In blends of highly polar NBRs, v_d^{rel} sharply decreases and the differences between the blends level out. This leveling is due to a decrease in the length of the interphase interaction zone as a result of an increase in the amount of polar acrylonitrile units in NBR. The breaking of adhesion contacts facilitates microphase separation during the deformation of a specimen and considerably increases the rate of growth of ozone micro cracks and their coalescence (Table 11.3).

There is no difference in v_d^{rel} between BKNS-40 co vulcanizates with DSM EPDMs at amounts of the ozone-resistant components of 15 and 30 phr. This indifference is explained by the low compatibility of the components, an increase in the particle size of the dispersed copolymer, and an extremely low strength of its network or its absence as such (Fig. 11.3). Whenever the network is formed in the co vulcanizates, its degradation takes place at strains of 30–50%, unlike in the case of Uniroyal EPDMs of the same compositions [5].

The amount of cross links between the phases in the blend of EPDM R521 is considerably greater, unlike the case of EPDM 712 of the same monomer composition with a low microtacticity of propylene units [5, 9–11]. As follows from the data presented in Table 11.3, for the BNKS-18 blend with EPDM R521 is three times that of the blend with EPDM 712.

A reduction in the compatibility of components is responsible for the formation of a dispersion structure with coarser particles of the dispersed phase and weak bonding between these particles. The topological structure of a blend is characterized by a large distance between neighboring elements of the EPDM network [5]. The growth and coalescence of micro cracks stop when a micro crack propagates to the interphase. The crack formation rate decreases as the distance between neighboring network elements becomes shorter. This decrease is responsible for the considerable difference in the efficiency of retardation of BNKS-18 degradation between the blends containing EPDM 778 and EPDM 712, EPDM R512, and EPDM R521, which is due to the specifics of interphase interaction in heterophasic blends.

An increase in polarity during the use of BNKS-28 does not lead to a decrease in the interphase density in the blends with both EPDM brands, a result that is due to the influence of the isomer composition of butadiene units on the compatibility of NBR with EPDM. It was shown that the interphase density and the amount of crosslinks in the interlayer increase with an increase in the total content of *cis*-1,4- and 1,2-butadiene units in BNKS (Table 11.1). Despite this circumstance, a considerable decrease in v_d^{rel} takes place, a result that is due to the decrease in the interphase volume with an increase in the amount of polar units in NBR. Correspondingly, has a minimal value v_d^{rel} in the blends with BNKS-40.

An increased amount of ENB units in EPDM R505 and EPDM 714 at the same ratio of ethylene to propylene units as in EPDM 521 and EPDM 712, respectively (Table 11.2), facilitates a decrease in the number of crosslinks between the phases if the propylene units have the isotactic configuration (R505) or an increase if the configuration is atactic. Nonetheless, a decrease in the efficiency of retardation of ozone degradation and, as will be shown below, in the EPDM network strength takes place in all cases. A large amount of the bulky diene impedes the formation of a dense bulk interlayer. A doubled amount of diene groups in R505 as compared to R521 leads to a twofold decrease in the amount of interphase ligaments in the BNKS-18 blend and to a drop in v_d^{rel} by a factor of 4.5. In the blends of all NBRs with EPDM 714, the amount of cross links increases by a factor of 1.6–1.8, and falls in the BNKS-18 and BNKS-28 blends by a factor of 1.3.

Thus, even an increase in the amount of cross-links primarily along the interface decreases the ozone protection effect. This result is due to a small depth of interpenetration of copolymer chain segments, a situation that leads to a reduction in the strength of interphase contacts. The most effective protection is provided by the compatibility of NBR with EPDM having a large

proportion of ethylene units at a high isotacticity of propylene sequences and a moderate amount of diene groups.

The character of the dependence of the quasi-equilibrium stress σ^* on strain in accordance with the specifics of adhesive interaction of phases reflects the ability of the EPDM network to withstand deformation [5]. Owing to the fact the EPDM phase is bonded to the matrix through inter layers containing chemical links between the components, the value of σ^* is the superposition of stresses in the matrix and the EPDM phase. Nonetheless, this quantity gives an insight into stresses that emerge in the EPDM vulcanization network during deformation of the specimen. The network strength is affected by the degree of development of the interlayer, its length, the amount of chemical links between the phases, and the degree of cross linking in the EPDM phase [5–7]. At a low compatibility of components (blends based on BKNS-40), the dependence of σ^* on the presence of microcrystalline domains of ethylene and isotactic propylene sequences in the copolymer manifests itself also [9–11, 23, 24].

All DSM EPDMs with propylene units of low stereo regularity form a less strong network in their cured blends with NBR as compared to Uniroyal EPDMs (Fig. 11.3). The networked structures of EPDMs in the blends with less polar BNKS-18 show the highest strength. In the covulcanizates with BKNS-28 the degradation of the EPDM 712 network with a high content of atactic propylene units occurs at $\varepsilon = 110\%$ and that of the EPDM 714 network with a higher proportion (8%) of diene groups takes place at $\varepsilon = 50\%$ [5].

At the same time, the values of σ^* for the Royalen EPDM blends of the same composition (R505 and R521) are retained at a sufficiently high level (0.10–0.15 MPa) [5]. The EPDMs of both brands with a large amount of ethylene units form the strongest three dimensional frame works.

EPDM R512 and EPDM R521 with a high proportion of ethylene or isotactic propylene units form networks that vary in strength; however, a decrease in the quasi-equilibrium stress σ^* on passing from NBR blends having a lower amount of acrylonitrile units to the blends with a higher amount of acrylonitrile units (e.g., from BNKS-18 to BNKS-28) is proportional to the increase in the amount of polar units in NBRs. Thus, the strength of the EPDM network depends on the interlayer length. The breaking of interphase interaction by polar units leads to a drop in the EPDM network strength [5].

In the case of blends of DSM EPDMs, a proportional decrease in the stress σ^* with a growth in the proportion of acrylonitrile units is observed only for EPDM 778 co vulcanizates with BNKS-18 and BNKS-28. Consequently, such a decrease in these materials is also the basic reason for the drop in the strength of the EPDM network structure. However, in the blend of BNKS-40

with EPDM 778, a decline in the EPDM network strength is nonproportionally strong at a rather high density and a large amount of cross links in the interlayer (Table 11.3). This behavior is explained in terms of a decrease in the volume of the interphase owing to its length and the depth of the interpenetration region of unlike chains. When the atactic fraction prevails in EPDM with a high ethylene contents, the compatibility with highly polar NBR is limited by the amount of butadiene units in the latter.

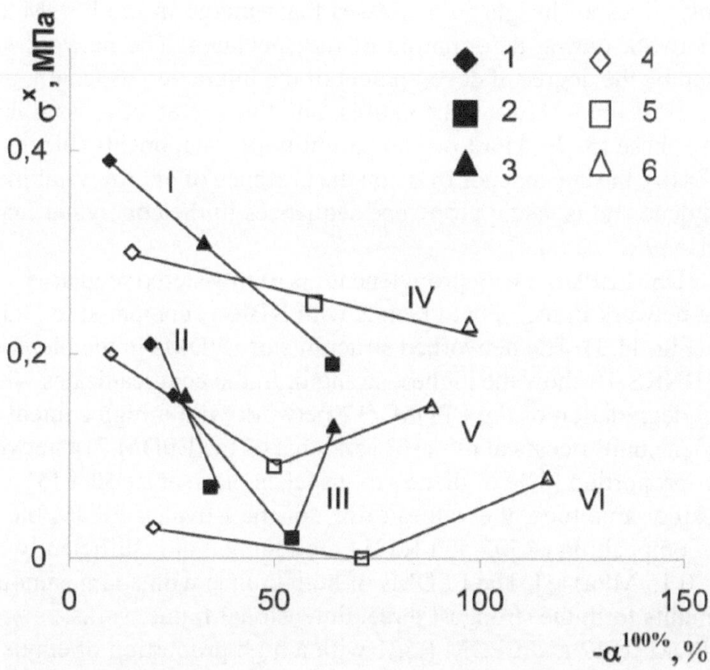

FIGURE 11.3 Quasi-equilibrium stress σ* as a function of the departure of equilibrium swelling from the additive value, –α: BNKS-18 (I, IV), BNKS-28 (II-V), and BNKS-40 (III-VI) covulcanizates with (*1–3*) EPDM R512, EPDM R505, and EPDM R521 and (*4–6*) EPDM 778, EPDM 714, and EPDM 712.

Figure 11.3 depicts the plots of the quasi-equilibrium stress versus $-\alpha^{100\%}$ for NBR co vulcanizates with EPDMs manufactured by different companies. As the interphase density and the amount of cross links between the components increase ($-\alpha^{100\%}$ decreases), the EPDM network strength in the blends with the most developed interlayer proportionally increases. The slope of the linear plot characterizes the relative contribution of the interphase length and

the depth of inters diffusion of unlike chains in an interlayer to the network strength. At the hypothetical maximal interphase density corresponding to – $\alpha^{100\%} = 0$, the network strengths of Royalen and Keltan EPDMs in these blends will be 4.3 and 3.1 MPa, respectively. The lower EPDM network strength in the latter case is due to a smaller interphase volume along the depth of the interphase interaction zone as well as to a lower concentration of microcrystalline domains [9–11, 23].

EPDM-40 (Table. 11.4) with a higher degree of isotacticity of propylene units (as compared with EPDM R 512) is characterized by a far more intensive penetration of propylene chain fragments in their rigid isotactic configuration into BNKS regions containing single polar groups [5, 6]. As the region of adhesion interaction of elastomers is increased, the efficacy of reduction in the rate of ozone degradation v_d^{rel} of the diene matrix dramatically increases in the blends with the less polar BNKS-18 (Tables 11.3, 11.4).

TABLE 11.4 Efficacy of Ozone Degradation Retardation of BNKS (V_d^{rel}) by Ethylene–Propylene–Diene Elastomers (30 Phr) and Σ* at Different Tensile Strains

ε, %	EPDM –40		Elastokam 6305		EPDM –60 (I)		EPDM –60 (II)		Elastokam 7505		$v_d\ 10^3$ min^{-1} BNKS
	v_d^{rel}	s^x, MPa	v_d^{rel}	s^x, MPa	v_d^{rel}	s^x, MPa	v_d^{rel}	s^x, MPa	v_d^{rel}	s^x, MPa	
BNKS-18											
30	48,0	0,27	11,6	0,34	12,1	0,30	4,5	0,24	1,8	0	240
50	43,0	0,30	7,4	0,43	5,7	0,31	3,4	0,25	2,7	0	256
70	3,5	0,23	2,9	0,35	4,0	0,25	2,1	0			200
90	3,0	0,23	4,1	0,34	4,0	0,30					190
110	2,2	0,19	2,4	0,34	2,8	0,26					122
130		0,17		0,36		0,26					
150		0,17		0,31							
BNKS-28											
30	1,8	0,09	2,2	0,11	2,8	0,16	1,8	0,13	1,9	0,10	165
50	3,0	0,21	2,0	0,15	4,0	0,22	2,4	0,12	2,5	0,13	192
70	1,4	0,17	1,6	0,11	3,2	0,23	1,7	0	2,3	0	163
90	1,3	0,17	1,3	0,16	1,8	0,15					134
110	1,2	0,18	1,0	0	1,5	0,13					100
130						0,14					

TABLE 11.4 *(Continued)*

ε, %	EPDM −40		Elastokam 6305		EPDM −60 (I)		EPDM −60 (II)		Elastokam 7505		ν_d 10^3 min^{-1} BNKS
	ν_d^{rel}	s^x, MPa	ν_d^{rel}	s^x, MPa	ν_d^{rel}	s^x, MPa	ν_d^{rel}	s^x, MPa	ν_d^{rel}	s^x, MPa	
BNKS-40											
30	2,9	0,15	2,0	0	3,2	0,08	1,3	0,08	2,2	0	115
50	1,4	0,13	1,2	0	1,6	0,09	0,8	0,04	1,3	0	94
70	1,2	0,10			1,7	0,07	1,0	0,07	2,2	0	99
90	1,3	0,08			2,0	0	1,3	0			117
110	2,1	0									130

This behavior can be explained by reduced dimensions of EPDM-40 particles and a small distance between network elements that are the consequences of high compatibility. In the blends with BNKS-40 with the increasing strain, the interfacial layer partially breaks down owing to its discontinuous structure; as a result, ν_d^{rel} decreases dramatically (Table 11.4). For the blend containing EPDM-40, this breakdown is observed at ε = 110%, while for the EPDM R 512-containing blend, this effect occurs at ε = 70% [5–7]. Therefore, in the case of the blends based on EPDM-40, the interfacial micro separation takes place at higher strains. With the increasing polarity of BNKS, ν_d^{rel} abruptly decreases owing to decreased length, depth, and density of the interfacial layer and cross-links between phases.

Comparing the efficacy of degradation retardation in BNKS −18 containing EPDM-40 or Elastokam 6305 with a high content of ethylene units (74%) and a very low degree of isotacticity of propylene sequences (9.5%) (Table 11.2), one can conclude that the properties of the latter blend appear to be much worse (Table 11.4). This copolymer is compatible only with butadiene chain fragments of BNKS containing preferably *cis*-1,4- and 1,2-isomers (Fig. 11.1). Hence, when propylene units (9.5%) are characterized by a very low microtacticity, compatibility between the components decreases. Even in the blends containing BNKS-18, ν_d^{rel} is not high: the depth of penetration of EPDM segments is limited by the butadiene part of the copolymer. The volume of the formed interfacial layer is smaller.

This blend is characterized by a high strength of the EPDM network. High values of the quasi-equilibrium stress are preserved up to a strain of 150% [6]. The dependence of σ* on the presence of microcrystalline domains of ethylene and isotactic propylene sequences [9–11, 23, 24] in the copolymer manifests itself also.

EPDM-60 (I) (ENB 4%) is characterized by a lower content of ethylene units and a somewhat higher degree of isotacticity of propylene chain fragments (13%). In the blends containing EPDM-60(I), the dependence of the number of cross links in the interfacial layer is similar to that as observed for the blends containing DSM 778 of close co monomer composition and stereo regularity of propylene units [5] (Fig. 11.1). Because, in EPDM-60 (I), higher propylene units exist in the isotactic configuration and the content of diene groups is lower, its segments are able to diffuse into non-polar SKN regions to a greater depth than segments of Elastokam 6305. As a result, with an increase in tensile strain, the protection efficacy of EPDM-60 (I) in the blends with BNKS-28 and BNKS-40 decreases to a lesser extent and the strength σ^* of EPDM-60 (I) network is higher than the strength of the Elastokam 6305 network (Table 11.4) [6].

EPDM-60 (II) with a higher content of ENB (6.7%) is compatibilized only with butadiene units (Fig. 11.2) and $(-\alpha^{100\%})$ increases by a factor of 1.8 as their content grows. This tendency suggests that this interfacial layer is characterized by an increased friability and the number of cross-links in this layer is low. The higher the volume of this layer (the lower the polarity of BNKS), the higher the $(-\alpha^{100\%})$ values. Therefore, the efficacy of ozone degradation retardation is low even for the blends based on BNKS −18 containing EPDM-60 (II) and the strength of the EPDM network is small (Table 11.4).

The ratio between ethylene and propylene units in Elastokam 7505 is similar to that in EPDM −60 (II), but the degree of isotacticity of propylene units is lower (9.5%) and the Mooney viscosity is very high (Table 11.3). Elastokam 7505 is characterized by the minimum compatibility with all SKN samples. Evidently, the efficacy of ozone degradation retardation of the diene matrix in the presence of Elastokam 7505 is low and the quasi-equilibrium stress σ^* in the cross linked blends of BNKS-18 and BNKS-40 is zero at the minimum strain (30%) [6] In the above blends, compatibility between components is so small that, as can be judged by the σ^* value, the copolymer exists in the dispersed state and does not form any continuous three-dimensional structure (Table 11.4). The fact that the strength of the EPDM network decreases to zero at low strains indicates the development of a coarse dispersed structure with weak bonds between particles; this behavior is likewise observed for the blends based on BNKS-40 and Elastokam 6305.

Thus, the reduction of the rate of ozone degradation of the diene elastomer is explained in terms of compatibility of the components, the volume and density of the interphase, and the amount of cross links in the interlayer, the EPDM network strength depending on the amount of chemical cross links between the phases and on the degree of cross linking of the EPDM phase or on

the presence of microcrystalline domains of ethylene and isotactic propylene sequences.

It has been found that not only the co monomer ratio in EPDM but also the stereo regularity of propylene sequences determines the specifics of formation of the interphase and its volume in cross linked heterophasic blends with NBRs of different polarities. The most dense and voluminous interlayer with the largest amount of cross links provides the most effective protection against ozone degradation to the diene matrix.

KEYWORDS

- **Butadiene–Nitrile Rubbers**
- **Compatibility Phase**
- **Ethylene–Propylene–Diene Elastomers**
- **Interphase Layers**
- **Isomeric Composition**
- **Isotacticity**
- **Ozone-Induced Degradation**
- **Ozone-Protective Action**
- **Stereo Regularity**
- **Stress Relaxation**
- **Structure**

REFERENCES

1. Khanin, S. E. (1984). *Abstract of C and. Sci. (Tech.) Dissertation*, Moskow: Tire Industry Research Inst.
2. Krisyuk, B. E., Popov, A. A., Livanova, N. M., & Farmakovskaya, M. P. (1999). *Polymer Science, Ser. A 41(94)*, [*Vysokomol. Soedin, Ser. A 41(102)*].
3. Livanova, N. M., Popov, A. A., Karpova, S. G., Shershnev, V. A., & Ivashkin, V. B. (2002). *Polymer Science, Ser.* A **44**,) [*Vysokomol. Soedin, Ser.* A *44(71)*].
4. Livanova, N. M. (2006). *Polymer Science, Ser. A 48(821)* [Vysokomol. Soedin, Ser. A *48(1424)* (2006)].
5. Livanova, N. M., Lyakin Yu. I., Popov, A. A., Shershnev, V. A., *Polymer Science, Ser.* A **49**, 63 (2007) [*Vysokomol. Soedin, Ser.* A *49*, 79].
6. Livanova, N. M., Lyakin Yu. I., Popov, A. A., & Shershnev, V. A. (2007). *Polymer Science, Ser. A 49*, 300 [*Vysokomol. Soedin, Ser. A 49*, 465].
7. Livanova, N. M., Lyakin, Yu. I., Popov, A. A., & Shershnev, V. A. (2006). Kauch Rezina *4*, 2.

8. Zateev, V. S. (1972). *Abstract of C and Sci. (Tech) Dissertation* Volgograd: Volgograd Polytechnical Inst.
9. Livanova, N. M., Karpova, S. G., & Popov, A. A. (2003). *Polymer Science, Ser A 45(238)* [*Vysokomol. Soedin, Ser A 45(417),* (2003)].
10. Livanova, N. M., Evreinov, Yu. V., Popov, A. A., & Shershnev, V. A. (2003). *Polymer Science, Ser. A 45,* 530 [*Vysokomol. Soedin, Ser A 45(903)*].
11. Livanova, N. M., Karpova, S. G., & Popov, A. A. (2005). Plast. Massy, *2,* 11.
12. Dechant J., Danz R., Kimmer W. & Schmolke R., *Ultrarotspektroskopische Untersuchungen an Polymeren* (Akademie, Berlin, 1972; Khimiya, Moscow, 1976).
13. Kozlova, N. V., Sukhov, F. F., & Bazov, V. P. (1965). *Zavod Lab 31 (968).*
14. Kissin, Yu. V., Tsvetkova, V. I., & Chirkov, N. M. (1968). *Vysokomol. Soedin, Ser.* A10, 1092.
15. Kissin, Yu. V., Popov, I. T., Lisitsin, D. M. et al. (1966). *Proizvod. Shin Rezino-Tekh. Asbesto-Tekh. Izdelii,* No. 7, 22.
16. Gould, R. F. *Multicomponent Polymer Systems*, (American Chemical Society, Washington, 1972.
17. Lednev, Yu. N., Zakharov, N. D., Zakharkin, O. A. et al. (1977). Kolloidn Zh *39(170).*
18. Kuleznev, V. N. (1980). *Polymer Blends* (Khimiya, Moscow) [in Russian].
19. Lipatov, Yu. S. *Physical Chemistry of Multi component Polymer Systems*, (Naukova Dumka, Kiev, 1986), 2 [in Russian].
20. Livanova, N. M., Karpova, S. G., & Popov, A. A. (2011). *Polymer Science, Ser. A53*, 1128 (2011) [*Vysokomol. Soedin Ser*A *53*, 2043].
21. Saltman, W. M. *The Stereo Rubbers*, (Wiley, New York, 1979), 2.
22. Nesterov, A. E. (1984). *Handbook on Physical Chemistry of Polymers* (Naukova Dumka, Kiev,) [in Russian].
23. Bartenev, G. M. (1979). *Structure and Relaxation Properties of Elastomers* (Khimiya, Moscow) [in Russian].
24. Bukhina, M. F. (1973). *Crystallization of Rubbers and Vulcanized Rubbers* (Khimiya, Moscow) [in Russian].
25. Vol'kenshtein, M. V. (1959). *Configurational Statistics of Polymer Chains* (Akad Nauk SSSR, Moscow) [in Russian].

CHAPTER 12

A RESEARCH NOTE ON INFLUENCE OF POLYSULFONAMIDE MEMBRANES ON THE PRODUCTIVITY OF ULTRAFILTRATION PROCESSES

E. M. KUVARDINA, F. F. NIYAZI, B. A. HOWELL, G. E. ZAIKOV, and N. V. KUVARDIN

CONTENTS

ABSTRACT

Article concerns questions of division of multi-component solutions by means of polyamide membranes in the course of ultra filtration. The question of influence of low-frequency fluctuations on a polyamide membrane for the purpose of increase of its productivity is considered.

12.1 INTRODUCTION

Pellicles semi permeable membranes, which are used for separation of multicomponent solutions, have a two-layer structure.

Polymeric membranes are formed mainly by methods based on processing polymer solutions (and less often melts). There are various methods for producing membranes such as irrigating of the solution, molding from the melt, forming pores in the pellicles of nuclear particles, followed by leach of the destruction products, pressing, rolling, etc. As the result, membranes have through system of pores, often forming a maze of interconnected channels. The pore diameter of the membrane varies in depth, that indicating their isotropy.

Such membranes have unequal structure in the thickness. The surface layer, which is responsible for the selectivity of separation, is dense (whose thickness is 20···40 microns), an inner layer is coarse-pored (the thickness of which is 100···200 microns), and it is used as a substrate and improves the mechanical strength of the membrane and has a big impact on the formation of the selective barrier layer.

Work layer is made mainly of synthetic polymers, copolymers, and mixtures thereof. Membranes, which working layer is based on cellulose and its derivatives (cellulose acetate), and polyamides, received most practical use in the ultrafiltration processes.

In this chapter, we used the membrane UPM-50P, which working layer is made of polysulfonamide.

12.2 EXPERIMENTAL PART

Physicochemical properties of the membrane are largely dependent on the nature of the material, from which they are formed, and methods for their preparation. They are determined by the chemical structure and composition of the polymer, which depend on such important characteristics as the flexibility of the polymer chains, the nature and the energy of intermolecular interactions, as well as interaction with components of shared solutions. The molecular

structure largely determines the physicochemical and mechanical properties of the membranes.

There are mechanical properties of polymers, which include mechanical strength, deformability, the ability to develop reversible and irreversible deformation, resistance to fatigue in multiple deformations and deterioration.

In the matter of determining the most rational parameters of membrane apparatus necessary to consider the physical condition of the membranes in the process of their work for the separation of multicomponent solutions, taking into account the major external factors, which are affecting the working surface of the membrane. During the operation of the membrane apparatus as external factors can be considered temperature, which values increases in direct proportion to the operation time, the chemical composition of the partial solution, operating overpressure and the mode of its change. In this article we consider the effects of temperature and pressure on the mechanical properties of the polysulfonamide membrane. As the mechanical properties of the membrane, we consider the value of the membrane deformations, arising under a vibration of the circulating solution.

It is known, that one of the factors, that influence on the mechanical properties of polymers is temperature.

Melting temperature of the polyamide lies within the range of 65–88°C· Values of entropy and heat of fusion of polyamides, which are calculated theoretically and confirmed experimentally, are described by the following relationship

$$\Delta H = T_m \cdot \Delta S; \tag{1}$$

where ΔH – heat of fusion kJ/mol, T_m - the melting temperature of polyamide, K; ΔS – entropy change kJ/(mol K) [1].

Melting temperature of the polysulfonamide lies within the range of 65 ± 20 °C.

In most cases, physical properties of the polysulfonamide don't vary linearly and depend on the composition polysulfonamide. They are characterized by the presence of a minimum in the curve of dependence of the melting temperature from the composition, often in the equimolar ratio of elementary units or nearly equimolar ratio.

Thus, the temperature of the ultrafiltration process is maintained in the range of 60 °C· The temperatures above 60 °C may negatively impact on the quality of solutions, so in our experiments the temperature was in the range of 4–60 °C.

As a partial solution were used a diffusion juice of sugar beets. Pressure generated in the apparatus was 0.25–0.3 MPA.

As a result, the pressure generated in the apparatus compresses the membrane. Conversation of mechanical compression energy into heat increases the temperature of the membrane [2]. However, its value disproportionately low compared with the temperature of the circulating solution, therefore, we don't consider it. The upper temperature limit is 60 °C, but it does not have great practical importance on the mechanical properties of the membrane, as it does not lead to a change in the physicochemical properties of the membrane material.

Curve 1 in the Fig. 12.1 illustrates the speed reduction of the separation process when the temperature of initial solution is 10 °C· From the point of view of the theory of concentration polarization, we certainty can say, that the largest pores of the working layer of the membrane become clogged, thereby forming a sediment layer on its surface without altering of the membrane's structural properties.

The interval of low temperatures (4–10 °C) caused the greatest interest during the experiment. Curve 2 in the Fig. 12.1 illustrates this interval. The performance of the membrane increases in the initial period. This can be explained by knowing the structural properties of the polymer, so that the membrane material behaves like a solid body, increasing the packing density of the molecules, and hence the density of the membrane. Large pores of the membrane are compressed, preventing the formation of a so-called "specula" [4]. So at this stage of the process we can see a phenomenon that suspended particles and macromolecular compounds remain on the surface of the membrane in the form of colloidal unstable sediments, which are easily washed off by the circulating solution and low molecular weight compounds are readily pass through the fine pores of the membrane without stopping. Under these conditions, the gel formation in the pores by sorption is practically eliminated.

Under such conditions of temperature deformation of the membrane, which is induced by the pressure, generated by the machine and vibration, is small.

With increasing temperature of the solution the membrane behaves like an elastic body, in other words, under the periodic influence of the low frequency vibrations, which are introduced into the circulating solution and create environmental fluctuations, we can see changes in the distance between the molecules of the polymeric membrane layer.

Effects of vibration on the mechanical properties of the membrane were investigated in the frequency range from 10 Hz to 1500 [4].

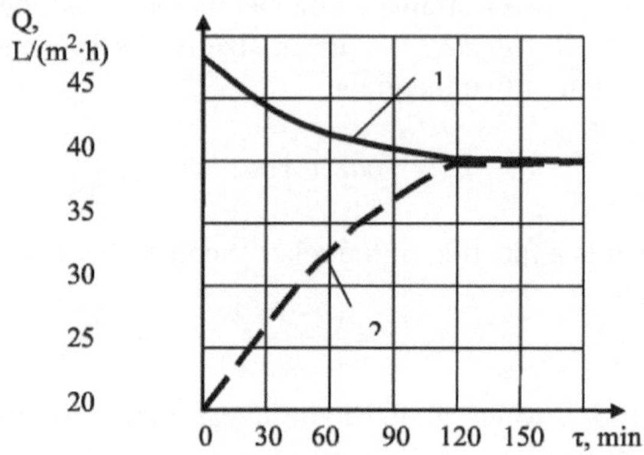

FIGURE 12.1 The dependence of the performance of the membrane in the first time period: 1 – initial solution's temperature above 10 °C; 2 – initial solution's temperature 4 °C.

The membrane's performance for all values of the specified frequency range hasn't changed, remaining within the performance of the membrane apparatus, which working under pressure and without introducing vibrations into the solution, but performance of the membrane significantly increased with exposure to low frequency in the range of 60–70 Hz. Maximum performance was equal to 67 Hz in this limit. This is shown at Fig. 12.2.

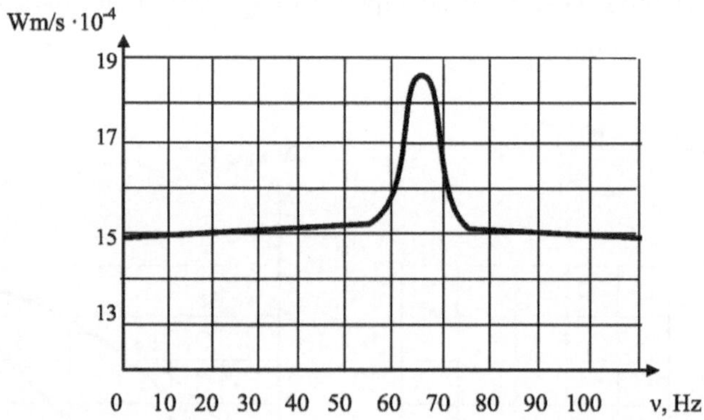

FIGURE 2 Filtration rate dependence from the oscillation frequency, which is introduced into the solution.

We try to analyze the obtained effect. It is known, that as a criterion for characterizing the viscoelastic properties of the membrane we can take a square of hysteresis loop, described by the curve G=f(P) at gradually increasing pressure from zero to a certain value, then the pressure change in the reverse sequence. [5]

A typical curve "load – strain" while membrane's working under constant pressure is shown at Fig. 12.3a. Current size of the hysteresis loop can be represented as a sum of two integrals:

$$S = \int_{0}^{\varepsilon_2} \sigma d\varepsilon + \int_{\varepsilon_2}^{\varepsilon^1} \sigma d\varepsilon \tag{2}$$

In this case, σ is a function of the relative compression ε,

$$\varepsilon = \frac{(h - h_0)}{h_0}$$

and a product (σdε), which is under the integral sign, has the dimension of work per unit volume.

$$\sigma d\varepsilon = \frac{F}{S} \cdot \frac{dh}{h_0} = \frac{Fdh}{v} \tag{3}$$

Thus, the area of the hysteresis loop is the difference between the specific work expended in compression (loading) of the membrane and energy received at the unloading of the membrane.

When we introduce vibrations in the flow than membrane's loading is uneven, thus load curves (compression and unloading) do not coincide and a graph describing these relationships will has a different character. This is shown at Fig. 12.3a.

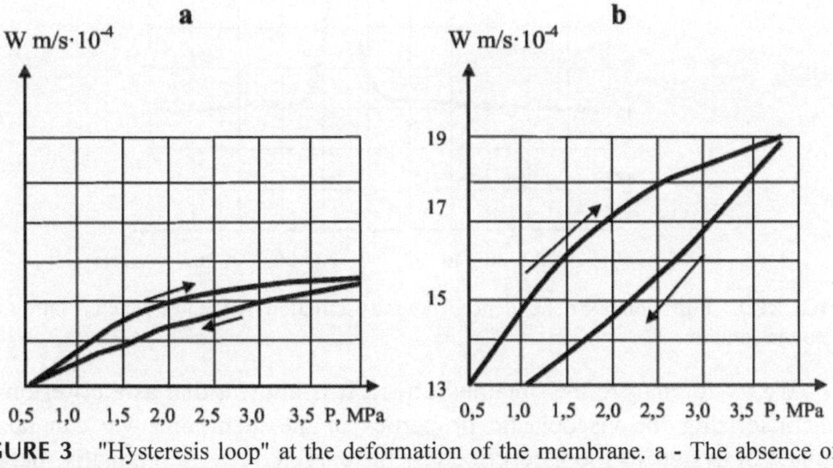

FIGURE 3 "Hysteresis loop" at the deformation of the membrane. a - The absence of vibrations introduced into the solution; b - The presence of 60-70 Hz vibrations introduced into the solution.

When we unload the membrane's material, the same load values correspond to larger values of compression than during loading, and the curve of unloading misses the origin, indicating the presence of residual deformation in the membrane. This is shown at Fig. 12.3b.

The hysteresis loop area depends on the rate of application of force and, in this case, on the frequency of the vibrations, which are introduced into the solution. We think that the value of the frequency of vibration below 60 Hz of values σ and ε are close to equilibrium and the loading and unloading curves are close to each other. This suggests that the area of the hysteresis loop has a small value. On the other hand, for values of the vibration more than 70 Hz, loading rate of the membrane material is very high and the elements of the structure of the polymer does not have time to regroup, and the material of the membrane is deformed not by regrouping kinetic units, but due to changes in the distances between the particles. In this case, the deformation of the membrane is small, its values are close to each other when loading and unloading, therefore, the area of the hysteresis loop also has a small value.

The gel layer is formed at the membrane surface at temperatures above 10 °C and along with it; the sorption takes place in the pores with the formation of additional gel layers.

Vibration frequency of 60–70 Hz helps to prevent gel formation in the pores of the polymer membrane.

12.3 CONCLUSIONS

The following conclusions are drawn from this chapter.

1. The hysteresis loop area is small both at very high and at very low rates of application of the force and its area is maximum for the time values of the force, which are comparable with the value of the relaxation time of the membrane.
2. The range of frequencies of vibrations, which are introduced into the circulating flow at the ultrafiltration of sugar beet raw juice, corresponds to values of 60–70 Hz. For such values, hysteresis loop area is maximum, as evidenced by an increase in productivity of the membrane apparatus.

KEYWORDS

- **Membrane**
- **Multicomponent Solution**
- **Polyamide**
- **Process**
- **Ultrafiltration**
- **Vibration**

REFERENCES

1. Kudryavtsev, G. I., Nosov, M. P., & Volokhina, A. V. (1976). Polyamide Fibers [Text]: Kudryavtsev, G. I., Nosov, M. P., Volokhina, A. V. M: "*Chemistry*" 260.
2. Gul, V. E. (1976). Structure and Mechanical Properties of Polymers [Text]: Gul, V. E., Kuleznev, V. N. M. Higher School. 314.
3. Cherkasov, A. N., & Pasechnik, V. A. (1991). Membranes and Sorbents in Biotechnology [Text]: A.N. Cherkasov, V.A., Pasechnik, L.: Chemistry. 240.
4. Kuvardina, E. M. PhD thesis (2003). Dynamics of the Ultra Filtration Apparatus for the Separation of Sugar Beet Diffusion Juice [Text]: Kuvardina, E. M. Kursk KSTU. 143.
5. Dytnersky, Y. I. (1986). *Boro Membrane Processes*. Theory and Calculation [Text]: Dytnersky, Y. I. M.: "Chemistry." 272.

CHAPTER 13

A RESEARCH NOTE ON ENVIRONMENTAL DURABILITY OF POWDER POLYESTER PAINT COATINGS

T. N. KUKHTA, N. R. PROKOPCHUK, and B. A. HOWELL

CONTENTS

ABSTRACT

For the first time ever durability of powder polyester paint coatings was determined by an express method. The method is based on empirical exponential dependence of coating durability on activation energy of thermal-oxidative degradation of paint filming agent. Quantitative evaluation of the impact of key destructive factors on coatings durability was made. The express method noticeably shortens time for certification of powder paints by such factor as "durability," and their manufacturing application in obtaining coatings for metal, concrete, asbestos boards.

13.1 INTRODUCTION

Technological advance in the field of paintwork materials (PWM) that requires enhancement of coating protective properties under severe service conditions, as well as solution of a number of ecological and economical problems, has resulted in the development of brand new PWM powder ones. Within relatively short period of time these materials proved themselves to be quite promising, with their formulation being one of the highest priorities of the current material science development. At the present time powder paints in terms of coatings manufacture technology, durability, as well as ecology and economics, have practically no alternatives. Absence of solvents in powder paints dramatically reduces environmental pollution, with absence of expenditures connected with organic solvents (30–70% of liquid PWM makeup), treatment of air and sewage waters. On top of that, the technology of powder paint coating fabrication is nonwaste (production waste is fully recyclable), less energy-consuming (no power is required for solvent evaporation, the costs of production premises ventilation drop), more automated (maintenance personnel and production floor are reduced), more manufacture efficient (several times).

Now the most wide-spread are powder paints on the basis of the following filming agents: epoxy resins, epoxy- polyester oligomers (combination of epoxy and polyether resins), hybrid filming agents; not-saturated polyester resins [1].

In selection of powder paints one of the most important properties is resistance to weather conditions. The influence of ambient environment leads to energy absorption in ultraviolet band of electromagnetic spectrum. This energy has negative effect both on film-forming polymer, and the pigment resulting in loss of glitter and change of color. Due to tendency to chalking epoxy and hybrid paint coatings are usually not recommended for use in the open air.

Polyether materials form coatings that provide good resistance to the influence of ultraviolet rays.

Composition of polyether powder paints contain curing agents that cross–link oligomer macromolecules and form low-molecular polymer spatial cross-linked structure. Powder paints with extensively used curing agents were selected for the research.

One of them is triglycidyl isocyanurate (TGIC), which was used for rather long period of time, but lately regarded as a harmful reagent in some European countries.

Another one is hydroxyalkylamide considered as harmless and known under the trade name "primide." Therefore, study of environmentally resistant polyether powder paint coatings cured with TGIC and primide, as compared to coatings of hybrid powder makeup is a crucial task.

Work objective is to quantitatively evaluate destructive factors that influence polymer protective coatings made of powder paints when in service; propose empiric exponential relationship between durability of metal tiles and asbestos cement boards coatings and activation energy of thermal oxidative breakdown of filming agent for polyether powder paints; develop express method for evaluation of these coatings durability with regard to influence of environmental factors.

13.2 EXPERIMENTAL PART

Subjects of the research were films of 0.3–0.4 mm in thickness, (10 ± 2) mm in width and 100 mm in length, and coatings of 0.1 mm in thickness fabricated of powder paints, samples of:

- 1, 2, 3, 4 – polyether paint, curing agent – primide, colors – red, white, green and black, correspondingly;
- 5, 6 – polyether paint, curing agent – TGIC, colors – green, black, correspondingly;
- 7, 8 – hybrid paint, colors – blue and black, correspondingly.

The films were formed on fluoroplastic sheet. Powder paint was deposited through the sieve.

The tests were performed in the climatic chamber "Feutron," type 3826/16 (Germany) according to the following cycle:

- moisture treatment of samples at temperature (40 ± 2)°C and relative humidity (97 ± 3)% for 2 h;
- moisture treatment free of heating at relative humidity (97 ± 3)% for 2 h;

- freezing at temperature minus (30 ± 3)°C for 6 h;
- irradiation of samples by creating light flux with surface density of total radiation energy (730 + 140) W/m^2 with surface density of UV radiation flux (30 + 5) W/m^2 and periodic sprinkling with water for 3 min in each 17 min for 5 h;
- freezing at temperature minus (60 ±3)°C for 3 h;
- conditioning at temperature 15–30°C and relative air humidity of 80% for 6 h.

Samples were taken each 25; 50; 75; 100 cycles.

The films were artificially aged under influence of UV and IR radiation by means of dummy emitter of sunlight SOL 1200S (Germany).

Artificial aging mode:

- temperature in the climatic chamber – 50°C;
- relative air humidity – 60%;
- UV radiation mode – 57.7 W/m^2; IR – 730 W/m^2;
- visible range – 320 W/m^2.

Total optical radiation flux from dummy emitter HSA 1200S at a distance of 60 cm from radiation source was 1107.7 W/m^2.

Magnitude of samples radiation energy from dummy emitter for 600 h amounted to 2393 MJ/m^2; 1200 h – 4786 MJ/m^2; 2,400 h – 9572 MJ/m^2.

Films porosity was determined by their specific area values calculated by BET method (Brunour, Emmet, Teller). Nitrogen adsorption isotherms were read out on instrument NOVA 2200. Gaseous nitrogen with operation temperature of 77 K was obtained by evaporation of liquid nitrogen. Measurements error did not exceed 10% of specific area values.

Mechanical tests were conducted on tensile testing machine T 2020 DC 10 SH (Alpha Technologies UK, USA).

Ambient air temperature – 18°C, speed of top grip motion – 10 mm/min, clamping length of samples – 54 mm, number of test samples – 10.

By diagrams "tensile strain σ (MPa) – tensile deformation ε (%)" using the instrument computer program, the rupture resistance (σ, MPa), relative elongation (ε, %), Young elasticity modulus (E, MPa) were computed as mean arithmetic of ten measurements.

The value of activation energy E_a was determined by Broido computational method on the basis of dynamic thermogravimetric data [2].

Morphology of films surface (in research film surface being in contact with air and not with fluoroplastic substrate was used in order to exclude influence of substrate) was studied on scanning electronic microscope JSM 5610LV (Jeol, Japan).

13.3 RESULTS AND DISCUSSION

Results of films mechanical tests before influence of artificial climatic factors are given in Fig. 13.1.

Data analysis shows that more durable and elastic films are formed from polyether films cured with primide in particular. Durability of 1–4 samples on the average is 24.6 MPa, whereas durability of 7, 8 samples is only 17.4 MPa. Relative elongation at rupture 2.35% and 1.66%, elasticity modulus 1610 and 1865 MPa correspondingly.

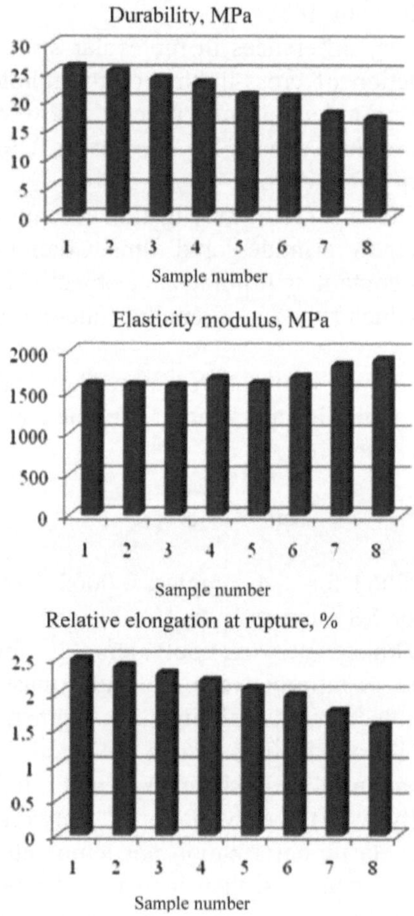

FIGURE 13.1 Deformation durability and elastic properties of films made of powder paints.

Therefore, films of polyether powder paints are more durable than films of hybrid powder compositions by 41%, and more elastic by 42%. Less than by 16% elasticity modulus of polyether films as against hybrid ones gives ground to believe that smaller inherent stresses will be developed in coatings of polyether powder paints due to their greater relaxation in film that is easier deformed. In its turn, small inherent stresses in the film provide its longer service life (service life of coatings without loss of their protective properties).

Comparing compositions of polyether powder, containing different curing agents, one may note that primide is more preferable than TGIC. The films cured by primide are more durable than films cured by TGIC by 14% on the average, and more elastic by 16%.

This is explained by differences in molecular structure of primide and TGIC. Chemical reaction of cross-linking of unsaturated polyether oligomeric molecules with primide molecules proceeds slower and requires higher temperatures and more time consuming. However, it forms more uniform, durable and elastic polymer network.

As a result, durability and relative elongation at rupture is higher, and elasticity modulus is lower for primide-cured films. Coatings made of polyether powder compositions containing primide have objectively to be more durable than TGIC cured coatings because of smaller inherent stresses that develop in them.

Since films made of powder paints represent chemically cross-linked spatial network patterns, their disintegration has fragile nature. For each sample the following ratio is observed:

$$\sigma = K \times E \times \varepsilon,$$

where K= 0.0066–0.0063 for 1–4 samples; 0.0062–0.0063 for 5,6 samples and 0.0057–0.0055 for 7,8 samples.

In spatially cross-linked (network) polymers their mechanical properties are greatly influenced by ratio between molecular mass of section between network nodes and molecular mass of kinematic segment [3].

If molecular weight of kinetic segment (MWk.s) << molecular weight of an interval between crosslinks (MWc) (kinematic linkage is flexible and network is sufficiently wide), then variation of network density practically affects neither highly elastic deformation nor temperature of polymer glass transition. But if MWc > MWk.s, then increase of network density (reduction of Mc) results in reduction of highly elastic deformation and rise of glass transition temperature [4].

At very high density of three-dimensional network the highly elastic deformation is impossible, and at room temperature material is in glass state. Variation of durability and increase of network density is, as a rule, expressed by a curve with maximum. Small amount of cross linkages does not hamper straightening of chains in deformation, resulting in the enhancement of durability. Still increase of density over optimal values hinders orientation processes during stretching of films, and their durability starts to degrade the more the larger density of formed three-dimensional network is.

According to the data obtained by DSC (differential scanning calorimetry) (Fig. 13.2), the lowest glass state temperature have nonadhesive films obtained from primide-cured polyether filming agent (glass-transition temperature of sample No 2 is 62°C). During solidification of this filming agent with TGIC curing agent the glass state temperature goes up by 6°C (samples No 5 and 6 in Fig. 13.4).

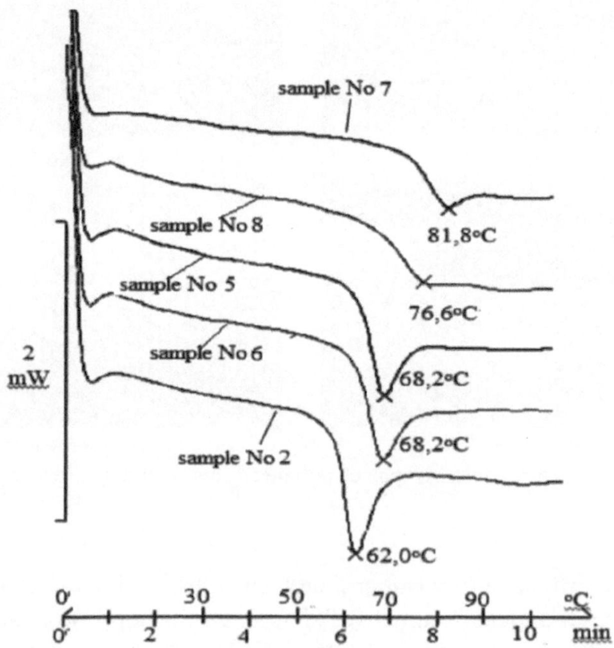

FIGURE 13.2 DSC of powder paint films.

Finally, with the highest glass transition temperature (77–82°C) is characterized by samples of films Nos. 7 and 8 obtained from hybrid filming agent

(combination of epoxy and polyester resins). Consequently, logical interrelationship is observed between values of glass transition temperature, durability, elasticity modulus, relative elongation upon rupture of films.

Electron micrographs (Fig. 13.3) give evidence of different nature of films surface, along with various size and shape of coloring agents particles.

Surface of films of primide-cured polyether powder paint (Sample No. 1) is smooth and uniform.

Network structure is formed free of considerable inherent stresses.

Surface of films of hybrid paint (Sample No. 7), and TGIC cured polyether paint (Sample No. 6) is texturized, as it was formed under conditions of great inherent stresses.

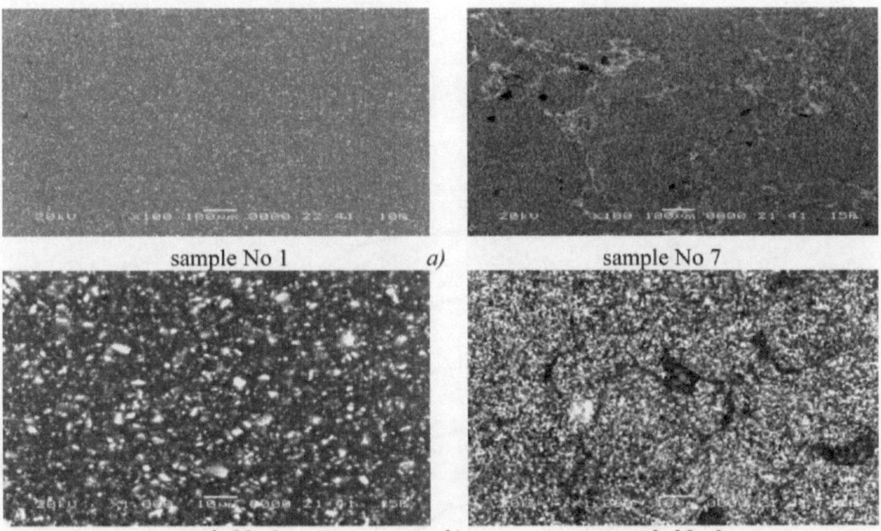

FIGURE 13.3 Electron micrographs of polymer films with magnification: a – 100×; b – 1000×.

Upon the influence of weather factor on films in climatic chamber rigidness of the three-dimensional network has small variation, so elasticity modulus varies within several percent (10% maximum). However, under impact of heat, UV-radiation and humidity sections of macromolecules between network nodes disintegrate and its elasticity drops. With practically linear rise of film stress upon its extension, and smaller relative elongation of upon rupture lower durability is achieved. Due to this with the increase of the number of cycles affecting films, both durability and relative elongation diminish.

Dependences of durability and relative elongation upon rupture on exposure time have similar S-shaped nature (Fig. 13.4).

As kinetics of radical reactions in solid polymers has strong dependence on the degree of macromolecules cross-linking and mechanical stress on their chemical bonds [5], the films under study differ in aging rate.

So films obtained from primide-cured polyester resin (Samples Nos. 1–4) retain 65% of initial durability and 72% of relative elongation upon rupture after impact of 100 cycles in the climatic chamber.

Invariance of values of these indices for TGIC cured polyester films is equal to 60 and 70% correspondingly, and for hybrid epoxy-polyester resin films – 50 and 67%.

The data we obtained for powder paint films comply with provision of the kinetic theory of strength [6] stating that durability of a solid, including that of polymer, decreases with the increase of mechanical stress σ that affects chemical carbon-to-carbon bonds in the backbone chain.

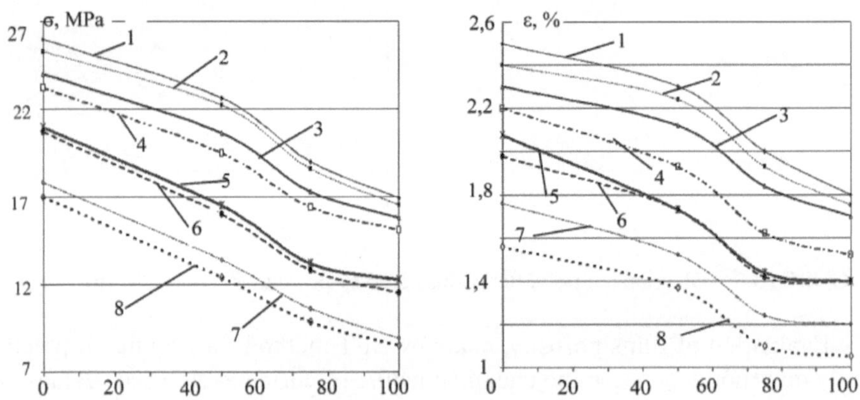

FIGURE 13.4 Dependence of stress-strain properties of films on time of exposure in the climatic chamber.

On trials in the climatic chamber polymer films were periodically subjected to 100% moistening. An important role in the reduction of durability and deformation characteristics of stressed composite materials is played by liquid media affecting them on a long-term basis [7].

Cross-linking reaction of polyester binder oligomeric molecules when cured by primide is exercised through chemical interaction of resin molecules terminal carboxyl with hydroxyl groups of curing agent (Fig. 13.5a).

Alongside with that, molecular structure of primide (aliphatic architecture of molecule matrix) provides comparably flexible mobile cross linkage of cured film three-dimensional network and, as a consequence, good mechanical properties and high weathering resistance.

With the same TGIC-cured powder polyester paint, terminal carboxyl of the binder interacts with three epoxy groups of the curing agent (Fig. 13.5b).

Furthermore, ester links are formed and water is exuded which, with large thickness of films, may form defects in the form of punctures. However, this negative effect can be brought to a minimum by using degassing additives in make up of powder paints, along with selection of optimal temperature-time mode of coatings formation.

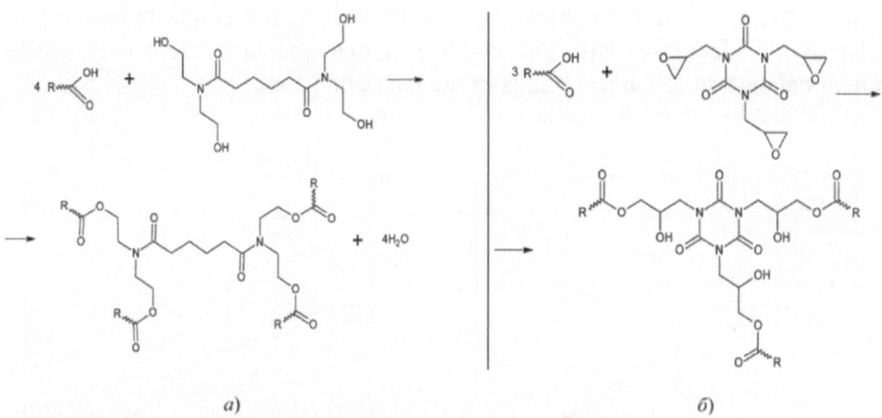

FIGURE 13.5 Diagrams of polyester resins cured by primide (a) and TGIC (b).

Evaluation of films porosity, made by BET method showed that it practically does not depend on the chemical nature of curing agent used. Values of specific surface area of film samples Nos. 1–6 varied from 10 to 11 m^2/g, that is, being within the measuring error. For this reason one may say that porosity has identical effect on all film samples during their tests. Whereas differences in mechanical properties and weathering resistance of powder polyester paint films are conditional on different nature of chemical cross-links in the three-dimensional network, that is, on differences in chemical architecture of curing agent molecules.

As a result, individual molecules of polyester resin are cross-linked into the three-dimensional network without emission of volatile low-molecular compound. Enhanced reacting capacity of TGIC epoxy groups provides more easy and quick cross-linking at lower temperatures and within shorter time of coatings cure. Meanwhile cyclic TGIC structure specifies higher rigidity of

network cross links, with inherent stress slow relaxation. Due to this complex of mechanical properties of films and their weathering resistance worsen.

Analysis of Fig. 13.4 shows that after impact of 100 cycles in the climatic chamber the films still retain on the average 65% of initial durability and 72% of relative elongation upon rupture. However, with regard to impact of other operational factors on protective films, like inherent stresses, external mechanical forces, it is reputed that 100 exposure cycles result in complete loss of films service life, as it is common knowledge that durability corresponds to time within which durability and/or elasticity drops twofold.

We have established that nonadhesive films of primide-cured powder polyester paint of different colors disintegrate with energy practically identical to activation energy of thermal-oxidative degradation E_d equal to 140±2 kJ/mol.

The key prevailing destructive factors that lower potential barrier of filming agent bond opening E_d are:

- ΔE^{clim} – lowering of E_d due to exposure in the climatic chamber;
- ΔE^{UV} – lowering of E_d under influence of sunlight-simulating radiation;
- $\Delta E^{in.str}$ – lowering of E_d by inherent stresses that occur due to differences in coefficients of thermal expansion of coating and protected surface;
- $\Delta E^{mech.\ in.}$ – lowering of E_d due to static and dynamic loads on coatings. The following values of destruction factors were obtained:

ΔE^{clim} =50 kJ/mol, ΔE^{UV} = 10 kJ/mol; $\Delta E^{in.str.}$= 7 kJ/mol; $\Delta E^{mech.\ in}$ = 3 kJ/mol.

Due to this calculated value of thermal-oxidative degradation activation energy of coatings that determines their durability equals to:

$$E_{calc.} = 140–50–10–7–3=70 \text{ kJ/mol.}$$

In our earlier works [8–10], we have proposed and implemented in the system of certification tests the express method for determination of rubber and thermoplastic articles durability, which is based on the interrelation between durability of polymer material τ and value of activation energy of its thermal-oxidative degradation E_d. It has been repeatedly demonstrated that E_d determines quality of polymer material and is reduced under influence of service factors.

In this work for the first time ever the method is proposed for evaluation of durability of powder paint coatings having tridimensional cross-linked structure.

Durability of coatings τ_{Ts} at set value of service temperature T_s is calculated by empirical formula we have established:

$$\tau_{T_э} = K \cdot (10^{-\alpha \cdot E_{calc.} - \beta} \cdot e^{Ecalc./R \cdot T_э}),$$

where K = 2.74×10^{-3} years, α = −0.1167 mol/kJ, β = 0.090.

Values of coefficients are established by mathematical treatment of experimental datasets obtained at prolonged (six months) aging of powder polyester paint films from different manufacturers differing considerably in values of E_d.

Calculated durability of polymer coating in years (τ_{total}) at variable values of article service temperature is determined by the formula:

$$\tau_{total} = [\sum_{i=1}^{n} \frac{m_i}{\sum m_i}] \cdot \tau_{T_s}$$

where m_i – number of hours of impact at particular values of service temperature; Σm_i – total number of hours of impact at variable values of service temperature; τ_{Ts} – durability of a polymer article in years at particular value of article service temperature.

For climatic conditions of Eastern Europe duration in hours was determined for service within one year and at temperatures that develop within materials upon impact on them of direct sun beams (Table 13.1).

TABLE 13.1 Climatic Conditions of Eastern Europe

Index name	Value				
Number of hours of temperature impact, h	2250	1400	440	200	100
Temperature, °C	20	30	40	50	60

Example of durability calculation for on-metal coatings from polyester powder paint produced by MAV PUPE is given below.

Calculated durability of coating in years at temperatures 20, 30, 40, 50 and 60°C is:

$$\tau_{20°C} = 2.74 \times 10^{-3} \times [10^{-0.1167 \times 70 - 0.090} \times e^{70/2.435}] = 46$$

$$\tau_{30°C} = 2.74\times10^{-3}\times[10^{-0.1167\times70-0.090}\times e^{70/2.518}] = 17.9$$

$$\tau_{40°C} = 2.74\times10^{-3}\times[10^{-0.1167\times70-0.090}\times e^{70/2.601}] = 7.4$$

$$\tau_{50°C} = 2.74\times10^{-3}\times[10^{-0.1167\times70-0.090}\times e^{70/2.684}] = 3.2$$

$$\tau_{60°C} = 2.74\times10^{-3}\times[10^{-0.1167\times70-0.090}\times e^{70/2.767}] = 1.5$$

Calculated durability of coating in years when in service in climatic conditions of Eastern Europe is:

$$\tau_{total} = \left[\frac{2250}{4392}\times\frac{1}{46}+\frac{1400}{4392}\times\frac{1}{17.9}+\frac{440}{4392}\times\frac{1}{7.4}+\frac{200}{4392}\times\frac{1}{3.2}+\frac{100}{4392}\times\frac{1}{1.5}\right]^{-1} = 14$$

13.4 CONCLUSIONS

Express method was developed for evaluation of durability of protective coatings made of powder paints of polyester class. The method is based on empirical exponential dependence of coatings durability on activation energy of thermal-oxidative degradation of powder paint filming agent. Evaluation of the impact of destructive factors acting on macromolecules of protective coatings when in service, and lowering potential barrier of empirical bonds opening of polymer macromolecules was made and, as a consequence, the coating deformation-durability properties and service life. For the first time data was obtained on real service time of powder paint coatings, notably made of local powder paint. Determination of coating durability by express method is performed in one working day, whereas the method established by standard takes four months. This makes it possible to promptly evaluate quality level of powder polyester paints from different manufacturing companies available on the market of paintwork materials of the Republic of Belarus. Besides, producers of powder paints of polyester class are permitted to noticeably shorten time for the development of new formulations of durable protective coatings.

KEYWORDS

- **Activation energy of thermal-oxidative degradation**
- **Climatic factors**
- **Coating**
- **Durability**
- **Powder polyester paint**

REFERENCES

1. Yakovlev, A. D. (1987). Powder paints Yakovlev, A. D. L.: *Chemistry*, 216 p.
2. Broido, A. (1969). Sensitive Graphical Method of Treating Thermogravimetrie Analysis Data Broido, A., *A Simple J. Polymer. Sci. Part A2. 7(10)*, 1761-1773.
3. Tager, A. A. (2007). Physics-chemistry of polymer TagerA. A.; edited by prof. Askersky, A. A. M: Nauchniy mir, 573 p.
4. Encyclopedia of polymers: in 3 v. Kabanov, V. A. (1977) (editor-in-chief) [et.al]. M: "Soviet Encyclopedia", 3: P-Y (1977). 1152 p.
5. Emanuel, N. M. M. (1982). Chemical physics of ageing and stabilization of polymers Emanuel, N. M., Bugachenko, A. L. M: Nauka, 360p.
6. Regel, V. R. (1974). Kinetic nature of solid durability Regel, V. R., Slutsker, A. I., Tomashevsky, E. M. M.: Nauka, 560p.
7. Shevchenko, A. A. (2010). Physics-chemistry and mechanics of composite materials: Textbook for HEE A.A. Shevchenko. SPb: COP "Profession" 224p.
8. Method for determination of elastomers durability: patent 1791753 USSR, IPC G01N318// G01N1700 Alexeev, A. G., Prokopchuk, N. R., Starostina, T. V., Kisel, L. O. (USSR). N4843144/08; appl.26.09.90; publ.30.01.93 Bull. N4. 8p.
9. Polymer articles for construction. Method of durability determination by activation energy of thermal-oxidative degradation of polymer materials. State Standard of the Republic of Belarus STB1333.0-2002.
10. Prokopchuk, N. R. (2008). Evaluation of durability of polymer articles Prokopchuk, N. R. *Standardization, 1*, 41–45.

CHAPTER 14

A RESEARCH NOTE ON ELASTOMERIC COMPOSITIONS BASED ON BUTADIENE-NITRILE RUBBER CONTAINING POLYTETRAFLUORETHYLENE PYROLYSIS PRODUCTS

N. R. PROKOPCHUK, V. D. POLONIK, ZH. S. SHASHOK, and E. M. PEARCE

CONTENTS

ABSTRACT

The effect polytetrafluorethylene pyrolysis products "Forum" on the properties of the elastomer compositions based on butadiene-nitrile rubber are investigated. It is established that introduction of modifying agent promote viscosity reduction, process acceleration of vulcanization and improve the technical properties of vulcanizate.

14.1 INTRODUCTION

Elastomers are one of the most important structural materials in modern engineering and occupy a unique place among a variety of polymeric materials. They are the only material capable of large reversible deformation in a wide range of temperatures; they possess high durability, wear-ability and water-resistance as well as a number of other valuable qualities [1].

At the same time, the products of elastomeric materials are determinative for different kinds of process equipment, automobiles, special purpose equipment. Failure of such determinative product leads to a loss of functional characteristics of machinery, mechanisms due to wear or destruction [2].

Despite the high importance of creating new formulations of rubber, it is advisable to carry out the modification of commercially produced rubber compounds. This would greatly save financial resources. Analysis of the literature data shows that the most common method of polymeric material modifying including rubber is the introduction of the modifying additive compositions in powder form.

The aim was to determine the influence of powdered polytetrafluoroethylene pyrolysis product on plasto-elastic properties and vulcanization kinetics of rubber compounds as well as on the basic operational properties of vulcanizates based on them.

14.2 EXPERIMENTAL PART

The objects of the research were filled elastomer compositions on basis of synthetic butadiene-nitrile rubber (NBR) with the dosage of bound acrylonitrile 17–23% intended for the production of compacting rubber products for different purposes. The modifying additive was added in dosage of from 0.1 to 0.5 parts per hundred of rubber (phr). The samples without the additive were used as objects of comparison.

Polytetrafluoroethylene pyrolysis product is produced by the Institute of Chemistry, Far East Branch of the Russian Academy of Science, under the

trade name "Forum." It's produced by thermal effects on the base polymer. The destruction of macromolecules of polytetrafluoroethylene in the most stressed areas of the sample followed by sublimation fragments of different molecular shape and mass is the most probable mechanism of this process. Macromolecular particles are the main products of the synthesis; they are characterized as "spray," which is formed by the interaction of molecular radicals – pyrolysis products of polytetrafluoroethylene and monomer molecules [3].

For the most part pyrolysis product PTFE contains sphere-like particles with an average diameter of 0.6 microns, which can be assembled into larger, easily breakable airflow conglomerates up to 15 μm. Particles can also sphere-like with the same dimensions as the diameters of the individual particles [4].

In accordance with studies [5–8], the powder particles contain the low- and high-molecular fractions of polytetrafluoroethylene. The molecules of low molecular weight part contain fluoroolefin end-groups with double bonds ($-CF=CF_2$) and side trifluoromethyl groups-CF_3. The material in these end-groups depends on the process conditions of block polytetrafluoroethylene pyrolysis.

The electronic picture of surface of the product "Forum" is presented in Fig. 14.1.

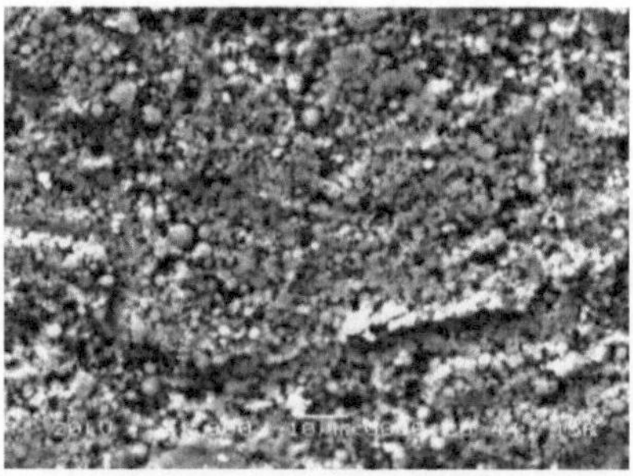

FIGURE 14.1 The electronic picture of surface of the product "Forum."

The Mooney viscosity test was carried out in viscometer MV2000 company "Alpha Technologies" according to ASTM D1646, and vulcanization kinetics parameters on vibroreometer ODR2000 Company "Alpha Technologies" by ASTM D2084. Physical and mechanical properties such as conditional tensile strength at break σ_p elongation at break ε_p were determined by tensiometer T2020DC company "Alpha Technologies" by ASTM D412. According to State Standard 9.024–74, State Standard 9.029–74 and State

Standard 9.030–74 tests to determine resistance to heat aging of rubbers in the unloaded and loaded conditions were carried out (the accumulation level of relative compression set) and the action of liquid hydro carbonaceous medium.

Influence of modifying additive on the vulcanization grid parameters was assessed by concentration values of cross links by the Flory-Rener equation based on data on the equilibrium swelling in toluene at temperature (23 ± 2)°C [8]:

$$\frac{1}{M_c} = -\frac{V_r + \chi \cdot V_r^2 + \ln(1 - V_r)}{\rho_K \cdot V_0 \cdot (V_r^{1/3} - 0,5 \cdot V_r)}$$

where M_c – average molecular weight of the chain segment enclosed between two cross links, kg/mol; V_r – volume fraction of rubber at swollen vulcanizate, м^3/mol; V_0 – molar volume of solvent, м^3/mol; χ – the Huggins constant which characterizes the interaction between the rubber and the solvent.

The viscosity of the material being processed determines the dynamics of the recycling process; it is a measure of the force, which must be applied to the material flow to implement it at a given speed for a particular stage of the process [9]. The results of identifying viscosity research of the elastomeric compositions are presented in Table 14.1.

TABLE 14.1 Viscosity and Kinetics of Vulcanization of the Elastomer Compositions Containing the Product of PTFE Pyrolysis

The modifying additive dosage, phr	The Mooney viscosity, conv.u.	Optimal time of vulcanization, min
0	85.1	15.5
0.1	84.5	15.3
0.2	83.3	15.1
0.3	82.8	14.9
0.4	83.1	14.4
0.5	83.5	14.5

The analysis of this data shows that the introduction of polytetrafluoroethylene pyrolysis product in the compositions leads to a decrease of viscosity. Thus, the minimum value of the elastomeric compositions viscosity based on NBR-18 is achieved at a dosage of 0.3 phr and is 82.8 conv. u Mooney, while

the viscosity of the unmodified compositions is 84.1 conv. u Muni. Reducing the viscosity of the elastomer compositions is probably due to segments orientation facilitation of macromolecules in the direction of load application. The particles oligomeric fraction of modifying additive acting as a plasticizer contributes to this process.

Vulcanization is hemoreological process, as a result of chemical transformations the material passes from plastic and viscous-flow into highly elastic status. During vulcanization rubber macromolecules dimensional cross-linking occurs to form a vulcanization space grid. The main parameter characterizing vulcanization process is optimal time of vulcanization.

The obtained data (see Table 14.1) indicate that using the product "Forum" decreases optimal time of vulcanization. The minimum value of this ratio is achieved with a dosage of modifying additive of 0.4 phr. Perhaps the additive particles contribute to the segmental mobility facilitation of rubber macromolecules and to more uniform distribution of vulcanizing group components in a volume of elastomeric composition, which led to their more intensive interaction with rubber macromolecules at the double bonds and reduce the formation of vulcanization grid.

Since the compacting rubber products were exposed to elevated temperatures during operation, the thermal stability of elastomeric compositions was determined. Thermal aging of the compositions in the unloaded state was performed in a heat chamber at 125 °C for 72 h, while in the loaded state (relative compression set) for 24 h. Change in elongation at break of δ_ε and the level of accumulation relative compression set are shown in Table 14.2.

TABLE 14.2 Change in Elongation at Break of Vulcanizates after Thermal Aging

The modifying additive dosage, phr	δ_ε, %	Accumulation level of relative compression set, %
0	–25.0	24.0
0.1	–15.0	19.8
0.2	–14.3	19.3
0.3	–13.6	19.6
0.4	–15.5	20.5
0.5	–16.2	21.1

It is seen that the imposition of polytetrafluoroethylene pyrolysis product promotes the thermal stability of the elastomeric compositions. Thus, the relative elongation of vulcanizates at break containing the product "Forum" is reduced by 13.6% at a dosage of 0.3 phr after thermal aging, whereas the decrease of the indicator in the unmodified samples is 25%. PTFE pyrolysis

product injection also reduces relative compression set savings at elevated temperatures. The minimum value of this ratio is achieved with a dosage of 0.2 phr, and it is 19.3%, while comparison sample-24%.

Vulcanizates thermal stability increasing in unloaded condition is probably due to the fact that the application of modifying additive helps to reduce the rate of oxygen diffusion in the bulk of vulcanizate. This is probably achieved by the migration of low molecular weight fractions of "Forum" in the surface layers of the samples. This process helps to reduce the degree of exposure to elevated temperatures. Specific compression set reduction is probably determined by the acceleration of relaxation processes due to the increasing mobility of macromolecules segments.

Liquid hydrocarbon media in relation to rubbers are physically active media. It does not give rise to profound structural changes with the destruction of chemical bonds. When liquid hydrocarbons contact rubber products, a series of simultaneously proceeding processes such as sorption of medium by surface and volume of the rubber, medium diffusion through the rubber and extraction with soluble ingredients occur.

Resistance to liquid hydrocarbon media is an important indicator of operational reliability of rubber used for the production of compacting and cushioning rubber applied in sealing and casketing products as swelling and erosion of ingredients occurring in contact with oils, greases and hydraulic fluids, reduces efficiency and durability of materials [10].

The data in Table 14.3 show the results of determining the concentration of cross-linking as well as the degree of swelling and degree of erosion of the elastomeric compositions.

Presented numbers show that "Forum" using in the elastomeric compositions based on butadiene nitrile rubber reduces diffusion of liquid medium deep into rubber products as well as reduces the bulk of extractable plasticizer.

TABLE 14.3 The Equilibrium Degree of Swelling of Compositions under Research Containing Product "Forum"

Modifying additive dosage, phr	Degree of swelling, %	Degree of erosion, %	Concentration of cross-linking $\times 10^{-19}$, mol/cm^3
0	111.0	14.0	7.9
0.1	108.8	12.6	8.4
0.2	107.5	11.8	8.5
0.3	106.8	11.6	8.6
0.4	104.9	11.3	8.6
0.5	104.5	10.8	8.7

Thus, the degree of swelling of the composition based on NBR-18 in toluene is reduced by 7%, and the erosion – 30% as compared with the sample not containing the modifying additive. At the same time, increasing dosage of the modifying additive increases the concentration of vulcanizate crosslinking: 8.72×10^{-19} mol/cm^3 in a sample containing 0.5 phr additives, whereas the comparison sample – 7.87×10^{-19} mol/cm^3.

The increased concentrations of crosslinking and, hence, the durability to the action of the liquid hydrocarbon medium, is apparently caused by the interaction of particles of the modifying additive with the components of vulcanizing systems. During vulcanization the formation of physical interaction additional linkages by the polar groups and the double bonds of rubber and additives active centers is possible.

Rubber compounds on the basis of NBR are used for the manufacture of rubber products operating under abrasion in harsh environments, so it seemed appropriate to determine the effect of modifying additives at the abrasion resistance of the studied elastomeric compositions.

Determination results of the abrasion resistance of the rubbers are shown in Fig. 14.2.

FIGURE 14.2 The abrasion resistance of the elastomeric compositions, containing PTFE pyrolysis product.

The presented data show that there is an increase in durability of the samples by using additives "Forum." Thus, the abrasion resistance index value for elastomer compositions on the basis of butadiene-nitrile rubber NBR-18 containing an additive in the maximum dosage is 14.2 J/mm^3, whereas the comparison sample – 8.4 J/mm^3.

The injection of the modifying additive into the elastomeric compositions helps to reduce the bulk of separated material during friction. This is probably

due to the formation of zones of plastic deformation caused by the migration of low molecular weight fractions of the PTFE pyrolysis products in the surface layers of samples.

14.3 CONCLUSION

The application of ultradisperse polytetrafluoroethylene as modifying additive to compositions on the basis of elastomeric butadiene-nitrile rubbers enables to reduce the viscosity of rubber compounds, to speed up the process of vulcanization grid formation, to improve the technical properties of the finished product such as heat resistance, resistance to attack by liquid hydrocarbon media, durability.

KEYWORDS

- **Polytetrafluorethylene pyrolysis product**
- **Processing behavior**
- **Rubber**
- **Technical characteristic**

REFERENCES

1. Kercha Y. Y. et al. (1989). Structural and chemical modification of elastomers [in Russian]. Sergeev, L. F., Ed. Navuk. Dumka, Kiev, 232 p.
2. Avdeychik, S. V. et al. (2006). Nanocomposite Engineering Materials: Development and Application Experience: Monograph [in Russian]. Struk, V. A. Ed. Grodno State University, Grodno, 403 p.
3. Struck, A. V. et al. (2011). Structure and properties of fluorine-containing nanocomposites based on vulcanized rubbers [in Russian] Weight Nat. Acad. Navuka of Belarus, Ser. fiz-TECH. Navuka No.1, 25–31.
4. Kuryavyi, V. G. et al. (2002). The morphological structure of the PTFE ultra disperse pyrolysis products [in Russian] *Advanced Materials, 6*, 71–73.
5. Vopilov, Y. Y. et al. (2012). Properties of ultradisperse PTFE fractions which are soluble in supercritical carbon dioxide [in Russian] High-molecular structures, V. 54, Series A, No 6, 842–850.
6. Bouznik, V. M. et al. (2002). Particle size and shape of ultradisperse polytetrafluoroethylene obtained in thermal gas-dynamic method [in Russian]. Advanced Materials, No 2, 69–72.
7. Avdeychik, S. V. et al. (2012). Engineering fluorocomposites: Structure, Technology, Applications: Monograph [in Russian]. Struk, V. A., Ed. Grodno State University, Grodno, 339 p.

8. Averko-Antonovich, I. Y., Bikmullin, R. T. (2002). Methods of Research of Structure and Properties of Polymers [in Russian]. KSTU, Kazan, 604 p.
9. Shutilin, Y. F. (2003). Handbook on properties and applications of elastomers: monograph [in Russian]. Voronezh state tehnol. acad., Voronezh, 871 p.
10. Uralsky, M. L. (1983). Control and Regulation of Technological Properties of Rubber Compounds [in Russian]. Uralsky, M. L., Gorelik, R. A., Bukanov, A. M. Khimiya, Moscow, 126 p.

CHAPTER 15

SPECTRAL-FLUORESCENT STUDY OF THE EFFECT OF COMPLEXATION WITH ANIONIC POLYELECTROLYTES ON CIS-TRANS EQUILIBRIUM OF OXACARBOCYANINE DYES

P. G. PRONKIN and A. S. TATIKOLOV

CONTENTS

ABSTRACT

The effect of interaction with anionic polymers (DNA, poly(sodium 4-styrenesulfonate)) was studied on *cis–trans* isomer equilibrium and spectral and fluorescent properties of a number of oxacarbocyanine dyes: 3,3′-diethyloxacarbocyanine iodide (K1), 3,3′-diethyl-9-methyloxacarbocyanine iodide (K2), 3,3′-dimethyl-9-ethyloxacarbocyanine iodide (K3), 3,3,'9-triethyl-6,6′-dimethoxyoxacarbocyanine iodide (K4), and 3,3,'9-triethyl-5,5′-dimethyloxacarbocyanine iodide (K5). The interaction of the dyes with anionic polymers leads to the formation of stable noncovalent complexes. A shift of the *cis-trans* isomer equilibrium toward the formation of the *trans*-isomer was observed for *meso*-substituted oxacarbocyanine dyes in the presence of anionic polymers, which determined in many respects the spectral effects observed upon the complexation of the oxacarbocyanine dyes. A steep rise of fluorescence (due to binding of the trans-isomer) in a complex with DNA is favorable for using oxacarbocyanine dyes to determine DNA.

15.1 AIM AND BACKGROUND

Cyanine dyes are among the most widely used noncovalent fluorescent probes for nucleic acid detection with various applications in research, as well as in clinical and medical analysis [1, 2]. For instance, indocyanine dyes Cy3 and Cy5 are frequently used in biomedical practice for DNA labeling and detection. Spectral and fluorescence properties of these dyes are sensitive to their nucleo base environment [3, 4]. Along with noncovalent probes, the properties of covalent conjugates of cyanine dyes (derivatives of thiazole orange and oxazole yellow) with DNA have been thoroughly investigated [5]. DNA conjugates of cyanine dyes SYBR Green, TO-PRO, TOTO have been used for molecular imaging [6], for ECHO probes [7].

Binding a dye to a biopolymer drastically changes the photo physical and photochemical properties of cyanine dyes (fluorescence and triplet quantum yields increase due to the growing "rigidity" of the molecule), thus forming the basis for their practical use in biochemistry and photo medicine. The unique properties of dyes of this class are determined by the presence of the polymethine chain, which is the most important part of the molecular structure of these compounds.

Studying the spectral and fluorescence properties of cyanine dyes in complexes with biopolymers (nucleic acids, proteins, peptides) attracts great interest, first, owing to the possibility of using these dyes as molecular probes [8, 9].

15.2 INTRODUCTION

Cyanine dyes are characterized by relatively narrow (of the order of 1000 cm^{-1}) and intense absorption bands in the visible and near-infrared spectral regions. The position of a maximum in the absorption spectra of polymethine dyes depends on the length of the polymethine chain, thus making it possible of bathochromic shift of the absorption band by increasing the number of –CH=CH– units. The short-wavelength side of cyanine dye absorption bands usually has a shoulder, 10–15 nm apart the principal maximum, attributed to the 0–1 vibronic transition, which is polarized, like the 0–0 transition, along the polymethine chain axis [10]. Due to complete π-π-electron conjugation in the polymethine chain, cyanine dyes generally have a high molar absorption coefficient. The structure of polymethine dyes is not rigid; therefore, a distinctive feature of these dyes is the fast dissipation of electronic excitation energy according to the mechanism of nonradiative deactivation, which efficiently competes with the processes of fluorescence and intersystem crossing to the triplet state $S_1 \rightarrow T$ [11, 12].

The lack of rigidity in molecules of these dyes provides a possibility of isomerization due to rotation about the polymethine chain, with the all-trans (EEEE) isomer being generally more stable thermodynamically for the unsubstituted dyes [10, 13], which occur in solution in this form. Introduction of substituent in the *meso*-position of the polymethine chain creates steric hindrances to the formation of the planar trans-isomer. For *meso*-substituted dyes, the steric effect of the substituent leads to the presence in dye solutions of the *trans* and *cis*-isomers, which are in equilibrium depending on polarity of the medium [13, 14].

The absorption spectra of the substituted thiacarbocyanine dyes are split into two bands: the long-wavelength band corresponds to the *trans*-isomer, whereas the short-wavelength band to the *cis*-isomer. The interaction of *meso*-substituted thiacarbocyanine dyes with ds-DNA shifts the isomeric equilibrium toward the *cis*-configuration, which can be clearly seen from the changes in the absorption spectra [15, 16]. The split of the bands is not observed in the absorption spectra of *meso*-substituted oxacarbocyanine dyes in both the absence and the presence of DNA.

In this chapter, we present the results of a comprehensive study by spectral and fluorescent methods of oxacarbocyanine dyes in solutions and in complexes with DNA and poly(sodium 4-styrenesulfonate). We chose as objects of the study a number of oxacarbocyanine dyes (K1-K5), including those with substituents (CH_3 and C_2H_5 groups) in the *meso*-position of the polymethine chain.

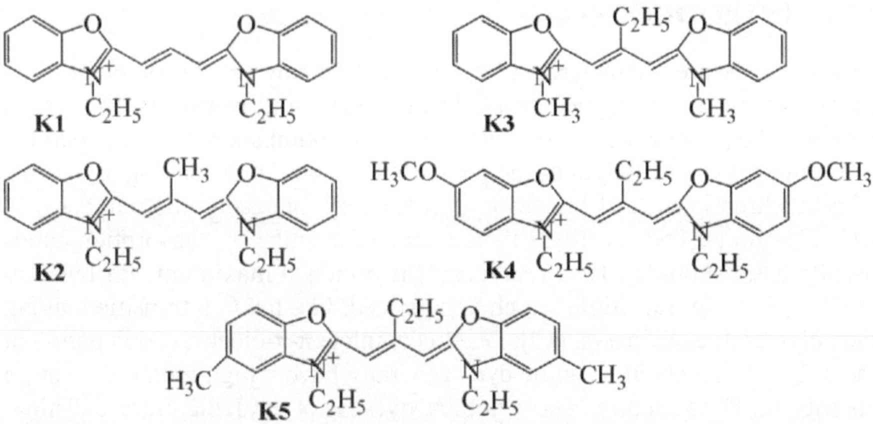

FIGURE 15.1 Structures of the oxacarbocyanine dyes: 3,3′-diethyloxacarbocyanine iodide (K1), 3,3′-diethyl-9-methyloxacarbocyanine iodide (K2), 3,3′-dimethyl-9-ethyloxacarbocyanine iodide (K3), 3,3,'9-triethyl-6,6′-dimethoxyoxacarbocyanine iodide (K4), and 3,3,'9-triethyl-5,5′-dimethyloxacarbocyanine iodide (K5).

15.3 EXPERIMENTAL PART

The absorption spectra of the dyes were measured with Shimadzu UV-3101 PC (Japan) and SF-2000 (Russia) spectrophotometers, and the fluorescent measurements with a Flyuorat-02-Panorama (Russia) spectrofluorimeter in standard quartz cells with the optical path length of 1 cm. The fluorescence and fluorescence excitation spectra were corrected for the spectral characteristics of the instrument (the signal of the reference channel) and the transmittance signal of a sample.

The dyes provided by the NIIKHIMFOTOPROEKT Research Center (B.I. Shapiro) and commercial chicken DNA (Reanal, Hungary) and poly(sodium 4-styrenesulfonate) (Sigma-Aldrich) were used. The concentrations of the dyes were constant; the experiments were carried out with the solutions having the absorbance at the maximums of the absorption spectra of about 0.1–0.15 (corresponding to the concentration of about $1–2 \times 10^{-6}$ mol l^{-1}).

The pH 7 phosphate buffer (at a concentration of 20 mmol l^{-1}), isopropanol, acetonitrile, chloroform, and 1,4-dioxane (reagent grade) were used as solvents.

The DNA concentration in the phosphate buffer solution was determined using the absorption coefficient of a base pair $\varepsilon = 13{,}200$ l mol^{-1} cm^{-1} at a wavelength of 250 nm [17]. The experiments were carried out at room temperature (22 ± 2)°C.

To characterize quantitatively the shape of the dye spectral bands, the method of moments was used [10]:

$$S_l = \int v^l \rho(v) dv \quad (1)$$

$$M_l = \frac{1}{S_0} \int v^l (v - <v>)^l \rho(v) dv \quad (2)$$

where S_l and M_l are the initial and central moments, respectively; l is the order of the moment ($l = 0, 1, 2$); $\rho(\nu)=\varepsilon(\nu)/\nu$, $W(\nu)/\nu^4$, $\varepsilon(\nu)$ is the absorbance of the sample; W is the fluorescence intensity of the sample; ν – is the wave number; and $<\nu>$ is the first moment of the normalized distribution of photons ($<\nu> = S_1/S_0$). The parameter $<\nu>$ characterizes "the gravity center" of the spectral band, and the value of $M_1 = 10^7/<\nu>$ the average position of the band on the wavelength scale [10]. The width of the spectral curve (σ) is determined by the value of the second central moment: $(\sigma) = M_2^{1/2}$.

The following simplified equations were used to estimate the equilibrium constant of the dye–polymer complex formation (binding constants K_{eq}) value:

$$F = F_{max} K_{eq} [Pol] / (1 + K_{eq} [Pol]), \quad (3)$$

where [Pol] is the concentration of the binding sites of the polymer (for DNA calculated using the base pair concentration), F_{max} is the intensity of the fluorescence at $[Pol] \rightarrow \infty$, and $K_{eq} [Pol] / (1 + K_{eq} [Pol])$ is the part of the dye molecules combined with biopolymers.

15.4 RESULTS AND DISCUSSION

In the absorption spectra of dyes K1–K5, single bands are observed. The position of the band maxima depends on the solvent: a bathochromic shift is observed on passing to the systems with lower polarity (Figs. 15.2a and b). Figure 15.2a shows the absorption spectra of K5 in different solvents (acetonitrile, isopropanol, dimethyl sulfoxide, and dioxane).

A growth of the refractive index of the media results in a long-wavelength shift of the dyes absorption spectra. Figure 15.2b shows the dependence of the positions of the absorption spectral maxima for the K1, K3–K5 on the refractive index of the solvent (n_D, at 20 °C). In the case of dye K5 this shift is about 12 nm on going from acetonitrile to dioxane.

Figure 15.3a shows the absorption, fluorescence, and fluorescence excitation spectra of K2 in acetonitrile and chloroform. The Stokes shift of the

emission bands of K1–K5 is 15–17 nm. For K1 without *meso*-substituents, the fluorescence excitation spectra correspond to the absorption spectra in both polar and nonpolar solvents (Fig. 15.3b), which indicates the presence of only one isomeric form of the dye in solutions of K1 irrespective of the solvent polarity.

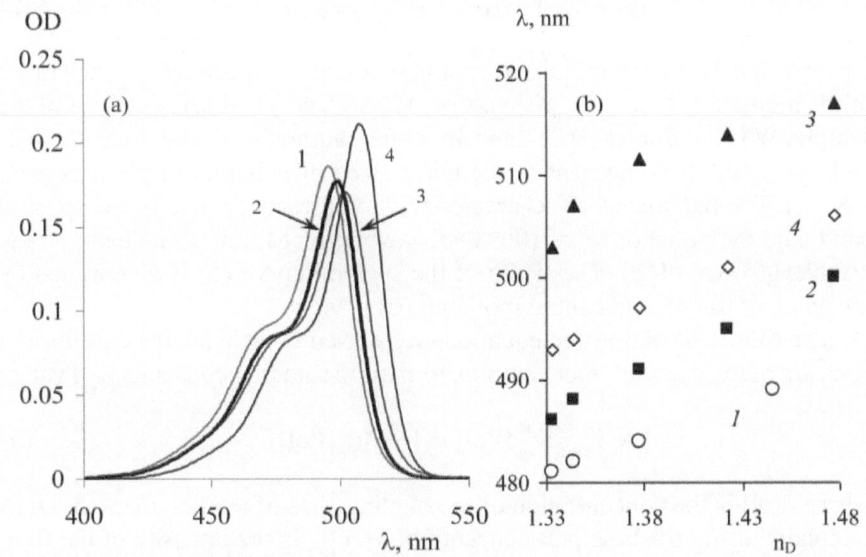

FIGURE 15.2 Absorption spectra of K5 (a; $c_{K5} = 2 \times 10^{-6}$ mol l^{-1}) in (*1*) acetonitrile, (*2*) isopropanol, (*3*) dimethyl sulfoxide, and (*4*) dioxane-1,4; (b) – the dependence of the position of the band maximum on the refractive index of the solvent (n_D, at 20°C) for (*1*) K1, (*2*) K3, (*3*) K4 and (*4*) K5.

In nonpolar solvents, the fluorescence excitation spectra of K2-K5 differ from the absorption spectra. In chloroform (Fig. 15.3a, curves 2, 6), the maximum of the fluorescence excitation spectrum of K2 (with the CH3 substituent) is bathochromically shifted by 7 nm with respect to the absorption spectrum (l_{max} = 492 nm). In more polar solvents (acetonitrile, isopropanol), the difference between the maxima of the bands in the spectra of K2 is 2–3 nm (Fig. 15.3a, curves 1, 5). In the spectra of dye K5 (with the C_2H_5 substituent), a similar shift of the maximums is also observed: ~7 nm in chloroform and ~4 nm in acetonitrile. However, an appearance of new bands belonging to dye isomers is not observed in the absorption spectra.

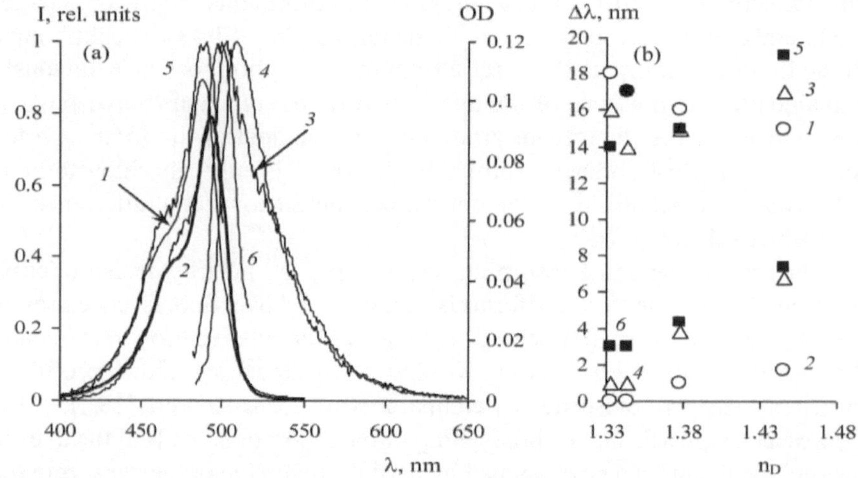

FIGURE 15.3 Absorption spectra (a) of dye K2 (c_{K2} = 1.09×10^{-6} mol l^{-1}) obtained in (*1*) acetonitrile and (*2*) chloroform; (*3, 4*) fluorescence (at λ_{ex} = 470 nm) and (*5, 6*) fluorescence excitation (at λ_{reg} = 510 and 525 nm, respectively) spectra of K2 in acetonitrile and chloroform solutions, respectively; (b) the dependences of the Stokes shifts of (*1*) K1, (*3*) K2, (*5*) K5, and the difference between the maxima of the excitation and absorption spectra D_2 ($D_2 = l_{ex.} - l_{abs}$.) of (*2*) K1, (*4*) K2, (*6*) K5 on the refractive index of the solvent (n_D, at 20°C). The fluorescence and fluorescence excitation spectra (3a – 6a) are normalized.

The substantial spectral differences between the absorption bands of *c is* and *trans*-isomers of thiacarbocyanines having bulky *meso*-substituents (SCH_3, OCH_3, Cl) permit unambiguous conclusion on the presence of a mobile isomeric equilibrium from the absorption spectra [13–15]. In the case of oxacarbocyanines, steep changes in the absorption spectra on passing from polar to nonpolar solvents are not observed [8]; hence, to characterize the spectra more accurately, we used the method of moments (Eqs. (1) and (2)) [10].

The gravity centers of the bands (the first moment of the normalized distribution of photons) ν (and their positions on the wavelength scale M_1) in the absorption and fluorescence excitation spectra of K5 in acetonitrile solution are rather close: 20,954 and 20,780 cm^{-1}, respectively; in the excitation spectrum, M_1 is bathochromically shifted by 4 nm with respect to the absorption spectrum. In chloroform the gravity centers of the bands of K5 are 20711 and 28,681 cm^{-1} in the absorption and fluorescence excitation spectra, respectively (M_1 is bathochromically shifted by 10 nm). In acetonitrile the band widths of the absorption (σ_{abs} = 1089 cm^{-1}) and fluorescence excitation (σ_{ex} = 1029

cm^{-1}) spectra of K5 differ little, whereas in chloroform this difference is larger (1311 and 889 cm^{-1}, respectively, with the ratio $\sigma_{abs}/\sigma_{ex} = 1.48$). The difference of the fluorescence excitation spectra from the absorption spectra is obviously explained by the presence of the dye in two forms (*cis*- and *trans*-isomers), which differ in the fluorescent properties. The content of the form with the long-wavelength absorbance (probably the *trans*-isomer) in chloroform is larger than in acetonitrile, which determines the great spectral differences in the former solvent.

The broadening of the absorption spectra upon the polarity decrease (passing from acetonitrile to chloroform) is accompanied by a fluorescence growth for dye K5 more than 40 times (for K1 and K2 the relative growth is 6.2 and 3.5, respectively). Since it is known that *c is*-isomers are characterized by extremely weak fluorescence as compared with *trans*-isomers [13, 14], the fluorescence growth and the broadening of the absorption bands of the dyes in less polar solvents can be explained by partial conversion of the dyes from the *cis*-form, in which they probably occur in polar solvents, to the *trans*-isomer.

Hence, the method of moments permits revealing changes in the absorption spectra upon the *cis*-*trans* transitions more clearly.

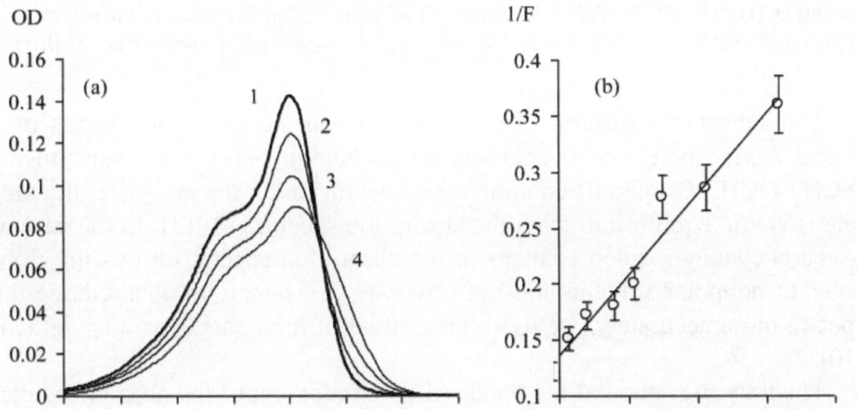

FIGURE 15.4 Absorption spectra (a) of dye K1 ($c_{K1} = 1.6\times10^{-6}$ mol l^{-1}) in the presence of DNA: c_{DNA} = (*1*) 0, (*2*) 2.0×10^{-5}, (*3*) 5.0×10^{-5}, (*4*) 1.5×10^{-4} mol l^{-1} respectively; (b) – dependences used to determine K_{eq} from the fluorescence spectra of the K1, in which the points show the experimental data, the line is the best fit to the experimental points ((Eq. (3)).

Dyes K1-K5 form noncovalent complexes with DNA [15], and the interaction with DNA drastically changes their photo physical properties. In the absorption spectra at low DNA concentrations (~ $0.5–1\times10^{-5}$ mol l^{-1}), a noticeable decrease in the apparent absorption coefficient of the dyes and broadening of the band are observed; a further increase in the biopolymer concentration leads to a long-wavelength shift and some growth in the intensity of the absorption band (Fig. 15.4a).

The interaction with DNA is accompanied by a bathochromic shift of the maximums of the bands in the dye fluorescence spectra (at $c_{DNA} \sim 2.5\times10^{-4}$ mol l^{-1} the shift ~ 7–10 nm).

Binding with DNA also leads to a fluorescence growth (Fig. 15.4b): at $c_{DNA} = 4.77\times10^{-4}$ mol l^{-1} the fluorescence intensity for K1 and K2 increases 6.2 and 5.2 times, respectively, and for K3, K4, and K5 fluorescence increases 55, 27, and 41 times, respectively (at $c_{DNA} = 2.5\times10^{-4}$ mol l^{-1}) [15].

The binding constants with DNA (K_{eq}) can be calculated from the fluorescence data (Eq. (3)). Eq. (3) can be transferred into linear function in coordinates 1/F is a function of 1/[DNA] with the slope $1/(F_{max} K_{eq})$. (Fig. 15.4b). K_{eq} equals to 5×10^4 l mol^{-1} for the unsubstituted dye K1 and $2.2–4.2\times10^4$ l mol^{-1} for the oxacarbocyanine dyes K3–K5.

The changes in the spectral and fluorescent properties of oxacarbocyanines observed upon complexation with DNA are traditionally explained in the context of the solvation effect of DNA [18], aggregation of the dye on the biomolecule (particularly noticeable at low DNA concentrations) [19], and an increase in the rigidity of the ligand structure (a growth in the fluorescence quantum yield at the expense of the spatial fixation of the dye on DNA). However, upon complexation with DNA, a shift of the isomeric equilibrium can also occur, which should contribute to the changes in the spectral and fluorescent properties of the dyes.

TABLE 15.1 The Value of the First Moment of the Normalized Distribution of Photons (N) of the Width of the Spectral Curve (Σ) is a Function of the DNA Concentration (in a Phosphate Buffer Solution) for the Absorption Spectra of Dyes K1, K3– K5

c_{DNA}, mol l^{-1}	K1		K3		K4		K5	
	ν, cm^{-1}	σ, cm^{-1}	ν, cm^{-1}	σ, cm^{-1}	ν, cm^{-1}	σ, cm^{-1}	ν, cm^{-1}	σ, cm^{-1}
0	21,539	1029	21,141	910	20,512	1003	20,882	921
$5–10^{-6}$	21,527	1036	21,061	927	20,487	1019	20,925	1058
$1–10^{-4}$	21,340	1133	20,732	921	20,241	1049	20,538	953
$2.5–10^{-4}$	21,243	1145	20,649	848	20,146	994	20,498	902

To reveal the effect of *cis-trans* equilibrium in the absorption spectra of the dyes, the spectra of K1 and K3–K5 in the presence of DNA were analyzed using the method of moments (Table 15.1). An increase in the DNA concentration is accompanied by long-wavelength shifts of the position of the band ν for all dyes; on the wavelength scale these shifts are 5–12 nm. With growing the DNA concentration, the width of the spectral curve (σ) increases and reaches a maximum at $c_{DNA} \sim (0.5\text{–}1.5)\times10^{-4}$ mol l^{-1}. At these DNA concentrations, a maximum drop in the apparent absorption coefficient of the absorption spectral bands is observed. The increase in σ is 3–5% of the initial width for K3-K5 and ~10% for K1. With the further increase in the DNA concentration, the widths of the absorption bands of K3-K5 decrease and at c_{DNA} = 2.5–10–^{4} mol l^{-1} become even somewhat less than the initial values. These facts indicate an appearance and a growth (with growing the DNA concentration) of the long-wavelength component in the absorption spectra [20].

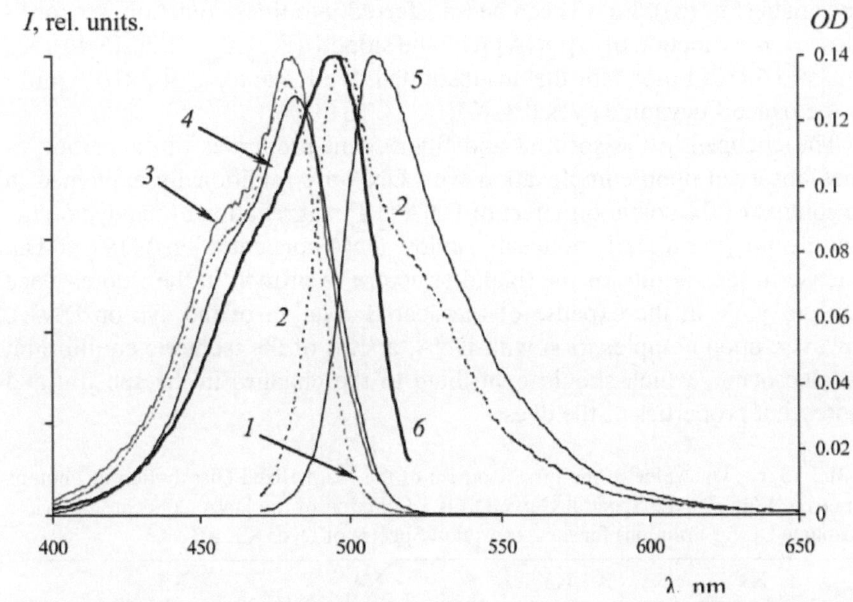

FIGURE 15.5 Spectra of dye K1 ($c_{K1} = 1.1\times10^{-6}$ mol l^{-1}) obtained in phosphate buffer solutions (*1–3*) in the absence of DNA and (*4–6*) at $c_{DNA} = 4.77\times10^{-4}$ mol l^{-1}: (*1, 4*) absorption, (*2, 5*) fluorescence (at λ_{ex} = 460 nm), and (*3, 6*) fluorescence excitation (at λ_{reg} = 530 nm) spectra. The fluorescence and fluorescence excitation spectra are normalized.

At c_{DNA} = 0 mol l^{-1}, the fluorescence excitation spectrum of K1 practically resembles the absorption spectrum (Fig. 15.5, curves 1, 2), which indicates

the presence of only one dye form (the *cis*-isomer) in the solution. At $c_{DNA} = 4.77\times10^{-4}$ mol l^{-1}, the maximum of the band of the fluorescence excitation spectrum (λ_{max} = 496 nm) is shifted toward the long-wavelength side with respect to the maximum of the absorption spectrum (λ_{max} = 483 nm; see Fig. 15.5, curves 4, 6). In the presence of DNA, the fluorescence band is also shifted bathochromically (curve 6). The fluorescence spectra were calculated using the method of the spectral moments. The increase in the DNA concentration is followed by the long-wavelength shifts of the position of the bands (on the wavelength scale $M_1 = 10^7/\langle\nu\rangle$) for all dyes. These shifts vary in the range of 3–6 nm (on the wavelength scale). The increase in the biopolymer concentration results in the reduction of the width of the spectral curve (σ) for K1-K5. The σ-value reduction is ~ 4–13% comparing to the initial width of the spectral bands for K2–K4 and 20% for K5.

For dyes K3–K5, the shifts of the maxima of the fluorescence excitation spectra with respect to the absorption spectra are larger (Fig. 15.6). At c_{DNA} = 0 mol l^{-1}, the maxima of the absorption and fluorescence excitation spectra of K5 differ by 3 nm (493 and 496 nm, respectively). In the presence of DNA, this shift increases to 4–5 nm (λ_{max} = 502 and 507 nm in the absorption and fluorescence excitation spectra, respectively).

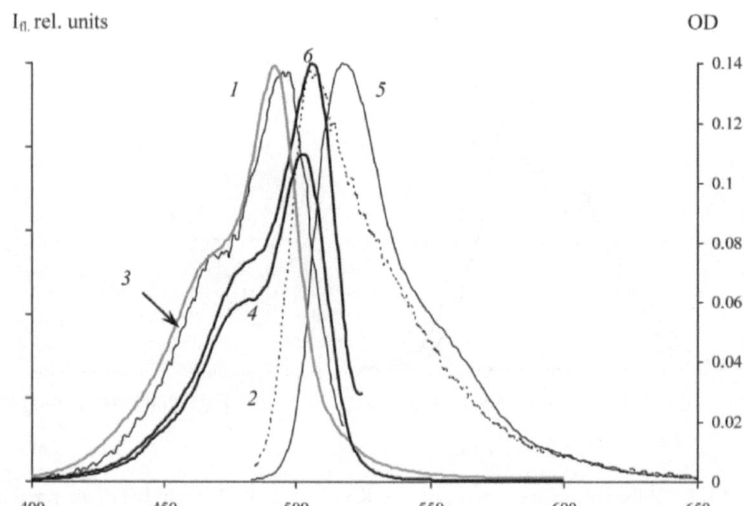

FIGURE 15.6 Spectra of dye K5 ($c_{K5} = 1.6\times10^{-6}$ mol l^{-1}) obtained in phosphate buffer solutions (*1–3*) in the absence of DNA and (*4–6*) at $c_{DNA} = 4.77\times10^{-4}$ mol l^{-1}: (*1, 4*) absorption, (*2, 5*) fluorescence (at λ_{ex} = 460 nm), and (*3, 6*) fluorescence excitation (at λ_{reg} = 530 nm) spectra. The fluorescence and fluorescence excitation spectra are normalized.

The difference between the maximums of the absorption and fluorescence excitation spectra (the long-wavelength shift of the excitation spectral bands observed in the presence of DNA) indicates the manifestation in the spectra of oxacarbocyanines of the band of the long-wavelength form of the dye: the *trans*-isomer of the dye bound to DNA, which fluoresces stronger than the *cis*-isomer (Figs. 15.5 and 15.6). The fluorescence excitation spectra of K1–K5 show that fluorescence of the dyes in a complex with DNA is determined mainly by fluorescence of the *trans*-isomers.

The difference in binding of oxacarbocyanine dyes with DNA from structurally similar thiacarbocyanine dyes (the latter bind with DNA in the *cis*-form [15]) can be explained by the possible specific interaction between terminal benzoxazole fragments oxacarbocyanine dyes and DNA bases upon the complex formation [20]. The molecules of oxacarbocyanine dyes are capable to interact with the other anionic polymers, one of these is poly(sodium 4-styrenesulfonate) (PSS). The complexation of cationic dyes seems to result from Coulomb attraction and hydrophobic interaction. The interaction of K1 and K5 with PSS has been studied for elucidation of possible effects of dye – DNA specific interaction on *cis-trans* isomeric equilibrium of oxacarbocyanines.

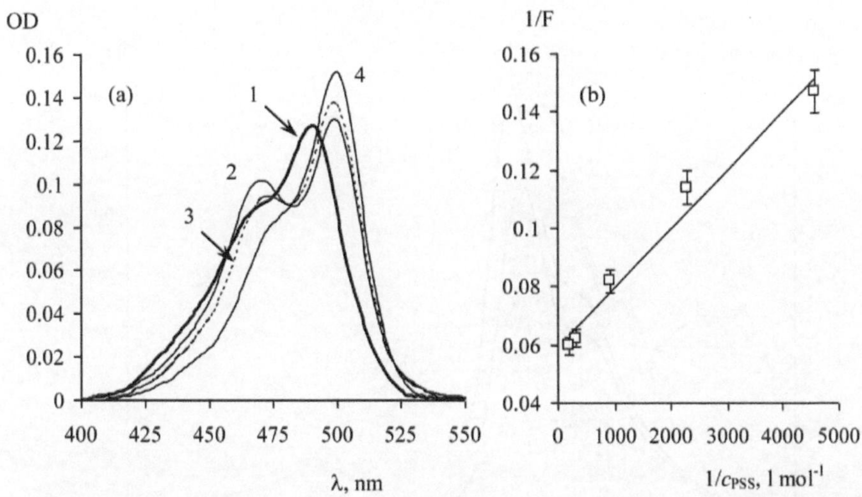

FIGURE 15.7 Absorption spectra (a) of dye K5 ($c_{K5} = 1.5\times10^{-6}$ mol l^{-1}) in the presence of PSS: c_{PSS} = (*1*) 0, (*2*) 2.19×10^{-4}, (*3*) 4.4×10^{-4}, (*4*) 5.46×10^{-3} mol l^{-1}; (b) – dependences used to determine K_{eq} from the fluorescence spectra of K5 in the presence of PSS: the points show the experimental data, the line is the linear fit to the experimental points.

Figure 15.7a shows the absorption spectra of K5 in presence of different concentration of PSS. In the absorption spectra in the presence of PSS the absorption band shifts to the long-wavelength side (λ_{max} = 490 nm for K1 and λ_{max} = 500 nm for K5), and its apparent molar absorption coefficient decreases. The absorption spectra shift also at a low concentration of the polymer (0.04 mg mL^{-1}). The further increase in the PSS concentration to 10 mg mL^{-1} does not lead to the greater shift. At low polymer concentrations in the solution, the dye-PSS interaction leads to an appearance of the band of dye aggregates in the spectra (λ_{max} = 472 nm for K5). At higher concentrations of PSS (0.3–0.5 mg mL^{-1}) these aggregates disappear and a complex of the dye monomer is formed. At high concentrations of the polymer the absorption spectra are broadened insignificantly and the equilibrium is shifted to the dye monomer complex formation. To characterize the absorption spectra more accurately, we used the method of moments. At PSS concentration of 1 mg mL^{-1} the width of the spectral curve (σ) for K5 increases by 5–7%.

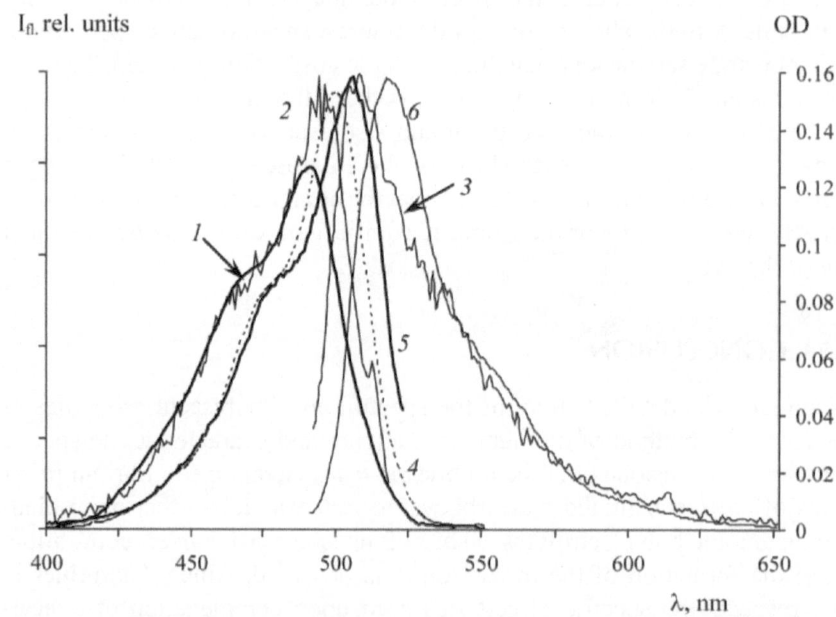

FIGURE 15.8 Spectra of dye K5 (c_{K5} = 1.5×10^{-6} mol l^{-1}) obtained in phosphate buffer solutions (*1–3*) in the absence of PSS and (*4–6*) at c_{PSS} = 5.46×10^{-3} mol l^{-1}: (*1, 4*) absorption, (*2, 5*) fluorescence (at λ_{ex} = 460 nm), and (*3, 6*) fluorescence excitation (at λ_{reg} = 530 nm) spectra. The fluorescence and fluorescence excitation spectra are normalized.

The complex formation results in an increase in the fluorescence quantum yield. The 16 times fluorescence growth has been observed at a PSS concentration of 1 mg mL^{-1} for K5. For K1 the fluorescence increases four times at a PSS concentration of 2 mg mL^{-1}. The equilibrium constant of the complex formation (K_{eq}) of K5 with PSS determined from the fluorescence data (Fig. 15.7b) was rather low $K_{eq\_PSS} = 2.0–10^3$ l mol^{-1}.

The interaction with PSS leads to long-wavelength shifts of the fluorescence bands of dyes K1, K5 (6 nm for K1 at 2 mg mL^{-1} and 13 nm for K5 at 1 mg mL^{-1}). The growth of the PSS concentration results in a decrease in the width of the fluorescence spectral curve (σ) for K1 and K5. At a PSS concentration of 2 mg mL^{-1} the σ value is reduced by 10% for K1 and 25% for K5 comparing to the initial values.

The maxima of the fluorescence excitation spectra ($\lambda_{max.\ fl.}$) of K5 do not match the absorption spectral maxima in the presence of PSS. At a PSS concentration of 1 mg mL^{-1} λ_{max} of the fluorescence excitation spectra is 506 nm, while λ_{max} of the absorption spectrum is 501 nm (i.e., the difference is 6 nm). For dye K1 the difference of the spectral maxima is less (D_2 is 4 nm at [PSS] = 1 mg mL^{-1}). In the absence of PSS the fluorescence excitation spectrum of K1 demonstrate satisfactory matching with the absorption spectrum. Thus the only one isomeric form of the dye exists in the solution.

The differences in the absorption and fluorescence excitation spectra of the dyes (the long-wavelength shifts of the fluorescence excitation spectra observed in the presence of PSS) show that the complex formation of the dyes with PSS leads to shifts of the isomeric equilibrium toward the *trans*-isomer form of the dyes.

15.5 CONCLUSION

In summary, the detailed study of the spectral and fluorescent properties of dyes using the method of moments to compare and characterize the spectra permitted a conclusion about the mobile *cis-trans* isomeric equilibrium in solutions of *meso*-substituted oxacarbocyanines. It was found that upon binding with anionic polyelectrolytes, a shift of the isomeric *cis-trans* equilibrium toward the formation of the *trans*-isomer is observed, which determines in many respects the spectral effects observed upon complexation of oxacarbocyanine dyes with anionic polymers (DNA, PSS). The steep fluorescence growth (due to binding of the *trans*-isomer) in a complex with DNA opens up the prospects to use oxacarbocyanine dyes as spectral and fluorescent probes for DNA.

ACKNOWLEDGMENTS

We thank Prof. B.I. Shapiro (NIIKHIMFOTOPROEKT Research Center) for providing the polymethine dyes. This work was supported by the Russian Foundation for Basic Research, projects no. 13-03-00863, 14-03-31196.

KEYWORDS

- ***Cis-trans*** **isomer equilibrium**
- **DNA, poly(sodium 4-styrenesulfonate)**
- **Oxacarbocyanine dyes**

REFERENCES

1. Goncharuk, E. I., Borovoy, I. A., Pavlovich, E. V., Malyukin, Yu. V., & Grischenko, V. I. (2009). Newly Synthesized Carbocyanine Fluorescent Probes Their Characteristics and Behavior In Proliferating Cultures Biopolymers and Cell, *25(N6),* 484–490.
2. Escobedo, J., Rusin, O., Lim, S., & Strongin, R. (2010). NIR dyes for bioimaging applications, *Curr Opin Chem Biol 14*, 64–70.
3. Harvey, B. J., & Levitus, M. (2009). Nucleobase-Specific Enhancement of Cy3 Fluorescence *Journal of Fluorescence 19,* 443–448.
4. Harvey, B. J., Perez, C., & Levitus, M. (2009). DNA sequence dependent enhancement of Cy3 fluorescence *Photochemical & Photobiological Sciences 8* 1105–1110.
5. Biancardi, A., Biver, T., Marini, A., Mennucci, B., & Secco, F. (2011). Thiazole Orange (TO) As A Light-Switch Probe: A Combined Quantum-Mechanical And Spectroscopic Study, *Phys. Chem. Chem. Phys., 13*, 12595–12602.
6. Bethge, L., Singh, I., & Seitz, O. (2010). Designed thiazole orange nucleotides for the synthesis of single labelled oligonucleotides that fluoresce upon matched hybridization, *Org. Biomol. Chem., 8*, 2439–2448.
7. Okamoto, A. (2011). ECHO probes: a concept of fluorescence control for practical nucleic acid sensing, *Chem. Soc. Rev. 40*, 5815–5828.
8. Tatikolov, A. S. (2012). Polymethine dyes as spectral-fluorescent probes for biomacromolecules, *J. Photochem. Photobiol C: Photochem. Rev 13,* 55–90.
9. Levitus, M., & Ranjit, S. (2011). Cyanine Dyes in Biophysical Research: The Photophysics of Polymethine Fluorescent Dyes in Biomolecular Environment, *Q. Rev. Biophys. 44,* 123–151.
10. Ishchenko, A. A. (1994). Stroenie i spektral'no-lyuminestsentnye svoistva polimetinovykh krasitelei (Structure and Spectral–Luminescent Properties of Polymethine Dyes), Kiev: Naukova Dumka, 232. ISBN 5–12-004649–5
11. Kreig, M., & Redmond, R. W. (1993). Photophysical properties of 3, 3'-dialkylthiacarbocyanine dyes in homogeneous solution *Photochem Photobiol 57*, 472.
12. Roth, N. J. L., & Craig, A. C. (1974). Predicted Observable Fluorescent Lifetimes of Several Cyanines. *J. Phys. Chem., 78(12*), 1154–1155.

13. West, W., Pearce, S., & Grum, F. (1967). Stereoisomerism in cyanine dyes meso substituted thiacarbocyanines. *J. Phys. Chem. 71(5)*, 1316–1326.
14. Khimenko, V., Chibisov, A. K., & Görner, H. (1997). Effects of Alkyl Substituents in the Polymethine Chain on the Photoprocesses in Thiacarbocyanine Dyes. *J. Phys. Chem. A. 101(39)*, 7304–7310.
15. Pronkin, P. G., Tatikolov, A. S., Anikovskii, M. Yu., & Kuz'min, V. A. (2005). The Study of cis trans Equilibrium and Complexation with DNA of meso Substituted Carbocyanine Dyes *High Energy Chem., 39(4)*, 237–243.
16. Akimkin, T. M., Tatikolov, A. S., & Yarmolyuk, S. M. (2011). Spectral and fluorescent study of the interaction of cyanine dyes Cyan 2 and Cyan 45 with DNA, *High Energy Chem.(2011) 45(3)*, 222–259.
17. Baguley, B. C., & Falkenhang, E. M. (1978). The Interaction of Ethidium with Synthetic Double Stranded Polynucleotides At Low Ionic Strength. *Nucl Acids Res 5*, 161–168.
18. Lober, G. (1981). The Fluorescence of Dye Nucleic Acid Complexes, *J. Lumin. 22(3)*, 221–265.
19. Lown, J. W. (1998). DNA sequence recognition altered bis-benzimidazole minor-groove binders, In: *Advances in DNA Sequence-Specific Agents*. Editors: Jones, G. B., Palumbo, M., Elsevier, *3*, 67–95.
20. Pronkin, P. G., & Tatikolov, A. S. (2012). Study of Cis Trans Equilibrium of Oxacarbocyanine Dyes in Solution and in a Complex with DNA, *High Energy Chemistry, 46(4)*, 252–257.

CHAPTER 16

OZONE DECOMPOSITION

T. BATAKLIEV, V. GEORGIEV, M. ANACHKOV, S. RAKOVSKY, and G. E. ZAIKOV

CONTENTS

ABSTRACT

Catalytic ozone decomposition is of great significance because ozone is toxic substance commonly found or generated in human environments (aircraft cabins, offices with photocopiers, laser printers, sterilizers). Considerable work has been done on ozone decomposition reported in the literature. This review provides a comprehensive summary of this literature, concentrating on analysis of the physicochemical properties, synthesis and catalytic decomposition of the ozone. This is supplemented by review on kinetics and catalyst characterization, which ties together the previously reported results. It has been found that noble metals and oxides of transition metals are the most active substances for ozone decomposition. The high price of precious metals stimulated the use of metal oxide catalyst and particularly the catalysts based on manganese oxide. It has been determined the kinetics of ozone decomposition to be first order. It was discussed a mechanism of the reaction of catalytic ozone decomposition, based on detailed spectroscopic investigations of the catalytic surface, showing the existence of peroxide and superoxide surface intermediates.

16.1 INTRODUCTION

In recent years, the scientific research in all leading countries of the world is aimed primarily at solving the deep environmental problems on the planet, including air pollution and global warming. One of the factors affecting negatively these processes is the presence of ozone in ground atmospheric layers. This is a result of the wide use of ozone in lot of important industrial processes such as cleaning of potable water and soil, disinfection of plant and animal products, textile bleaching, complete oxidation of waste gases from the production of various organic chemicals, sterilization of medical supplies, etc. [1].

The history of ozone chemistry as a research field began immediately after its discovery from Schönbein in 1840 [2]. The atmospheric ozone is focused mainly in the so-called "ozone layer" at a height of 15 to 30 km above the earth surface wherein the ozone concentration ranges from 1 to 10 ppm [3]. The ozone synthesis at that altitude runs photochemical through the influence of solar radiation on molecular oxygen. The atmospheric ozone is invaluable to all living organisms because it absorbs the harmful ultraviolet radiation from the sun. The study of the kinetics and mechanism of ozone reactions in modern science is closely related to solving the ozone holes problem that

reflect the trend of recent decades to the depletion of the atmospheric ozone layer.

Ozone has oxidation, antibacteriological and antiviral properties that make it widely used in the treatment of natural, industrial and polluted waters, swimming pools, contaminated gases, for medical use, etc. Catalytic decomposition of residual ozone is imperative because from environmental point of view the release of ozone in the lower atmosphere has negative consequences.

The presence of ozone in the surrounding human environment (airplane cabins, copiers, laser printers, sterilizers, etc.) also raises the issue of its catalytic decomposition as ozone is highly toxic above concentrations of 0.1 mg/m^3 [4, 5] and can damage human health. The uses of catalysts based on transition metal oxides are emerging as very effective from an environmental perspective and at the same time practical and inexpensive method applied in the decomposition of residual ozone.

16.2 SOME PHYSICO-CHEMICAL PROPERTIES OF OZONE

Ozone is an allotropic modification of oxygen that can exist in all three physical conditions. In normal conditions, ozone is colorless gas with a pungent odor while at very low concentrations (up to 0.04 ppm) it can be feeling like a pleasant freshness. A characteristic property of ozone odor is the fact that it is easily addictive and at the same time hazardous for men who work with ozone in view of its high toxicity at concentrations above the limit (> 0.1 mg/m^3).

When the ozone concentration is more than 15–20% it has blue color. At atmospheric pressure and temperature of 161.3 K, the ozone becomes fluid in deep blue color. It cures at 80.6 K by acquiring dark purple color [6]. Ozone is explosive in all three physical conditions. The work with ozone concentrations of 0 to 15% can be considered safe [4]. The danger of ozone explosion is a function of its thermodynamic instability ($\Delta G°298$=-163 kJ/mol) and ozone decomposition to diatomic oxygen is a thermodynamically favorable process with heat of reaction $\Delta H°298$=-138 kJ/mol [7]. The ozone molecule consists of three oxygen atoms located at the vertices of obtuse-angled triangle with a central angle of 116°8'±5' and length of O-O-bond: $r_{o\text{-}o}$ = 0.1278±0.0003 nm (Fig. 16.1).

In gas phase the ozone molecule is a singlet biradical while in liquid phase it reacts generally as a dipolar ion. The homogeneous reaction of pure gaseous ozone decomposition is characterized by certain velocity at sufficiently high temperatures. The kinetics study of the ozone thermal decomposition is also complicated by the fact that above a certain critical temperature the steady

kinetic decomposition is transformed into explosion, subsequently passing to detonation [1]. According to Thorpe [8] the detonation of ozone is observed above 105 °C. The gaseous ozone is characterized by different time of half-life depending on the temperature (Table 16.1):

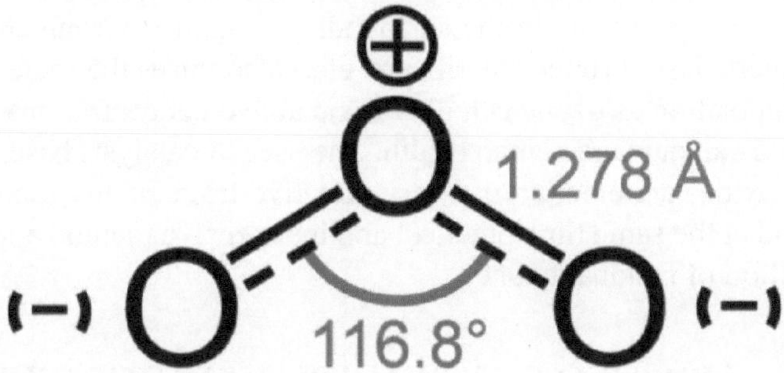

FIGURE 16.1 Structure of ozone molecule.

TABLE 16.1 Half-life of Ozone

Temperature (°C)	Half-life *
–50	3-months
–35	18-days
–25	8-days
20	3-days
120	1.5-hours
250	1.5-seconds

The ozone structure is resonance stabilized which is one of the reasons for its resistance against decomposition at low temperatures (Fig. 16.2).

In most reactions of ozone with inorganic compounds it reacts with participation of one atom oxygen and the other two are separate as O_2. Typically, the elements are oxidized to their highest oxidation states. For example,

manganese is oxidized to MnO^{4-}, halogen oxides to metals (ClO_2, Br_2O_5), ammonia-to NH_4NO_3, nitrogen oxides pass into N_2O_5 [9–11].

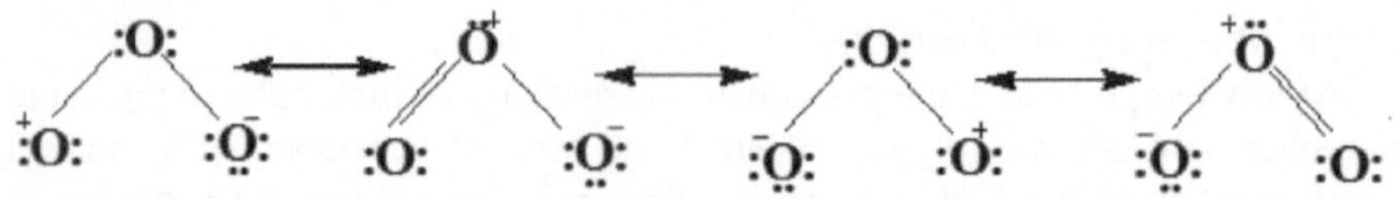

FIGURE 16.2 Resonance structures of ozone.

Ozone has spectral characteristics from the IR-region to the vacuum UV-region, which are present in a significant number of works, such as the majority of them are carried out with gaseous ozone [12–18]. It should be noted that most researchers, who use the spectrophotometric method for analysis of ozone, work in the main area of the spectrum 200÷310 nm, where is the wide band with maximum at ~255 nm for gaseous ozone [13, 15–17]. That maximum is characterized with high value of the coefficient of extinction (Table 16.2):

TABLE 16.2 Coefficient of Extinction (e, $L.Mol^{-1}.Cm^{-1}$) of Gaseous Ozone in UV-Region [19]

l, nm	[35]	[33]	[36]	[47]	l, nm	[35]	[33]	[36]
253.6*	2981	3024	2952	3316	296.7**	150.5	153.4	156.8
270	-	-	-	2302	302.2	74.4	-	74.4
289.4	383	337.5	387.5	-	334.2	1.50	-	1.46

* The values of e are 1830 $l.mol^{-1}.cm^{-1}$ at 254.0 nm [4] and 253.6 nm at 3020 $l.mol^{-1}.cm^{-1}$ [14].
** e = 160 $l.mol^{-1}.cm^{-1}$ at 295 nm [18].

16.3 OZONE SYNTHESIS AND ANALYSIS

Ozone for industrial aims is synthesized from pure oxygen by thermal, photochemical, chemical, electrochemical methods, in all forms of electrical discharge and under the action of a particles stream [1]. The synthesis of ozone is carried out by the following reactions shown in the scheme:

$$O_2 + (e^-, hn, T) \rightarrow 2O\ (O_2^*) \quad (1)$$

$$O_2 + O + M \leftrightarrow O_3 + M \tag{2}$$

$$O_2^* + O_2 \rightarrow O_3 + O \tag{3}$$

where: M is every third particle.

At low temperatures, the gas consists essentially of molecular oxygen and at higher – by atomic oxygen. There is no area of temperatures at normal pressure wherein the partial pressure of ozone is significant. The maximum steady-state pressure which is observed at temperature of 350 K is only 9.10–7 bar. The values of the equilibrium constant of reaction (2) at different temperatures are presented in Table 16.3 [20].

TABLE 16.3 Equilibrium Constant (K_e) of Reaction (2) Depending on the Temperature

T, K	1500	2000	3000	4000	5000	6000
K_e, M	1662×10^{-11}	4413×10^{-7}	1264×10^{-2}	2.104	48.37	382.9

At high temperatures, when the concentration of atomic oxygen is high, the equilibrium of reaction (2) is moved to the left and the ozone concentration is low. At low temperatures, the equilibrium is shifted to the right, but through the low concentration of atomic oxygen the ozone content is negligible. For synthesis of significant concentrations of ozone it is necessary to combine two following conditions: (i) low temperature, and (ii) the formation of super equilibrium concentrations of atomic oxygen. It is possible to synthesize super equilibrium atomic concentrations of oxygen at low temperatures by using nonthermal processes, such as dissociation of oxygen with particles flow, electrons, hn, electrochemical and chemical influences. Ozone will be formed at low temperatures always, when there is a process of oxygen dissociation. Higher concentrations of ozone may be obtained by thermal methods providing storage ("quenching") of super equilibrium concentrations of atomic oxygen at low temperatures. Photochemical synthesis of ozone occurs upon irradiation of gaseous or liquid oxygen by UV radiation with wavelength $\lambda < 210$ nm [21]. It is assumed [22], that the formation of ozone in presence of radiation with wavelength in the range of $175 < \lambda < 210$ may be associated with the formation of excited oxygen molecules:

$$O_2 + h\nu \rightarrow (O_2^*) \tag{4}$$

$$O_2^* + O_2 \rightarrow O_3 + O \quad (5)$$

$$O_2 + O + M \leftrightarrow O_3 + M \quad (6)$$

The photochemical generation of ozone has an important role in atmospheric processes, but it is hard to apply for industrial application because of the high energy costs (32 kWh/kg ozone) for the preparation of high-energy shortwave radiation. Nowadays the industrial synthesis of ozone happens mainly by any of the electric discharge methods by passing oxygen-containing gas through high-voltage (8–10 kV) electrodes. The aim is creation of conditions in which oxygen is dissociated into atoms. This is possible in all forms of electrical discharges: smoldering, silent, crown, arc, barrier, etc. The main reason for the dissociation is the hit of molecular oxygen with the accelerated in electric field electrons [1, 6], when part of the kinetic energy is transformed in energy of dissociation of the bond O–O and excitation of oxygen molecules. In general, molecule oxygen dissociates in two oxygen atoms, then on final stage two molecules of oxygen form one oxygen atom and one ozone molecule:

$$O_2 + e \rightarrow 2O\ (O_2^*) \quad (7)$$

$$O_2 + O + M \rightarrow O_3 + M \quad (8)$$

$$O_2^* + O_2 \leftrightarrow O_3 + O \quad (9)$$

Technically, the synthesis of ozone is carried out in discharge tubes – ozonators. One of the most commonly used type ozonators are Siemens (Fig. 16.3).

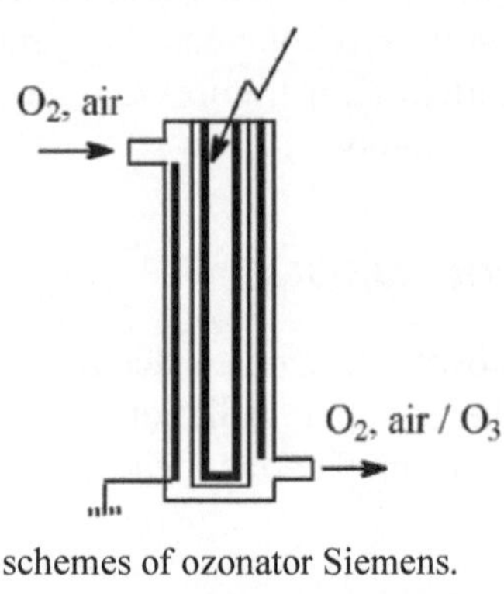

FIGURE 16.3 Principal schemes of ozonator Siemens.

This ozonator represents two brazed one to another coaxial pipes with supported inside of the inner side and outside of the outer side electrically conductive coating of aluminum, silver, copper and the like. It is submitted of high voltage up to 20 kV and dry and pure oxygen is left to pass through the ozonator. The ozone is synthesized on the other end of the ozonator. Ozone concentration depends on the parameters of the current – voltage, frequency and power properties of the ozonator thickness, length and type of glass tube, the distance between the electrodes and temperature. The ozone concentration depends on the current parameters – voltage, frequency and power, on the properties of ozonator – thickness, length and type of glass tube, on the distance between the electrodes and also on the temperature.

The analysis of the ozone concentration is developed by various physical and chemical methods discussed in details in monograph [23]. It should be noted the preference for the spectrophotometric method compared with the iodometric method which is determined by the lack of need of continuous pH monitoring during the ozone analysis and also by the possibility of direct observation of the inlet and outlet ozone concentrations in the system, allowing precise fixing of the experimental time and its parameters. For determination of the ozone amount can be used not absolute values of concentration but the ratios of the proportional values of optical densities which excludes the error influence at expense of the inaccuracy in calibration of the instrument.

16.4 OZONE DECOMPOSITION

The reaction of ozone on the surface of solids is of interest from different points of view. Along with the known gas cycles depleting the atmospheric ozone, a definitive role plays also its decomposition on aerosols that quantity already is growing at expense of the anthropogenic factor. The growth of ozone used in chemical industries set up the task for decomposition of the residual ozone on heterogeneous, environmentally friendly catalysts. In humid environments and under certain temperatures and gas flow rates this subject has not been yet entirely understood.

16.4.1 STRATOSPHERIC OZONE

The importance of the stratospheric ozone is determined by its optical properties- its ability to absorb the UV solar radiation with a wavelength of 220 to 300 nm. The absorption spectrum of ozone in the ultraviolet and visible region is given in Fig. 16.4.

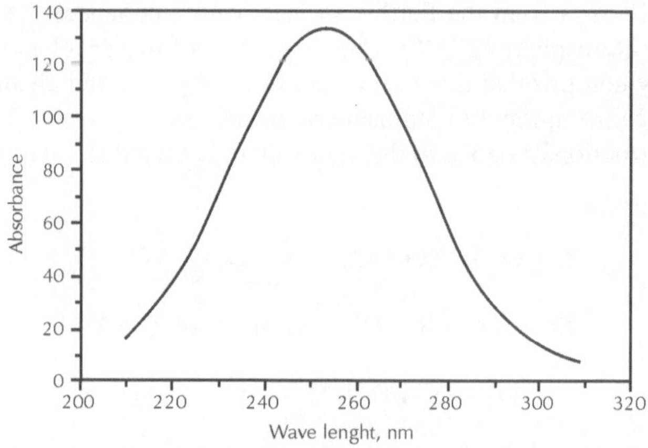

FIGURE 16.4 UV–absorption spectrum of ozone with maximum at 254 nm.

The main absorption band in the name of Hartley is in the range 200–300 nm with a maximum at a wavelength of 254 nm. Stratospheric ozone layer has a thickness of 3 mm and its ability to absorb UV rays protects the earth's surface of biologically active solar radiation which destroys the most important biological components, proteins and nucleic acids.

The fundamentals of the photochemical theory of stratospheric ozone have been made by the English chemist Chapman [24, 25], according to whom the photochemical decomposition of ozone occurs to the following reactions:

$$O_3 + O \rightarrow O_2 + O_2 \quad (10)$$

$$O_3 + h\nu \rightarrow O_2 + O\ (^3P) \quad (11)$$

$$O_3 + h\nu \rightarrow O_2 + O\ (^1D) \quad (12)$$

The proposed mechanism leads to formation of oxygen atoms in ground and excited state. This cycle of decomposition is called the cycle of "residual oxygen." The photo dissociation of ozone (11) and (12) takes place under the action of solar radiation with wavelength less than 1134 nm. Decomposition of ozone following these reactions (11) and (12) can be observed at any altitude near to ground level. The photochemical synthesis of ozone in the stratosphere requires its degradation in reaction (10). It was found that 20% of the stratospheric ozone decomposes following this reaction [26]. The effect of ozone layer on climate changes is related to the absorption of radiation which occurs not only in the UV-region, but also in the IR-region of the spectrum.

Absorbing IR rays from the Earth's surface, ozone enhances the greenhouse effect in the atmosphere [27, 28]. The connection of this problem with climate is extremely complicated due to the variety of physical and chemical factors that influence the amount of atmospheric ozone [29].

The destruction of ozone in the atmosphere is linked to catalytic cycles of the type:

$$X + O_3 \rightarrow XO + O_2 \quad X + O_3 \rightarrow XO + O_2$$

$$XO + O \rightarrow X + O_2 \quad XO + O_3 \rightarrow X + 2O_2$$

$$O_3 + O \rightarrow O_2 + O_2 \quad 2O_3 \rightarrow 3O_2$$

where: X is OH, NO or Cl, formed by dissociation of freons in the atmosphere [30].

The stratospheric cycles of ozone decomposition are discussed in a lot of works but fundamental contributions have most [26] and [31]. The effective action of the molecules-catalysts in the cycles of ozone decomposition is determined by their concentrations in the atmosphere, which depends on the rates of regeneration and exit from the respective cycle [32]. The ratio between the rate of decomposition of residual oxygen and the rate of catalyst outgoing from the cycle determines the length of the chain reaction and corresponds to the number of O_3 molecules destructed by one catalytic molecule. The number of stages of ozone decomposition in a single catalytic center can reach 106. The reduction of the ozone amount over Antarctica, that is, the thinning of ozone layer is mainly due to the action of the chlorine cycle [33–35]. During this catalytic cycle, the presence of one chlorine atom in the stratosphere can cause the decomposition of 100–000 ozone molecules.

16.4.2 CATALYSTS FOR DECOMPOSITION OF OZONE

The use of ozone for industrial aims is related with the application of effective catalysts for its decomposition, since as it was already mentioned; the release of ozone in the atmosphere near the ground level is dangerous and contaminates the air [1, 36, 37]. The ozone has a great number of advantages as oxidizing agent and, in that capacity, it has been used in different scientific investigations doing with neutralization of the organic contaminants [38, 39], as in presence of catalyst the ozonation efficiency increases [40, 41]. Due to its antibacterial and antiviral properties, ozone is one of most used agents in water treatment [42]. This facts result in considerable interest from the

researchers in study of homogeneous and heterogeneous reactions of ozone decomposition, as well as in participation of ozone in multiple oxidation processes.

16.4.2.1 OZONE DECOMPOSITION ON THE SURFACE OF METALS AND METAL OXIDES

The first studies on the catalytic decomposition of gaseous ozone [43] showed that the degradation of the ozone is accelerated in the presence of Pt, Pd, Ru, Cu, W, and the like. Catalytic activity of metals in the decomposition of ozone is studied by Kashtanov et al. [44], which highlight that the silver (Ag) show higher catalytic activity, compared to copper (Cu), palladium (Pd), and tin (Sn). From earlier works on decomposition of ozone, it should be noted the study of Schwab and Hartmann [45]. They have investigated catalysts based on metals from I-IV group and their oxides in different oxidation levels, and they have found out that the catalytic activity of these oxides increases with increase in the oxidation state of the metal. Below is presented the relationship between the catalytic activity degree of a number of elements and their oxides, in the reaction of the catalytic decomposition of ozone:

$$Cu < Cu_2O < CuO;\ Ag < Ag_2O < AgO;\ Ni < Ni_2O_3;\ Fe < Fe_2O_3;\ Au < Au_2O_3;\ Pt < Pt$$

These results can be explained by the higher activity of the ions of the respective elements relative to the activity of the elements themselves, as well as the importance of ionic charge in the catalytic reaction. Other early studies are presented in the article by Emelyanov et al. [46–48] in which has being discussed the catalytic decomposition of gaseous ozone at temperatures from −80 °C to +80 °C and an ozone concentration of 8.8 vol.%. For catalysts were used elements of the platinum group: Pt, colloidal Pt, Pd, Ir and colloidal Ru, NiO, and Ni_2O_3. Other early studies have been presented in the article by Emelyanov et al. [46–48] in which was being discussed catalytic decomposition of gaseous ozone at temperatures from −80°C to +80°C and an ozone concentration of 8.8 vol.%. For catalysts have been used elements of the platinum group: Pt, colloidal Pt, Pd, Ir and colloidal Ru, NiO, and Ni_2O_3. Experiments have been carried out in a tube reactor at a gas velocity of 5 L/h. Studies indicate identical activity of the nickel oxide and colloidal platinum at temperatures of from 20°C to 80 °C. At −80 °C, the activity of the nickel oxide falls to zero, while

the platinum remains active. This is related to the low activation energy of decomposition of O_3 on the metal surface (~ 1 kcal/mol). It was also found out that the process of decomposition of the ozone in the presence of colloidal Rh and Ir at +3 °C and +20 °C slowed down after four hours, and the catalysts lost activity. Sudak and Volfson's silver-manganese catalyst [49] for the decomposition of ozone was used at low temperatures in different gas mixtures. Commercial reactors operating with this catalyst have a long life of performance and maintain a constant catalytic activity. The preparation of other high performance metal catalyst for the decomposition of ozone, including the development of special technology for synthesis, determination the chemical composition, the experimental conditions and the catalytic activity have been noted in a number of studies in Japanese authors published in the patent literature [50–60]. The main metals used are Pt, Pd, Rh and Ce, as well as metals and metal oxides of Mn, Co, Fe, Ni, Zn, Ag and Cu. The high price of precious metals stimulates the use of metal oxide catalyst supporters with a highly specific surface such as γ-Al_2O_3, SiO_2, TiO_2, ZrO_2 and charcoal. Hata et al. [50] have synthesized a catalyst containing 2% Pt as active component on a supporter composed of a mixture of 5% SiO_2 and 95% γ-Al_2O_3 having a specific surface area 120 m^2/ g. They have impregnated Chloroplatinic acid onto the carrier at 80 °C then the sample has been dried successively at 120 °C and 400 °C and atmospheric pressure. At an operating temperature of 20 °C, the gas velocity of 20,000 h^{-1} and 10 ppm initial concentration of ozone in the fed gas, this catalyst shows 95% catalytic activity. Another catalyst comprising TiO_2, SiO_2 and Pt [52] degrades 94, 97 and 99% of the ozone in the air stream at temperatures of 20, 50 and 100 °C, respectively. Therui et al. [54] have deposited metals Mn, Co, Fe, Ni, Zn, Ag or their oxides in the weight ratio in regard to carrier in the range of 0 to 60%, as well as Pt, Pd and Rh of the 0 to 10 wt%, and also mixed oxides such as TiO_2-SiO_2; TiO_2-ZrO_2 and TiO_2-SiO_2-ZrO_2, supported on colloidal polyurethane with 400 m^2/g specific surface. The catalysts have been placed in tube reactors and their catalytic activity, measured in the decomposition of ozone at a concentration of 0.2 ppm has been 99%. In published patents may be noted that for catalysts with identical chemical composition have been used various precursors and methods for synthesis. For example, TiO_2 and MnO_2 [56] are obtained from aqueous solutions of H_2SO_4 + $TiOSO_4$ and Mn (NO_3).6 H_2O, while in another case [57], TiO_2 was purchased from the manufacturer, and the MnO_2 is prepared by precipitation of aqueous

solutions of $MnSO_4$ and NH_3 in an atmosphere of oxygen followed by calcination. The application of the proposed system conditions makes it possible to assess the effectiveness of the catalysts and may be used for development of new catalytic systems. It can be seen that many of the catalysts operate at ambient temperatures (293–323 K), high space velocities (> 20,000 h^{-1}), and exhibit high catalytic activity (conversion 95%). In recent years, more and more researchers are focused on the development, study and application of catalysts for decomposition of ozone based on supported or native metal oxides, in regard with the already mentioned fact – the high price of metals of the platinum group. In recent years, more and more researchers are focused on the development, study and application of catalysts for decomposition of ozone based on supported or unsupported metal oxides, because of the already mentioned fact – the high price of metals of the platinum group. In this regard, the most widely used are the oxides of Mn, Co, Cu, Fe, Ni, Si, Ti, Zr, Ag and Al [61–69]. It has been found that the high catalytic activity in the decomposition of ozone exhibit the oxides of transition metals [70], and particularly the catalysts based on manganese oxide [71]. The manganese oxide is used as a catalyst for various chemical reactions including the decomposition of nitrous oxide [72–74] and isopropanol [74, 75], oxidation of methanol [76], ethanol [77,] benzene [78, 79], CO [74, 75, 80] and propane [81] as well as for reduction of nitric oxide [82] and nitrobenzene [83]. In terms of technology for the control of air pollution, manganese oxides are used both in the decomposition of residual ozone, and the degradation of volatile organic compounds [84, 85]. Oxides of Mn, Co, Ni, Cr, Ag, Cu, Ce, Fe, V and Mo supported on γ-Al2O3 on cordierite foam (60 pores/cm2 and geometry 5.1 cm × 5.1 cm × 1.3 cm) have been prepared and tested in the reaction of decomposition of ozone [71]. Experiments have been carried out at a temperature of 313 K, a linear velocity of 0.7 m/s, inlet ozone concentration of 2 ppm, a relative humidity of 40% and a total gas flow rate of 1800 cm^3/s. It has been found out that the catalyst activity decreases over time and the measurements are made only when the speed of decomposition is stabilized. After comparison of the conversion degree of ozone, conclusions on their catalytic activity have been drawn, according to which they have been arranged as follows: MnO_2 (42%) > Co_3O_4 (39%) > NiO (35%) > Fe_2O_3 (24%) > Ag_2O (21%) > Cr_2O_3 (18%) > CeO_2 (11%) > MgO (8%) > V_2O_5 (8%) > CuO (5%) > MoO_3 (4%). The high dispersibility of the material is confirmed by X-ray diffraction analysis. TPR profile of supported

and unsupported MnO_2 has been done. It has been found that the reduction temperature of the supported MnO_2, Co_3O_4, NiO, MoO_3, V_2O_5 and Fe_2O_3 were in the range of 611–735 K, while that one of CeO_2 was above 1–000 K. It has turned out that from all examined metal oxide samples, MnO_2 had lowest reduction temperature hence it exhibited higher reduction ability in comparison with the others. On the basis of experiments it has been proposed mechanism on decomposition of ozone on the catalytic surface which includes formation of intermediate ionic particles possessing either superoxide or peroxide futures:

$$O_3 + 2* \leftrightarrow *O_2 + O* \tag{13}$$

$$*O_2 + * \rightarrow 2O* \tag{14}$$

$$2O* \rightarrow O_2 + 2* \tag{15}$$

where * denotes active center.

A distinguishing peculiarity of the proposed scheme is the assumption that the key intermediate particle $*O_2$ is not desorbed immediately. This could happen if she has a partial ionic character (O^{2-}, O_2^{2-}). The catalytic activity of iron oxide Fe_2O_3 in decomposition of ozone has been studied also in earlier work of Rubashov et al. [86, 87]. It was found that Fe_2O_3 showed a catalytic activity only when has been used in form of particles with small diameter while catalyst that was consisted from aggregated particles was not efficient. Furthermore, the stability of the catalyst is not high and its poisoning is due to the formation of oxygen directly associated with the surface. Reactivation of the catalyst is accomplished by its subjecting to the vacuum thermal treatment to remove oxygen. The same catalyst was also studied in terms of the fluidized bed, wherein the rate constant for the catalytic reaction is higher than in the normal gas flow. The same catalyst was also studied in terms of the fluidized bed, wherein the rate constant for the catalytic reaction is higher than in the normal gas flow. By determining the constant of the decomposition of ozone on the surface of Fe_2O_3 at various temperatures can be calculated activation energy of the process: $E = (12.5 \pm 1)$ kcal/mole. This value differs from that calculated by Schwab [45] Activation energy of decomposition of ozone on compact sample of iron oxide, which is of the order of 2–3 kcal/mol. The difference can be explained with the regime in which the process is carried out: kinetic, in the first case, and diffusion in the second. Ellis et al. [88] have investigated the activity of 35 different materials with respect to decomposition of ozone as the best catalysts have turned out nickel oxide and charcoal. Measurements have been carried out at room temperature, the gas velocities between 250–2000 L/h, atmospheric pressure and ozone concentra-

tions (1.1–25) × 10^{12} molecules/cm^3. The measured rate of decomposition of ozone corresponds to a first order reaction. The authors ignoring the decomposition of ozone in the bulk have proposed a homogeneous decomposition on the surface:

$$O_3 + M \rightleftharpoons O_2 + O + M \tag{16}$$

$$O_3 + O \rightarrow 2O_2 \tag{17}$$

In this case "M" is an active center on the surface which reduces the energy of activation of the reaction and accelerates the decomposition. Assuming that the portion of the surface area occupied by ozone q is proportional to the partial pressure of ozone in the volume, the kinetic region of ozone leaving from the gas phase is equal to:

$$-\frac{dp_{O_3}}{dt} = k\theta = k'p_{O_3} \tag{18}$$

It is obvious that first order of reaction with respect to the decomposition of ozone have been established, which has been confirmed experimentally. The decomposition of ozone in gradient-free reactor is considered in Ref. [89]. As a catalyst has been used thin film of nickel oxide coated on the walls of the reactor. Turbulence of gas flow in such type of reactor would avoid the physical transfer and will provide conditions for studying the kinetics only on the catalytic surface. In this case, the turbulent diffusion is superior to the rate of reaction and the process is conducted in the kinetic regime. It has been found that the rate constant in the range 10–40 °C is calculated by the expression:

$$k = 10^{2.54} p^{-0.5} \exp(-1250/T) \text{ cm/s} \tag{19}$$

where p is in atm.

The experiments which have been made with different pressures of the oxygen and ozone-helium mixtures have shown that the rate constant is changes depending on the total pressure and not by the partial pressure of ozone. Hence the authors have concluded that the catalyst operates in inner diffusion mode, that is, the reaction rate is limited by molecular diffusion in the pores of the catalyst. These and other assumptions made it possible to explain the resulting experimental dependence of the rate constant on the total pressure of the mixture in the gas phase. The kinetics of decomposi-

tion of ozone in gradient-free reactor was investigated in the work [90]. It has been measured activities of γ-Al_2O_3, hopkalite (MnO_2, CuO, bentonite), silver-manganese oxide, alumina-palladium and alumina-platinum catalysts. The experiments have been conducted at 60 °C and initial ozone concentration of 2.7×10^{-5} mol/L. The most active of the tested catalysts have turned out silver-manganese oxide and the hopkalite. It has been determined kinetics of decomposition of ozone to be first order. It have also been defined the rate constants and energies of activation, as the latter being fluctuated in the range of 6–10 kcal/mol. Of interest are conclusions of authors on the active surface of the catalyst. It has been found that the activation energy does not depend on the dispersibility, but influences the catalytic activity. On the other hand, dispersibility does not increase the inner surface of the porous material. Hence it has been concluded that the decomposition of ozone occurs in the outer kinetic region, i.e. the heterogeneous reaction is limited mainly by the outer surface of the catalyst. During the study of the decomposition of ozone in the presence of nickel oxide [91] is established that the reaction to be first order. It has been proposed that the limiting stage appear to be adsorption of ozone to the catalytic surface. The EPR spectra revealed the presence of ozonide ion radical O^{3-}. The formation of O^{3-} is explained by the process of electron transfer from the catalyst to the ozone, so that the surface Ni^{2+} ions are oxidized to the Ni^{3+}. Further studies of the authors have shown that in this way takes place approximately 5% of the total reaction. The remaining amount of the ozone has been decomposed by molecular mechanism:

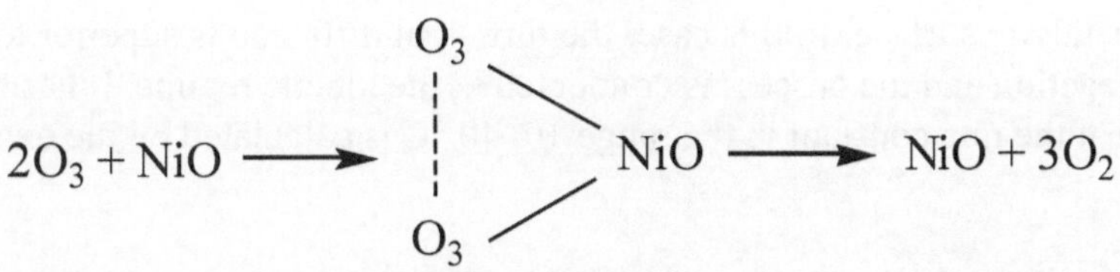

(20)

In some cases, the ozone does not remain in the molecular state and dissociate to atomic or diatomic oxygen species. Ozonide particles on catalytic surface are generated by the reaction:

$$O^- + O_2 \rightarrow O^{3-} \qquad (21)$$

Figure 16.5 shows the characteristic bands of the EPR spectrum at g_1 = 2.0147, g_2 = 2.0120 and g_3 = 2.0018, due to the formation of the ion-radical O^{3-}.

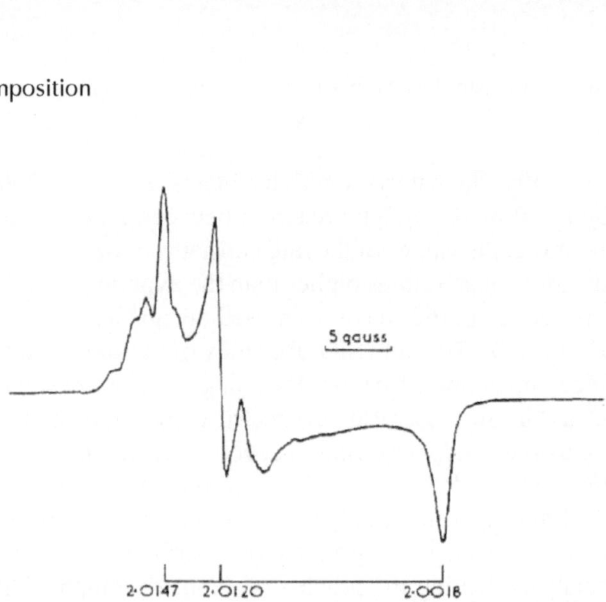

FIGURE 16.5 EPR spectrum of O^{3-} at temperature 77 K.

Martynov et al. [94] have studied the influence of the CuO, CoO and NiO on the process of ozone decomposition. The catalysts have been prepared using aqueous solutions of the respective nitrates and have supported on the walls of the tubular reactor (l=25 cm, D=1.6 cm). The reactor has been heated to 370 °C for 5 h, after which each of the catalysts was treated with ozone at a concentration of 0.5 vol. % for 6 h at a gas velocity of 100 L/h. After solving the diffusion-kinetic equation of the ozone decomposition reaction, the diffusion and the kinetic constants can be defined as well as the coefficients of ozone decomposition. The recent ones characterize the decomposition rate of ozone molecules on the catalyst surface toward the rate of the hits of molecules with the surface. The results of these calculations for different samples are presented in Table 16.4 wherein w is space velocity, g-coefficient of ozone decomposition and k_{exp} k_{kin} are respectively experimental and theoretical rate constants.

TABLE 16.4 Kinetic Parameters of the Reaction of Ozone Decomposition on Some Oxides

Catalyst	w, L/h	k_{exp}, s^{-1}	g.10^{-5}	k_{kin}, s^{-1}
CuO	20	1.0	4.8	2.8
CoO	100	0.9	4.1	2.4
CoO	200	1.3	6.3	3.6
NiO	20	0.9	4.4	2.6
NiO	100	1.0	4.8	2.8
NiO	200	1.7	8.1	4.4

It is evident that the catalysts activity increases at w = 200 L/h. This is explained by the flow of catalytic reaction between kinetic and diffusion regions. Furthermore, the values of the rate constants of ozone decomposition in kinetic mode are almost 3 times higher than the experimental.

Radhakrishnan et al. [68] have been used manganese oxide catalysts supported on Al_2O_3, ZrO_2, TiO_2 and SiO_2 for study of the support influence on the kinetics of decomposition of ozone. By using different physical methods of analysis such as "in situ" laser Raman spectroscopy, temperature-programmed desorption of oxygen and measuring of specific surface area (BET), is shown, that the manganese oxide is highly dispersed on the support surface. The Raman spectra of the supported catalysts reveal the presence of Mn–O bands as result from well-dispersed manganese oxide particles on the Al_2O_3 and SiO_2 supported catalysts. During the process of ozone decomposition on the catalytic surface, a signal from adsorbed particles appears at Raman spectra in the region 876–880 cm^{-1}. These particles were identified as oxygen particles from peroxide type (O_2^{2-}), which disappear upon catalysts heating to 500 K. Using temperature-programmed desorption of oxygen is calculated the number of active manganese centers on the catalyst surface. After integration of the TPD peaks area corresponding to desorption of adsorbed oxygen, besides the density of active sites, the corresponding dispersion values of the catalysts are also identified. These results, as the calculated specific surface areas of the catalytic samples are shown in Table 16.5. The catalysts were tested in reaction of ozone decomposition to determine their activity, and it has been found that the rate of decomposition increases with increasing the ozone partial pressure and the temperature. It has been calculated the kinetic parameters of the reaction. It has been found that the activation energy is in the ranges 3–15 kJ/mol, depending on the catalyst sample, as it is the lower (3 kJ/mol) in the case of ozone decomposition on MnO_x/Al_2O_3. It was suggested that this is related to the structure of this catalyst which only of the tested samples has mononuclear manganese center coordinated by five oxygen atoms. This was demonstrated using absorption fine-structure X-ray spectroscopy. Using this method, it was also found that the other three supported catalysts possess multicore active manganese centers surrounded as well by five oxygen atoms.

TABLE 16.5 Densities of Active Centers and Specific Surfaces

Catalyst	S_g, $m^2\ g^{-1}$	Density, mmol g^{-1} (O_2/O_3 TPD)	dispersion, %
MnO_x/Al_2O_3	92	40	12
MnO_x/ZrO_2	45	163	47
MnO_x/TiO_2	47	31	9
MnO_x/SiO_2	88	13	4

A mechanism of ozone decomposition on MnO_x/Al_2O_3 catalyst has been proposed:

$$O_3 + Mn^{n+} \rightarrow O^{2-} + Mn^{(n+2)+} + O_2 \quad (22)$$

$$O_3 + O^{2-} + Mn^{(n+2)+} \rightarrow O_2^{2-} + Mn^{(n+2)+} + O_2 \quad (23)$$

$$O_2^{2-} + Mn^{(n+2)+} \rightarrow Mn^{n+} + O_2 \quad (24)$$

The presented mechanism consists of electronic transfer from the manganese center to ozone, where after the manganese is reduced by desorption of peroxide particles to form oxygen ($O_2^{2-} \rightarrow O_2 + 2e-$).

Study of MnO_x/Al_2O_3 with absorption fine-structure X-ray spectroscopy is also presented in publication [95]. The aim of the work is to receive important information on the catalytic properties of the supported manganese oxides in oxidation reactions using ozone. The structural changes in the manganese oxide supported on alumina have been detected in the process of catalytic ozone decomposition at room temperature. It has been found that during the ozone decomposition in presence of water vapor, the manganese is oxidized to higher oxidation state. At the same time the water molecule combines to manganese active center due to cleavage of Mn-O-Al bond. The catalyst is completely regenerated after calcination in oxygen at 723 K.

In Ref. [96], it has been studied the influence of nickel oxide addition on the activity of cement containing catalyst for ozone decomposition. The activity of the samples was measured by calculating the rate of decomposition g that shows the degree of active interactions (leading to decomposition) of ozone molecules with catalytic surface. According to [6] the expression for g is:

$$\gamma = \frac{4\omega . \ln\left(C_0 / C\right)}{V_t . S}, \quad (25)$$

where: V_t – specific heat velocity of ozone molecules, S – geometric surface of the catalyst, w – space velocity of the gas stream, C_0 and C – inlet and outlet ozone concentrations.

It has been found that the addition of nickel oxide in the catalyst composition improves its catalytic properties. Upon decomposition of wet ozone is observed a decrease in the activity of all tested samples, as the calculated values for g are 23 times lower compared with the value obtained after decomposition of dry ozone. On the basis of the measured values for energy of

activation (E_a = 5.9 ± 0.3 kJ/mol) of ozone decomposition of in the region of high temperatures (300–400 K), and the corresponding values of E_a (15.2 ± 0.4 kJ/mol) in the region of low temperatures, it has been made a conclusion, that in the first case the process goes to the outer diffusion region, whereas in the second–into the inner diffusion region. The factor of diffusion suspension or the accessible part of the surface is estimated to make clear the role of the internal catalytic surface. This has permitted to do calculation that at room temperature the molecules of ozone enter into the pores of the catalyst at a distance not more than ~ 10^{-4} cm.

Lin et al. [97] have studied the activity of a series oxide supports and supported metal catalysts with respect to decomposition of ozone in water. From the tested in reaction conditions supports the activated carbon has showed relatively high activity, while the zeolite support (HY and modernite), Al_2O_3, SiO_2, SiO_2.Al_2O_3 and TiO_2 have showed zero or negligible activity. From all supported metal catalysts submitted to ozone dissolved in water the noble metals have highest activity in ozone decomposition excepting gold. The metals are deposited on four types of supports (Al_2O_3, SiO_2, SiO_2.Al_2O_3 and TiO_2). Highest activity was measured for the catalysts deposited on silica. It has been found that the catalyst containing 3% Pd/SiO_2 is most effective in the reaction of ozone decomposition from all tested samples. A comparison of some indicators for Pd catalysts deposited on different supports is presented in Table 16.6.

TABLE 16.6 Comparison of the Specific Surface Areas, the Size of Metal Particles and the Average Rates of Ozone Decomposition in Water Over Palladium-Containing Catalysts

Catalyst	Specific surface area (m^2 g^{-1})	Size of metal particle (Å)	Average rate ($mg_{(O_3)}$ min^{-1} $g^{-1}_{(cat.)}$)
Pd/SiO_2	206	90	0.77
Pd/SiO_2.Al_2O_3	221	70	0.54
Pd/Al_2O_3	139	75	0.39
Pd/TiO_2	34	109	0.35

It has been determined first order of reaction of ozone decomposition on 3% Pd/SiO_2 regarding the concentration of ozone. In presence of the same catalyst the calculated activation energy is about 3 kcal mol^{-1} and it is assumed that the reaction goes in the diffusion region. The proposed mechanism of catalytic decomposition of ozone in water is similar to the mechanism of decomposition of gaseous ozone. Table 16.7 shows two possible reaction paths

of ozone decomposition on metals or oxides, depending on the fact whether oxygen is adsorbed on the catalytic surface.

In the literature there are also data for the study of ozone decomposition in water in presence of aluminum (hydroxyl) oxide [98]. It has been suggested that the surface hydroxyl groups and the acid-basic properties of aluminum (hydroxyl) oxides play important role in catalytic decomposition of ozone.

TABLE 16.7 Possible Mechanisms of Catalytic Ozone Decomposition in Water [97]

Case of O_2 not adsorbed on metal	Case of O_2 adsorbed on metal
$O_3 \rightarrow O_{3(a)}$	$O_3 \rightarrow O_{3(a)}$
$O_{3(a)} \rightarrow O_{(a)} + O_2$	$O_{3(a)} \rightarrow O_{(a)} + O_{2(a)}$
$O_{(a)} + O_3 \rightarrow 2O_2$	$O_{(a)} + O_3 \rightarrow O_2 + O_{2(a)}$ $O_{2(a)} \rightarrow O_2$

The environmental application of ozone in catalysis has been demonstrated in article [99], devoted to the ozonation of naproxen and carbamazepine on titanium dioxide. The experiments were carried out in aqueous solution at T = 25 °C and in pH range of 3–7. The results have been indicated that the naproxen and the carbamazepine are completely destructed in the first few minutes of the reaction. The degree of mineralization during the noncatalytic reaction flow is measured up about 50% and it is formed primarily in the first 10–20 min. The presence of catalyst is increased to more than 75% the degree of mineralization of the initial hydrocarbon. Furthermore, it has been found that the catalyst increases the mineralization in both acid and neutral solution, as the best results are obtained at slightly acidic media. This effect is related with possible adsorption of intermediate reaction products on Lewis acidic catalytic sites. It is also reflected that the titanium dioxide catalyzes the ozone decomposition in acidic media, whereas in neutral solution the ozone destruction is inhibited. This is precluded the flow of mechanism based on surface formation of hydroxyl radicals followed by their migration and complete reaction with the organic compounds. The variation of the quantity total organic carbon is modeled as a function of the integral of the applied amount of ozone. On this basis, it is assumed that the reaction between organic compounds and ozone is of second order. The calculated pseudo homogeneous catalytic rate constants are $7.76 \times 10^{-3} \pm 3.9 \times 10^{-4}$ and $4.25 \times 10^{-3} \pm 9.7 \times 10^{-4}$ L mmol^{-1} s^{-1} for naproxen and carbamazepine, respectively, at pH 5 and catalyst amount of 1 g/L. The products of ozonation are investigated with a specific ultravio-

let absorption at 254 nm. The wide application of metal oxide catalysts in ozone decomposition necessitated the use of different instrumental methods for analysis. Based on the results of X-ray diffraction, X-ray photoelectron spectroscopy, EPR and TPD it has been found that during the destruction of ozone on silver catalyst supported on silica, the silver is oxidized to complex mixture of Ag_2O_3 and AgO [100]. This investigation of catalytic ozone decomposition on Ag/SiO2 is carried out in temperature range −40 °C – +25 °C, it is determined first order of reaction, and the calculated activation energy is 65 kJ/mol. X-ray diffraction has been also used to determine the phase composition of manganese oxide catalysts supported on γ-Al_2O_3 and SiO_2 [101]. Moreover, the samples are characterized by Raman and IR spectroscopy. The supported catalysts are prepared by nitrate precursors using the impregnation method. In addition to the Raman spectral bands of b-MnO_2 and a-Mn_2O_3 phases, other signals also are registered and attributed to isolated Mn^{2+} ions present in tetraedrical vacations on the support surface, and in some epitaxial layers, respectively on g-Mn_2O_3 and manganese silicate. The data from the IR spectra was not so much useful due to the fact that the supporter band overlaps the bands of manganese particles formed on the surface, and make it difficult to identify them.

Eynaga et al. [102] are carried out catalytic oxidation of cyclohexane with ozone on manganese oxide supported on aluminum oxide at temperature of 295 K. It has been performed "in situ" IR studies for taking information on the intermediates formed on the catalytic surface at the time of oxidation. It has been found that the intermediates are partially oxidized alcohols, ketones, acid anhydrides and carboxylic acids. These compounds are subsequently decomposed by ozone. In the beginning, the activity of the catalyst gradually decreases whereupon it reaches a steady state with mole fractions of CO (90%) and CO_2 (10%). It has been found that high resistant particles, containing C=O, COO-and CH groups, remain on the catalyst surface that caused the slow deactivation of the catalyst. The C=O groups are decomposed at relatively low temperatures (<473 K), while the COO- and CH groups are dissociated at temperatures >473 K.

The kinetics of gas-phase ozone decomposition was studied by heterogeneous interactions of ozone with aluminum thin films [103]. The ozone concentrations were monitored in real time using UV absorption spectroscopy at 254 nm. The films were prepared by dispersion of fine alumina powder in methanol, and their specific surface areas are determined by "in situ" adsorption of krypton at 77 K. It has been found that the reactivity of alumina decreases with increasing the ozone concentration. As consequence of multiple exposures to ozone of one film, it has been found that the number of

active sites is greater than 1.4×10^{14} per cm^2 surface or it is comparable with the total number of active sites. The coefficients of ozone decomposition are calculated (g) depending on the initial concentration of ozone in the reaction cell, using the expression:

$$g = (4 \times k^{I} \times V) / (c \times SA) \quad (26)$$

where c is the average velocity of gas phase molecules of ozone, k^I is the observed initial rate constant of ozone decomposition from first order (for the first 10 s), SA is the total surface area of the aluminum film, and V is volume of the reactor. From the results it was concluded that the coefficients of ozone decomposition on fresh films depend inversely on ozone concentration, ranging from 10^{-6} to ozone concentration of 10^{14} molecules/cm^3 to 10^{-5} for 10^{13} molecules/cm^3 ozone. It has been observed also that the coefficients of ozone decomposition do not depend on the relative humidity of gas stream. It was discussed a mechanism of the reaction of catalytic ozone decomposition, based on detailed spectroscopic investigations of the catalytic surface [104, 105], where there is evidence for the formation of peroxide particles on the surface:

$$O_3 + * \longrightarrow O_2 + O* \quad (27)$$

$$O_3 + O* \longrightarrow O_2* + O_2 \quad (28)$$

$$O_2* \longrightarrow O_2 + * \quad (29)$$

It has been suggested that molecular oxygen could also be initiated by the reaction:

$$O* + O* \rightarrow 2* + O_2 \quad (30)$$

It has been found that the reactivity of the oxidized aluminum film can be partially restored after being placed for certain period in a medium free of ozone, water vapor and carbon dioxide.

16.4.2.2 DECOMPOSITION OF OZONE ON THE SURFACE OF THE CARBON FIBER AND INERT MATERIALS

The use of carbon material in adsorption and catalysis is related to their structural properties and surface chemical groups. Their structural properties are determined by the specific surface area and porosity, while the chemical groups on the surface of the catalyst mainly are composed of oxygen-containing functional groups. Ozone reacts with various carbon materials such as activated carbon, carbon black, graphite, carbon fiber, etc. The character of these interactions depends both on the nature of the carbon surface and temperature [106]. It has been tested two types of interactions: the complete oxidation leading to formation of gaseous carbon oxides and partial oxidation, producing surface oxygen-containing functional groups [106, 107]. With increase of the temperature, the ratio between these two processes is displaced in the direction of complete oxidation. The latter is accompanied by an oxidative destruction on the surface with formation of gaseous carbon oxides. In addition, on the surface of the carbon takes place catalytic decomposition of ozone. In a number of publications [108–111] have been studied the physicochemical properties of activated carbon treated with ozone, as well as and the kinetics of the process related to the release of CO and CO_2. Subrahmanyam et al. [112] have examined the catalytic decomposition of ozone to molecular oxygen on active carbon in the form of granules and fibers at room temperature. It was found that the dynamics of the activity of the carbons is characterized by two distinct zones. The first one is the observed high activity with respect to the decomposition of ozone, which is mainly due to the chemical reaction of ozone with carbon. As a result of this interaction on the carbons, oxygen-containing surface groups are formed. Then sharp drop the conversion is registered and transition of the catalyst to a low active zone takes place. In this zone, the decomposition of the ozone to molecular oxygen takes place in a catalytic way. The activities of the carbons in dry environment on the one hand and in presence of water vapor and NO_X on the other have been compared. The presence of water vapor has reduced catalytic activity, while the presence of the NO_X has improved activity, due to the change in the carbon surface functional groups. They can be modified in two ways: boiling in dilute nitric acid or thermal treatment at 1273 K in helium media. Positive results were obtained only in the first treatment. The decomposition of ozone with respect to the gasification of carbon with forming of CO_x proceeds with a selectivity of less than 25%. The catalysts were characterized by means of temperature-programmed decomposition of surface functional groups, IR and

X-ray photo-electron spectroscopy. Mechanism of decomposition of ozone on activated carbon has been proposed (Fig. 16.6).

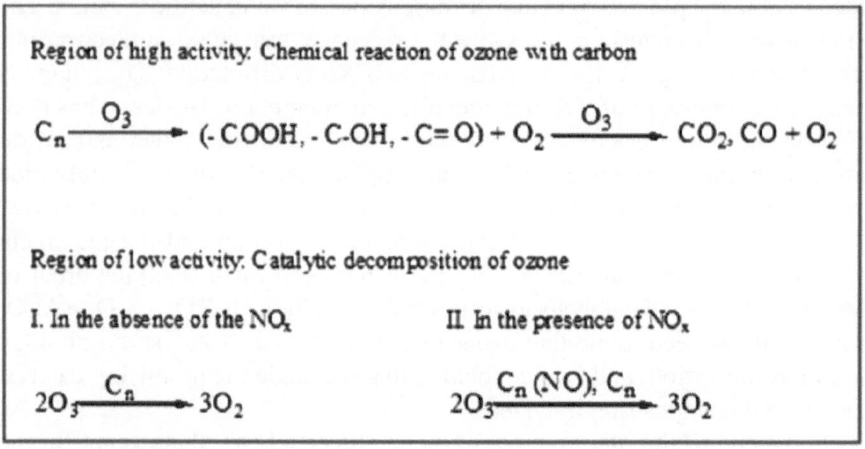

FIGURE 16.6 Simplified scheme of ozone decomposition on carbon.

In ozone decomposition process it is difficult to draw a line between relatively inert materials and heterogeneous catalysts. However, the separation can be made on the base of two evident signs: (i) the catalysts for ozone decomposition are specifically synthesized; and (ii) they have higher catalytic activity (higher values for the coefficient of decomposition g) compared to the inert materials. In the literature there are investigations on the reaction of ozone destruction on the surface of silica [113], glass [114], volcanic aerosols [115, 116] and ammonium hydrogen sulfate [117]. The published values for the coefficient of ozone decomposition g for quartz and glass are in the order of (1–2) $\times$ 10^{-11}, and those for aerosols and ammonium sulfate–from 1.61 $\times$ 10^{-6} to 7.71 $\times$ 10^{-8}. It is interesting to mark the studies of ozone decomposition on the surface of Saharan dust [118] as well as the application of slurry treated water as catalyst for ozone destruction in aqueous medium [119].

The heterogeneous reaction between O_3 and authentic Saharan dust surfaces [118] was investigated in a Knudsen reactor at ~296 K. O_3 was destroyed on the dust surface and O_2 was formed with conversion efficiencies of 1.0 and 1.3 molecules O_2 per O_3 molecule destroyed for unheated and heated samples, respectively. No O_3 desorbed from exposed dust samples, showing that the uptake was irreversible. The uptake coefficients for the irreversible destruction of O_3 on (unheated) Saharan dust surfaces depended on the O_3 concentration and varied between 4.8×10^{-5} and 2.2×10^{-6} for the steady-state

uptake coefficient. At very high O_3 concentrations the surface was deactivated, and O_3 uptake ceased after a certain exposure period.

New, effective and stable ecological catalyst based on slurry [119] has been used in the process of ozone decomposition in water acidic medium. The catalyst was characterized by X-ray fluorescence, transmission electron microscopy, scanning electron microscopy and X-ray diffraction. The sludge is essentially composed of different metallic and nonmetallic oxides. It has been investigated the effect of various experimental parameters such as catalyst amount, initial ozone concentration and application of different metal oxide catalysts. The decomposition of dissolved ozone is significantly increased with the enhancement of the initial ozone concentration and the increment of catalyst amount from 125 to 750 mg. It has been established the order of activity of the tested catalysts as follows: ZnO ≈ sludge > TiO_2 > SiO_2 > Al_2O_3 ≈ Fe_2O_3. It has been found that ozone does not affect the catalyst morphology and its composition, and it is concluded that the sludge is promising catalyst for ozone decomposition in water.

In the end of the literature review it can be concluded that except for the metals of platinum group, characterized by its high price, the metal oxide catalysts containing manganese oxide have the highest activity in decomposition of gaseous ozone and also in catalytic oxidation of pollutants. It is important to mention, that unlike the inert materials, for the oxide catalysts there is not exist strong dependence of catalytic activity from ozone concentration in gas phase. It should be emphasized that despite the great number of publications on the subject, the kinetics and the mechanism of ozone decomposition on the surface of heterogeneous metal oxide catalysts are not cleared up sufficiently.

KEYWORDS

- **Catalysts**
- **Decomposition**
- **Kinetics**
- **Mechanism**
- **Ozone**
- **Synthesis**

REFERENCES

1. Rakovsky, S. K., Zaikov, G. E.: "Kinetic and Mechanism of Ozone Reactions with Organic and Polymeric Compounds in Liquid Phase," monograph (second edition), *Nova Sci. Publ.*, Inc. New York, 1–340, (2007).
2. Schonbein, C. F., *Pogg. Ann.*, 49, 616 (1840); *Helb. Seances Acad. Sci.*, 10, 706 (1840).
3. Ulmann's Encyclopedia of Industrial Chemistry, Vol A18, 1991, 349.
4. Razumovskii, S. D., Rakovsky, S. K., Shopov, D. M., Zaikov, G. E. Ozone and Its Reactions with Organic Compounds (in Russian). *Publ. House of Bulgarian Academy of Sciences*, Sofia (1983).
5. Brown, Theodore, L.; H. Eugene Le May Jr., Bruce, E. Bursten, Julia, R. Burdge [1977]. "22," in Nicole Folchetti: http://en.wikipedia.org/w/index.php?title=Template:Cite_book/editor&action=editChemistry: The Central Science, 9th Edition (in English), *Pearson Education*, 882–883 (2003).
6. Lunin, V. V., Popovich, M. P., Tkachenko, S. N., *Physical Chemistry of Ozone* (in Russian), *Publ. MSU*, 1–480 (1998).
7. R. H. Perry and, D. Green, *Perry's Chemical Engineer's Handbook*, McGraw-Hill, New York, 1989, 3–147.
8. Thorp, C., *Bibliography of Ozone Technology*, 2, 30 (1955).
9. Kutsuna, S., Kasuda, M., Ibusuki, T., Transformation and Decomposition of 1, 1, 1-Trechloroethane on Titanium Dioxide in the Dark and Under Photo-illumination, *Atmospheric Environment*, 28 (9), 1627–1631 (1994).
10. Naydenov, A., Stoyanova, R., Mehandjiev, D., Ozone decomposition and CO oxidation on CeO_2, *Journal of Molecular Catalysis A-Chemical*, 98 (1), 9–14 (1995).
11. Rakitskaya, T. L., Vasileva, E. K., Bandurko, A. Yu., Paina, V. Ya., Kinetics of Ozone Decomposition on Activated Carbons, *Kinetics and Catalysis*, 35 (1), 90–92 (1994).
12. Tanaka, V., Inn, E. C. J., Watanabe, K. J., *Chem. Phys., 21(10),* 1651 (1953).
13. De More, W. B., Paper, O. J., *Phys. Chem.*, 68, 412 (1964).
14. Beitker, K. H., Schurath, U., Seitz, N., *Int. J. Chem. Kinet.*, 6, 725 (1974).
15. Inn, E. C. J., Tanaka, V. J., *Opt. Soc. Amer., 43(10),* 870 (1953).
16. Griggs, M. J., *Chem. Phys.*, 49 (2), 857 (1968).
17. Taube, H., *Trans. Faraday Soc.*, 53, 656 (1957).
18. Galimova, L. G., Komisarov, V. D., Denisov, E. T., *The Reports of Russian Academy of Sciences (in Russian), Ser. Chem.*, 2, 307 (1973).
19. Alexandrov, Y.A., Tarunin, B. I., Perepletchikov M. L., *J. Phys. Chem. (in Russian)*, LVII (10), 2385–2397 (1983).
20. Hon, Y. S., Yan, J. L., Ozonolytic Cleavage of Cycloalkenes in the Presence of Methyl Pyruvate to Yield the Terminally Differentiated Compounds. *Tetrahedron, Lett., 34(41),* 6591–6594 (1993).
21. Deninno, M. P., McCarthy, K. E., The C-14 radiolabelled synthesis of the cholesterol absorption inhibitor CP-148,623. A novel method for the incorporation of a C-14 label in enones. *Tetrahedron, 53(32),* 11007–11020 (1997).
22. Claudia, C., Mincione, E., Saladino, R., Nicoletti, R., Oxidation of Substituted 2-Thiouracils and Pyramidine-2-Thione with Ozone and 3,3-Dimethyl-1,2-Dioxiran. *Tetrahedron, 50(10),* 3259–3272 (1994).
23. Razumovskii, S. D., Zaikov, G. E., Ozone and its reactions with organic compounds, Publ.: Science (in Russian), Moscow (1974).
24. Chapman, S., *Phil. Mag.*, Ser. 7, 10, 369 (1930).
25. Chapman, S., *Met. Roy. Soc.*, 3, 103 (1930).

26. Johnston, H., *Rev. Geoph. Space Phys.*, 13, 637 (1975).
27. NASA Reference Publication, 1162, N. Y. Acad. Press (1986).
28. Gerchenson Yu., Zvenigorodskii, S., Rozenstein, V., *Success Chemistry (in Russian)*, 59, 1601 (1990).
29. Johnston, H., *Photochemistry in the Stratosphere*, UCLA, Berkley, 20 (1975).
30. Crutzen, P., Smalcel, R., *Planet Space Sci.*, 31, 1009 (1983).
31. Farmen, J., Gardiner, B., Shanklin, J., *Nature*, 315, 207 (1985).
32. Solomon, S., Garcia, R., Rowland, F., Wueblles, P., *Ibid*, 321, 755 (1986).
33. Vupputuri, R., *Atm. Environ.*, 22, 2809 (1988).
34. Kondrat'ev, K., *Meteorology and Climate (in Russian)*, 19, 212 (1989).
35. Kondrat'ev, K., *Success Chemistry (in Russian)*, 59, 1587 (1990).
36. Heisig, C., Zhang, W., Oyama, S. T., Decomposition of Ozone using Carbon Supported Metal Oxide Catalysts. *Appl. Catal. B: Environ.*, 14, 117 (1997).
37. Rakitskaya, T. L., Bandurko, A. Yu., Ennan, A. A., Paina, V. Ya., Rakitskiy, A. S., Carbon-fibrous-material-supported Base Catalysts of Ozone Decomposition. *Micro. Meso. Mater.*, 43, 153 (2001).
38. Skoumal, M., Cabot, P. L., Centellas, F., Arias, C., Rodriguez, R. M., Garrido, J. A., Brillas, E., *Appl. Catal. B Environ.*, 66, 228–240 (2006).
39. Bianchi, C. L., Pirola, C., Ragaini, V., Selli, E., *Appl. Catal. B Environ.*, 64, 131–138 (2006).
40. Ma J., Sui, M. H., Chen, Z. L., Li N. W., *Ozone Sci. Eng.*, 26, 3–10 (2004).
41. Zhao, L., Ma J., Sun, Z. Z., *Appl. Catal. B Environ.*, 79, 244–253 (2008).
42. Von Gunten, U., *Water Res.*, 37, 1443–1463 (2003).
43. Monhot, W., Kampschulte, W., *Ber.*, 40, 2891 (1907).
44. Kashtanov, L., Ivanova, N., Rizhov, B., Zh. *Applied Chemistry (in Russian)*, 9, 2176 (1936).
45. Schwab, G., Hartman, C., *J. Phys. Chem. (in Russian)*, 6, 72 (1964).
46. Emel'yanova, G., Lebedev, V., Kobozev, N., *J. Phys. Chem. (in Russian)*, 38, 170 (1964).
47. Emel'yanova, G., Lebedev, V., Kobozev, N., *J. Phys. Chem. (in Russian)*, 39, 540 (1965).
48. Emel'yanova, G., Strakhov, B., *Advanced Problems Physical Chemistry (in Russian)*, 2, 149 (1968).
49. Sudak, A., Vol'fson, V., Catalytic Ozone Purification of Air, *Scientific Notion (in Russian)*, Kiev, 87 (1983).
50. Hata, K., Horiuchi, M., Takasaki, T., Jap. Pat., *CA*, 108, 61754u (1988).
51. Tchihara, S., Jap. Pat., *CA*, 108, 192035h (1988).
52. Kobayashi, M., Mitsui, M., Kiichiro, K., Jap. Pat., *CA*, 109, 175615a (1988).
53. Terui, S., Sadao, K., Sano, N., Nichikawa, T., Jap. Pat., *CA*, 112, 20404p (1990).
54. Terui, S., Sadao, K., Sano, N., Nichikawa, T., Jap. Pat., *CA*, 114, 108179b (1991).
55. Oohachi, K., Fukutake, T., Sunao, T., Jap. Pat., *CA*, 119, 119194g (1993).
56. Kobayashi, M. and Mitsui, K., Jap. Pat. 63,267,439, Nov 4, (1988), to Nippon Shokubal Kagaku Kogyo Co., Ltd.
57. Kuwabara, H. and Fujita, H., Jap. Pat. 3016640 Jan 24, (1991), to Mitsubishi Heavy Industries, Ltd.
58. Hata, K., Horiuchi, M., Takasaki, K. and Ichihara, S., Jap. Pat. 62,201,648, Sep 5, (1987), to Nippon Shokubai Kagaku Kogyo Co., Ltd.
59. Terui, S., Miyoshi, K., Yokota, Y. and Inoue, A., Jap. Pat. 02,63,552, Mar 2, (1990).
60. Yoshimoto, M., Nakatsuji, T., Nagano, K. and Yoshida, K., Eur. Pat. 90,302,545.0, Sep 19, (1990), to Sakai Chemical Industry Co., Ltd.
61. Oyama, S. T., Chemical and Catalytic Properties of Ozone. *Catal. Rev. Sci. Eng.*, 42, 279 (2000).

62. Einaga, H., Futamura, S., Comparative Study on the Catalytic Activities of Alumina-supported Metal Oxides for Oxidation of Benzene and Cyclohexane with Ozone. *React. Kinet. Catal. Lett.*, 81, 121 (2004).
63. Tong, S., W. Liu, W. Leng, Q. Zhang, Characteristics of MnO_2 Catalytic Ozonation of Sulfosalicylic Acid Propionic Acid in Water. *Chemosphere*, 50, 1359 (2003).
64. Konova, P., M. Stoyanova, A. Naydenov, ST. Christoskova, D. Mehandjiev, Catalytic oxidation of VOCs and CO by ozone over alumina supported cobalt oxide. *J. Appl. Catal. A: Gen.* 298, 109 (2006).
65. Stoyanova, M., P. Konova, P. Nikolov, A. Naydenov, ST. Christoskova, D. Mehandjiev, Alumina-supported nickel oxide for ozone decomposition and catalytic ozonation of CO and VOCs. *Chem. Eng. Journal*, 122, 41 (2006).
66. Popovich, M., Smirnova, N., Sabitova, L., Filipov Yu., *J. of Moskow Univerity (in Russia)*, ser. chem., 26, 167 (1985).
67. Popovich, M., *J. of Moskow Univerity (in Russian)*, ser. chem., 29, 29 (1988).
68. Radhakrishnan, R., S. T. Oyama, J. Chen, A. Asakura, Electron Transfer Effects in Ozone Decomposition on Supported Manganese Oxide. *J. Phys. Chem. B*, 105 (19), 4245 (2001).
69. Zavadskii, A. V., S. G. Kireev, V. M. Muhin, S. N. Tkachenko, V. V. Chebkin, V. N. Klushin, D. E. Teplyakov, Thermal Treatment Influence over Hopcalite Activity in Ozone Decomposition. *J. Phys. Chem. (in Russian)*, 76, 2278 (2002).
70. Imamura, S., Ikebata, M., Ito, T., Ogita, T., *Ind. Eng. Chem. Res.*, *30*, 217 (1991).
71. Dhandapani, B., Oyama, S. T., *J. Appl. Catal. B: Environmental*, 11, 129 (1997).
72. Lo Jacono, M., Schiavello, M., The influence of preparation methods on structural and catalytic properties of transition metal ions supported on alumina. *In Preparation of catalysts I*; B. Delmon, P. Jacobs, G. Poncelet (Eds.), *Elsevier*, New York, 473 (1976).
73. Yamashita, T., Vannice, A., *J. Catalysis, 161*, 254, (1996).
74. Ma J., Chuah, G. K., Jaenicke, S., Gopalakrishnan, R., Tan, K. L., *Ber.Bunsenges. Phys. Chem*, *100*, 585, (1995).
75. Ma J., Chuah, G. K., Jaenicke, S., Gopalakrishnan, R., Tan, K. L., *Ber.Bunsenges. Phys. Chem*, *99*, 184, (1995).
76. Baltanas, M. A., Stiles, A. B., and Katzer, J. R., *Appl. Catal.*, *28*, 13 (1986).
77. Li Wei, Oyama, S. T., in *Heterogeneous Hydrocarbon Oxidation*, B. K. Warren and, S. T. Oyama (Eds.), *ACS Symp. Ser.* 638, ACS: Washington, DC, 364 (1996).
78. Naydenov, A., Mehandjiev, D., *Appl. Catal. A. : General.*, *97*, 17 (1993).
79. Einaga, H., Ogata, A., Benzene oxidation with ozone over supported manganese oxide catalysts: Effect of catalyst support and reaction conditions, *J. Hazard. Mater.* (2008).
80. Boreskov, G. K., *Adv. Catal.*, *15*, 285 (1964).
81. Baldi, M., Finochhio, E., Pistarino, C., Busca, G., *J. Appl.Catal. A: General*, 173, 61 (1998).
82. Kapteijn, F., Singoredjo, L., Andreini, A., Moulijn, J. A., *J. Appl.Catal. B: Environmental*, *3*, 173 (1994).
83. Maltha, A., Favre, L. F. T., Kist, H. F., Zuur, A. P., Ponec, V., *J. Catal.*, 149, 364 (1994).
84. Hunter, P., Oyama, S. T., *Control of Volatile Organic Compound Emissions*, John Wiley & Sons, Inc (2000).
85. Subrahmanyam Ch., Renken, A., Kiwi-Minsker, L., Novel catalytic non-thermal plasma reactor for the abatement of VOCs, *Chemical Engineering Journal*, 134, 78 (2007).
86. Rubashov, A. M., Pogorelov, V. V., Strahov, B. V., *J. Phys. Chem. (in Russian)*, 46 (9), 2283 (1972).
87. Rubashov, A. M., Strahov, B. V., *J. Phys. Chem. (in Russian)*, 47 (8), 2115 (1973).
88. Ellis, W. D., Tomets, P. V., *Atmospheric Environment Pergamon Press.*, 6 (10), 707 (1972).
89. Houzellot, J. Z., Villermaux, J., J. de Chemie Physique, 73 (7–8), 807 (1976).

90. Tarunin, B. I., Perepletchikov, M. L., Klimova, M. N., *Kinetics and Catalysis*, 22 (2), 431 (1981).
91. Rakovsky, S., Nenchev, L., Cherneva, D., *Proc. 4th Symp. Heterogeneous Catalysis*, Varna, 2, 231 (1979).
92. Che, M. and Tench, A. J., *Adv. Catal.*, 31, 77 (1982).
93. Tench, A. J. and Lawson, T., *Chem. Phys. Lett.*, 459 (1970).
94. Martinov, I., Demiduk, V., Tkachenko, S., Popovich, M., *J. Phys. Chem. (in Russian)*, 68, 1972 (1994).
95. Einaga, H., Harada, M., Futamura, S., Structural Changes in Alumina-supported Manganese Oxides during Ozone Decomposition, *Chem. Phys. Lett.*, 408, 377 (2005).
96. Martinov, I. V., Tkachenko, S. N., Demidyuk, V. I., Egorova, G. V., Lunin, V. V., NiO Addition Influence over Cement-containing Catalysts Activity in Ozone Decomposition. *J. of Moscow Univ. (in Russian)*, Ser. 2, Chemistry, 40, 355 (1999).
97. Lin, J., Kawai, A., Nakajima, T., Effective Catalysts for Decomposition of Aqueous Ozone, *Applied Catalysis B: Environmental*, 39, 157 (2002).
98. Qi F., Chen, Z., Xu B., Shen, J., Ma J., Joll, C., Heitz, A., Influence of Surface Texture and Acid-Base Properties on Ozone Decomposition Catalyzed by Aluminum (Hydroxyl) Oxides, *Applied Catalysis B: Environmental*, 84, 684 (2008).
99. Rosal, R., Rodriguez, A., Gonzalo, M. S., Garcia-Calvo, E., Catalytic Ozonation of Naproxen and Carbamazepin on Titanium Dioxide, *Applied Catalysis B: Environmental*, 84, 48 (2008).
100. Naydenov, A., Konova, P., Nikolov Pen., Klingstedt, F., Kumar, N., Kovacheva, D., Stefanov, P., Stoyanova, R., Mehandjiev, D., Decomposition of ozone on Ag/SiO_2 catalyst for abatement of waste gases emissions, *Catalysis Today*, 137, 471 (2008).
101. Buciuman, F., Patcas, F., Craciun, R., Zhan, D. R. T., Vibrational spectroscopy of bulk and supported manganese oxides, *Phys. Chem. Chem. Phys.*, 1, 185 (1998).
102. Einaga, H., Futamura, S., Oxidation behavior of cyclohexane on alumina-supported manganese oxide with ozone, *Applied Catalysis B: Environmental*, 60, 49 (2005).
103. Sullivan, R. C., Thornberry, T., Abbatt, J. P. D., Ozone decomposition kinetics on alumina: Effects of ozone partial pressure, relative humidity and repeated oxidation cycles, *Atmos. Chem. Phys.*, 4, 1301 (2004).
104. Li W., Gibbs, G. V., Oyama, S. T., Mechanism of Ozone Decomposition on Manganese Oxide: 1. In situ Laser Raman Spectroscopy and ab initio Molecular Orbital Calculations, *J. Am. Chem. Soc.*, 120, 9041 (1998).
105. Li W., Oyama, S. T., The Mechanism of Ozone Decomposition on Manganese Oxide: 2. Steady-state and Transient Kinetic Studies. *J. Am. Chem. Soc.*, 120, 9047 (1998).
106. Atale, Hitoshi, Kaneko, Taraichi, Yano, Jap. Pat., *CA*, 123, 121871 (1995).
107. Mori, Katsushiko, Hasimoto, Akira, Jap. Pat., *CA*, 118, 153488v (1993).
108. Kobayashi, Motonobu, Mitsui, Kiichiro, Jap. Pat., *CA*, 110, 120511d (1989).
109. Rakitskaya, T. L., Vasileva, E. K., Bandurko, A. Yu., Paina, V. Ya., *Kinetics and Catalysis*, 35, 103 (1994).
110. Aktyacheva, L., Emel'yanova, G., *J. Phys. Chem. (in Russian)*, *ser. chem.*, 31, 21 (1990).
111. Valdes, H., Sanches-Polo, M., Rivera-Utrilla, J., Zaror, C. A., Effect of ozone treatment on surface properties of activated carbon, *Langmuir*, 18, 2111 (2002).
112. Subrahmanyam, C., Bulushev, D. A., Kiwi-Minsker, L., Dynamic behaviour of activated carbon catalysts during ozone decomposition at room temperature, *Applied Catalysis B: Environmental*, 61, 98 (2005).
113. Tkalich, V. S., Klimovskii, A. O., Lissachenko, A. A., *Kinetics and Catalysis*, 25 (5), 1109 (1984).

114. Olshina, K., Cadle, R. D., DePena, R. G., *J. Geoph. Res.*, 84 (4), 1771 (1979).
115. Popovich, M. P., Smirnova, N. N., Sabitova, L. V., *J. Phys. Chem. (in Russian)*, *ser. Chem.*, 28 (6), 548 (1987).
116. Popovich, M. P., *J. Phys. Chem. (in Russian)*, *ser. chem.*, 29 (5), 427 (1988).
117. Egorova, G. V., Popovich, M. P., Filipov Yu. V., *J. Phys. Chem. (in Russian)*, *ser. Chem.*, 29 (4), 406 (1988).
118. Hanisch, F., Crowley, J. N., Ozone decomposition on Saharan dust: an experimental investigation, *Atmos. Chem. Phys. Discuss.*, 2, 1809 (2002).
119. Muruganadham, M., Chen, S. H., Wu J. J., Evaluation of water treatment sludge as a catalyst for aqueous ozone decomposition, *Catalysis Communications*, 8, 1609 (2007).

CHAPTER 17

A TECHNICAL NOTE ON DESIGNING, ANALYSIS AND INDUSTRIAL USE OF THE DYNAMIC SPRAY SCRUBBER

R. R. USMANOVA, M. I. ARTSIS, and G. E. ZAIKOV

CONTENTS

ABSTRACT

Struggle against an atmospheric pollution represents a multidisciplinary problem. Modern problems of designing of gas-cleaning installation buildings are observed. The analysis of organizational-technical and scientific bases of the solution of a problem of clearing of gas emissions is made. Design on modernization of system of an aspiration of smoke fumes of kilns of limestone with use of the new scrubber which novelty is confirmed with the patent for an invention. Efficiency is increased and power inputs of spent processes of clearing of gas emissions and power savings at the expense of modernization of installation of clearing of gas emissions are lowered.

17.1 INTRODUCTION

Last years in Russia a row of the legislative deeds sharply raising the demands to protection of a free air [22–24] is published. Conditions of designing of buildings for clearing and sterilization of plant emissions have accordingly changed. Customers of designs – the industrial factories are aimed thereto that these buildings were highly effective, failsafe, fade-resistant and inexpensive, that is would possess high technical and economic indexes.

Parameters of any industrial structures are defined first of all by a technological level of their designs. Meanwhile, objectively it is necessary to recognize that designing of gas-cleaning installation buildings in Russia as a whole essentially loses world level and elimination of this defect occurs is inadmissible slow rates. Struggle against an atmospheric pollution and protection against aftereffects of this pollution represents a multidisciplinary problem. Organizational-technical and scientific bases of its solution are still far from perfect. It is possible with sufficient basis to tell that creation of basic new operating procedures with dust-laden gas streams business concerning the remote prospect. In the future it is necessary to be oriented on such ways of protection of air basin which are realizable means of modern techniques.

Main is a decrease in volumes of the flying emissions during the basic operating procedure. Designers of gas-cleaning installation buildings should pay attention that complexity of problems solved by them is quite often aggravated with weak responsibility of the persons designing the basic manufacture [1, 19, 21]. Not too seldom economic benefit attained in sphere the basic manufacture, is completely recoated by costs on clearing of huge volumes of the flying emissions. There are situations when building of gas-cleaning installation buildings in general appears almost impossible. At last, it is necessary to pay attention once again that gas-cleaning installation buildings cannot

be observed as panacea which will correct all flaws of the specialists devising production engineering of the basic manufacture. Unfortunately, such point of view existed throughout many years and as a result on a row of the factories the gas-cleaning installation buildings which still fairly have not attained term of the operational deterioration, have ceased to cope with the increased technological loading. The role and a place of gas-cleaning installation buildings in system of provisions on aerosphere protection consist in liquidating and neutralizing those emissions which formation cannot be prevented any preventive measures. Such statement of a question is dictated by elementary economic reasons. Under the world data, cost of gas-cleaning installation buildings makes from 10 to 40–45% (in certain cases even to 50%) in relation to cost basic dust-laden the equipment and, in connection with toughening of sanitary demands, tends to the further growth.

It is necessary to note defects, to correct which it is necessary the proximal years.

1. Insufficiency of the nomenclature a gas-cleaning installation and its lag from growing powers of the industry.
2. Weakness of design baseline in which predominates empiric the analysis.
3. Absence of strict scientific criteria for designing of gas-cleaning installation buildings with number of steps of clearing two and more. For the specified reason at designing of such buildings the big role is played by purely heuristic factor.
4. The weakest and not an authentic level of scrutiny of the questions connected with drawing by the flying emissions of a damage to a circumambient and, accordingly, with definition of economic benefit of liquidation of this damage.

These and other unresolved problems should be solved that who initiates today to master designing of gas-cleaning installation buildings.

17.2 PROBLEMS OF DESIGNING OF GAS-CLEANING INSTALLATION BUILDINGS

The gas-cleaning installation building consisting of a complex of consistently working apparatuses and communications represents the aerodynamic system possessing a row of features, namely:

1. In most cases, through a channel of clearing of gas emissions gas, and an aerosol (firm, liquid, mixed) moves not. Industrial aerosols always the unequigranular. If the hydrodynamics of dispersion streams in

general is difficult enough, in an unequigranular stream [12] it repeatedly becomes complicated.

2. At passage through a channel of clearing of gas emissions the aerosol stream continuously undergoes changes of speed and a traffic route, flows round obstructions and overcomes channels of a various configuration. Veering can be smooth, sudden and on any angle. Specific types of channels are foam layers on lattices of foamy apparatuses. As special aspects of obstructions cyclone separators where on a short piece of a way and for very small time the stream undergoes many affecting serve.
3. In gas-cleaning installation buildings often there are so-called *free flooded streams* [13]. They originate, for example, at an entry of a stream from a gas pipeline in the apparatus of much larger cross-section already filled with gas. Thus between an inducted stream of a basin (aerosol) and a mix in the apparatus there are the difficult contacts which result is quite often expressed in not design fall of aerosol corpuscles. The specified phenomenon not always is desirable. For example, in the scrubber of full transpiration used as the air conditioner, but not as the dust precipitator, the bottom part turns to "the dry" dust-collecting chamber.
4. Special complexity in aerodynamics a gas-cleaning installation a building consists that in very many cases it is necessary to observe dynamics of gas and dispersion phases of an aerosol separately [3, 10]. Aerosol corpuscles, except for the smallest (2–3 microns) owing to the time lag and long relaxation time have not time to follow behind gas phase motion, and deviate it, sometimes so sharply that it is in hardly probable not solving image influences parameters of dust removal apparatuses. Not taking into account of the given factor leads to coarse design errors and quite often forces to bring in expensive alterations to already working buildings.

At designing of gas-cleaning installation buildings it is necessary to analyze and solve three basic hydro aerodynamic problems: calculation of a water resistance of a channel, sampling of fan devices and definition of places of their arrangement; provision on all channel of clearing of gas emissions of a regime of motion of the gas (aerosol) in the best way answering to set conditions; provision uniform (within pass-off standards) distributions of gas and dust loading between apparatuses and in them.

17.3 FEATURES OF DESIGNING OF WET GAS-CLEANING INSTALLATIONS

A standard row of wet apparatuses is rather extensive and odd. Its only small part is issued in the form of the normalized rows, and serial exhaustion is restricted by only several types. Many apparatuses are made as the nonstandard equipment at the factories where it is possible to place the order, or is direct on an assembly site. In the latter case the workmanship appears low. Wet apparatuses can carry out following functions:

- to chill gas (aerosol); as alternative – with salvaging of warmth of an irrigation water which in process heats up;
- to moisten to (condition) – gas (aerosol) before its supply on clearing; this process is often carried out in a regime of full transpiration of a chilling liquid;
- to absorb gas or steam blending agents from gas emission or from an aerosol disperse medium;
- to trap firm and liquid corpuscles of a dispersoid of an aerosol.

These functions in many cases are carried out simultaneously though it can be not provided the clearing purposes. The account of collateral functions, including contradicting the designing purposes, is absolutely obligatory. They can change all kinetics of process and change properties of a circulating opening to such extent that it can demand its additional difficult machining.

In wet apparatuses occur heat-and-mass transfer and their completeness depends on character and intensity of contact of phases. Technological calculation of the wet apparatus in the fullest aspect develops of three independent calculations: heat exchange between an irrigation water and medium in the apparatus, a mass transfer (absorption of gases and steams a liquid, liquid transpiration) and trapping of aerosol corpuscles.

The product trapped in gas-cleaning installations, can be in two conditions:

- in the form of a liquid – if during clearing there was only an absorption of components of a gas phase of emission or if the fog, d.h. a dispersoid liquid an aerosol was trapped;
- in the form of sludge if in the wet apparatus there was a dust trapping, d.h. a dispersoid firm an aerosol.

The liquid or is accepted the factory which uses it at own discretion, or goes to manufacturing system of clearing of flows, or passes local clearing within a gas-cleaning installation building and again is fed on apparatus irrigation (the closed cycle of an irrigation).

Sludge is carried on a mud space where parches and can be used, or is passed through system of settlers and fine gauge strainers; after a filtering the liquid is refunded on an irrigation, and the filtered off weight (in the form of so-called cakes) is reclaimed.

17.4 CIRCULATING WATER SYSTEM

Process flow sheets of preparation of an opening are so various, how much various the problems solved by wet clearing of gas emissions. The complicating factor is discrimination of an irrigation (on regimes and a chemical compound of irrigation waters), both on separate apparatuses, and in one apparatus on different knots of an irrigation.

Irrigation of wet apparatuses without an opening reuse is applied now seldom as conducts to unfairly big charge of a liquid and reagents containing in it. In designs cyclicity of irrigation, d.h. n-fold use of the same opening with its deductions from a cycle and the additive of a fresh opening is usually provided. If hot gas and a circulating opening is exposed to clearing heats up, a cycle is built in heat-transfer apparatus.

At designing of sprinkling systems the momentous role is played by concept about a limiting condition of an opening. If to irrigate the wet apparatus with a circulating opening without removal of its part and the additive fresh after a while the opening condition expels its further use. The limiting condition can be defined by following factors:

1. If the opening traps a dispersoid firm an aerosol the suspended matter should not exceed concentration above which work of sprinklers is broken. Other criterion of a limiting condition in this case is inadmissible decrease in extent of the trapping, called by the big removal of a trapped product with splashes of the concentrated suspended matter.
2. At accumulation in an opening of some components, their crystallization on an internal surface of pipes, apparatuses, armatures is under certain conditions initiated, and it is accompanied also by sedimentation of inert suspended matters. The crystallization beginning means that there has stepped a limiting condition of an opening; its further use will lead to rapid driving down of elements of an irrigation system.
3. At absorption of steams or gases a limiting condition is such saturation of an opening at which its further use loses meaning: between an opening and an absorbed component balance is installed, and absorption stops.

It is necessary to pay attention of designers to condition which is specific to clearing of gas emissions. The removal in an aerosphere of a trapped blending agent (d.h. its reentrainment) occurs for two reasons: first, a blending agent quantity is not entrained by a liquid; secondly, the blending agent entrained by a liquid partially is taken out from the apparatus with a carryover.

17.5 WORKING OUT A BUILD OF A SCRUBBER AND THE RECOMMENDATION ABOUT DESIGNING

On the basis of the analysis of builds of modern apparatuses for gas clearing the dynamic spray scrubber build is devised and proprietary. The apparatus is supplied by twirled air swirler and the central pipe for supply of an irrigating liquid. A centrifugal force originating at twirl of a rotor, secures with liquid crushing on micro fogs that causes intensive contact of gases and trapped corpuscles to a liquid. Thanks to act of a centrifugal force, intensive mixing of gas and a liquid and presence of the big interface of contact, there is an effective clearing of gas in a bubble column [12].

At research operating conditions changed: speed of a gas stream and quantity of the liquid fed on an irrigation of the apparatus (Fig. 17.1). In the capacity of design data values of number of blades and air swirler critical bucklings varied.

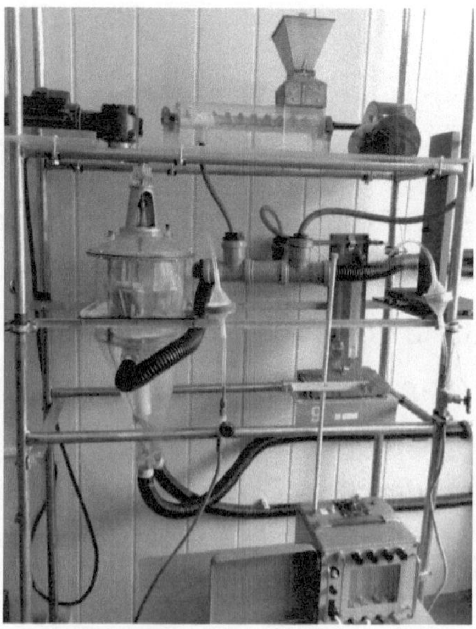

FIGURE 17.1 Experimental assembly.

Mechanically each of such apparatuses consists of contact channel fractionally entrained in a fluid and the drip pan merged in one body. The principle of performance of the apparatuses is based on intensive wash down of gases in contact channels of various configurations with the subsequent separation of a water gas flow in the drip pan. The used fluid is not discharged and recirculates several times for dust removal process.

Recommendations are resulted only on packaging a twirled air swirler:

1. Diameter of the apparatus is designed proceeding from productivity on gas.

$$D = 1,26\sqrt{\frac{Q}{W}}$$

2. Air swirler outside diameter is accepted equal $D' = (0.75 \div 0.85)\,D$.
3. The number of guide vanes is computed proceeding from diameter of an air swirlerz = (10÷25) D' and rounded off to the number convenient for staggered pitch length of a round on equal parts.
4. The length of blades is designed on a relationship.

$$l_\Lambda = \frac{0,5D_i \sin\dfrac{360}{z_i}}{\sin(\alpha_y + \dfrac{360}{z_i})}.$$

   And the angle of installation of blades a_y is recommended to be accepted 35÷45.
5. For the set size of corpuscles it is defined critical angular speed of twirl of an air swirler.

$$\omega_{opt} = 397.38 \cdot w^{1.65}(m \cdot 10^6)^{0.31}\breve{D}^{-0.31} \cdot \breve{z}^{1.05}\,\text{exp}\cdot$$
$$\cdot\exp\left[-0.018 \cdot 10^6 d_p - (1,06 + 0,034w) \cdot \cos\alpha - 2,18 \cdot \cos^2\alpha\right]$$

6. The air swirler direction of rotation is recommended such, at which 90°.
7. The air swirler water resistance is defined by formula

$$\Delta\rho_{\omega>0}=\frac{0,5\rho\omega D_1(0,5\omega D_1-W_1\cos\alpha)}{1+\dfrac{1,5+1,1\alpha/90^{\circ}}{z(1-d_1^{\ 2})}}$$

8. Power of the drive is roughly sized up on dependence:

$$P=Q\cdot\Delta\rho$$

17.6 CLEARING OF GAS EMISSIONS IN THE INDUSTRY

Results of researches have been implemented in practice. In manufacture of roasting of limestone at conducting of redesign of system of an aspiration of smoke gases of baking ovens. The devised scrubber is applied to clearing of smoke gases of baking ovens of limestone as secondary absorber.

Temperature of gases of baking ovens in main flue gas breeching before a exhaust-heat boiler 500–600°C, after exhaust-heat boiler 250°C. An average chemical compound of smoke gases (by volume): 17%$C0_2$; 16%N_2; 67% CO. Besides, in gas contains to 70 mg/m^3 SO_2; 30 mg/m^3 H_2S; 200 mg/m^3 F and 20 mg/m^3 CI. The gas dustiness on an exit from the converter reaches to 200/m^3 the Dust, as well as at a fume extraction with carbonic oxide after-burning, consists of the same components, but has the different maintenance of oxides of iron. In it than 1 micron, than in the dusty gas formed at after-burning of carbonic oxide contains less corpuscles a size less. It is possible to explain it to that at after-burning CO raises temperatures of gas and there is an additional excess in steam of oxides. Carbonic oxide before a gas heading on clearing burn in the special chamber. The dustiness of the cleared blast-furnace gas should be no more than 4 mg/m^3. The circuit design shown in Fig. 17.2 is applied to clearing of the blast-furnace gas of a dust.

Gas from a furnace mouth of a baking oven 1 on gas pipes 3 and 4 is taken away in the gas-cleaning plant. In raiser and duct gas is chilled, and the largest corpuscles of a dust, which in the form of sludge are trapped in the inertia sludge remover are inferred from it. In a centrifugal scrubber 5 blast-furnace gas is cleared of a coarse dust to final dust content 5–10/m^3 the Dust drained from the deduster loading pocket periodically from a feeding system of water or steam for dust moistening. The final cleaning of the blast-furnace gas is carried out in a dynamic spray scrubber where there is an integration of a finely divided dust. Most the coarse dust and drops of liquid are inferred from

gas in the inertia mist eliminator. The cleared gas is taken away in a collecting channel of pure gas 9, whence is fed in an aerosphere. The clarified sludge from a gravitation filter is fed again on irrigation of apparatuses. The closed cycle of supply of an irrigation water to what in the capacity of irrigations the lime milk close on the physical and chemical properties to composition of dusty gas is applied is implemented. As a result of implementation of trial installation clearings of gas emissions the maximum dustiness of the gases, which are thrown out in an aerosphere, has decreased with 3950 mg/m³ to 840 mg/m³, and total emissions of a dust from sources of limy manufacture were scaled down about 4800 to/a to 1300 to/a.

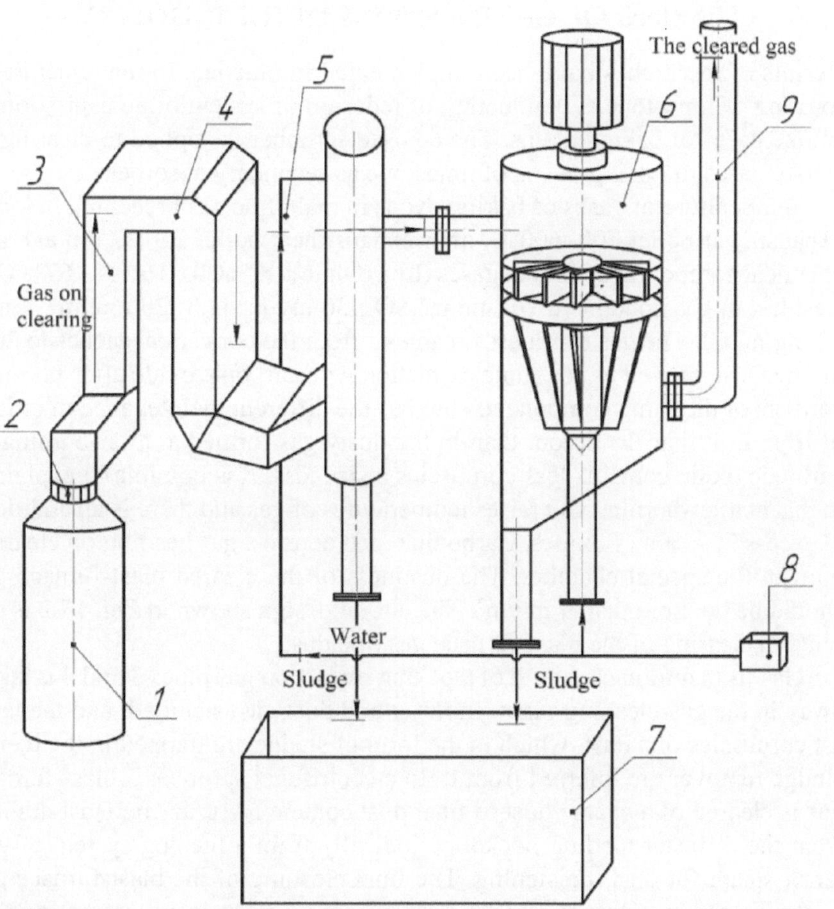

FIGURE 17.2 Process flowsheet of clearing of gas emissions.

Such method gives the chance to make gas clearing in much smaller quantity, demands smaller capital and operational expenses, reduces an atmospheric pollution and allows to use water recycling system.

17.7 CONCLUSION

1. Modern problems of designing of gas-cleaning installation buildings are observed. Specific features of process of clearing of gas emissions are in detail presented.
2. The solution of an actual problem on perfection of complex system of clearing of gas emissions and working out of measures on decrease in a dustiness of air medium of the industrial factories for the purpose of betterment of hygienic and sanitary conditions of work and decrease in negative affecting of dust emissions on a circumambient is in-process given.
3. Designs on modernization of system of an aspiration of smoke gases of calcinations of limestone with use of the new scrubber which novelty is confirmed with the patent for the invention are devised. Efficiency is raised and power inputs of spent processes of clearing of gas emissions and power savings at the expense of modernization of a process flow sheet of installation of clearing of gas emissions are lowered.

KEYWORDS

- **Smoke fumes**
- **The Closed cycle**
- **The Dynamic spray scrubber**
- **The Industry**
- **The Refire kiln**
- **Water recycling**

REFERENCES

1. Berezhinsky, A. I., & Homutinnikov, I. C. (1980). Salvaging, Cooling and Clearing of Smoke gases. M: Metallurgy.
2. Bogatich, C. A. (1978). Cyclonic and Foamy Apparatuses. L: Engineering industry.

3. Valdberg, A. J., Isjanov, L. M., & Tarat, E. J. (1985). Dust Separation Production Engineering. L.: Engineering Industry.
4. Effect of air pollutions on vegetation. The Reasons, Affecting, Retaliatory Measures. Under the Editorship of X. Dessler. M: The Forest industry, 1971.
5. The gas-cleaning installation Equipment. Bag Hoses: Catalogue. M: 1985.
6. Guderian, R. (1979). Pollution of air medium: Transfer with English M: The World.
7. Dubtchik, R. V. (1978). Rehash of a Waste of Aluminum Manufacture Abroad. M.
8. Idelchik, I. E. (1983). Fluid Kinetics of Industrial Apparatuses. M: Engineering Industry.
9. Idelchik, I. E., Alexanders, V. P., & Kogan, E. I. (1968). Research of direct-flow cyclone separators of system of an ash collection of a state district power station. *Heat Power Engineering, 8,* 45–48 (in Russian).
10. Lebeduk, K. E., et al. (1971). Methods of Clearing of Kiln Gases from a Dust. Chemical and Oil Engineering Industry. M.
11. Pazin, L. M, & Libina, V. L. (1977). Industrial and sanitary clearing of gases. *A Chemical and Oil Engineering Industry.* M, 2–3 (in Russian).
12. The patent 2339435 Russian Federations. Dynamic Spray Scrubber.R.R.Usmanova.27.11.2008.
13. Rakhmonov, T. Z., Salimov, S. C., & Umirov, R. R. (2005). Wet Clearing of Gases in Apparatuses with a Mobile Nozzle. T: The Fan.
14. Rusanov, A. A., Urbah, I. T., & Anastasiadi, A. P. (1969). Clearing of smoke gases in industrial power engineering. M: Energy.
15. Sagin, B. S., & Gudim, L. I. 1984 Dedusters with the Counter Twirled Streams. *The Chemical Industry*. Moscow, *8*, 50–54 (in Russian).
16. The directory after a heat – and to an ash collection. Under A. A. Rusanov's edition. M; Energy, 1975.
17. Stark, S. B. (1977). Dust Separation and Clearing of Gases in Metallurgy. M: Metallurgy.
18. Staritsky, V. A. (1973). The gas equipment of Factories of Ferrous Metallurgy. M: Metallurgy.
19. Ugov, V. N., & Valdberg, A. J. (1975). Preparation of Industrial Gases for Clearing. M: Chemistry.
20. Ugov, V. N., & Myagkov, B. I. (1970). Clearing of Industrial Gases by Fine Gauge Strainers. M: Chemistry.
21. Ugov, V. N., Valdberg, A. J., Myagkov, B. I., & Rashidov, I. K. (1981). Clearing of Industrial Gases of a Dust. M: Chemistry.
22. Directions and norms of technological designing and technical-and-economic indexes of a power equipment of the factories of ferrous metallurgy. Metal works. T. 18. Aerosphere Protection. Clearing Gases of a Dust. Sanitary code № 1-41-00. SSSR, 2001.
23. Directions for to dispersion calculation in an aerosphere of the harmful substances containing in emissions of the factory. Sanitary code 369-04. M: Building, 2005.
24. Economy of sterilization of gas emissions. *Chemical and Oil Engineering Industry.* M, 1979, *6*, 25.

CHAPTER 18

ENGINEERED NANOPOROUS MATERIALS: A COMPREHENSIVE REVIEW

AREZOO AFZALI and SHIMA MAGHSOODLOU

CONTENTS

ABSTRACT

Nanoporous solids are an important class of materials that have been studied extensively. Porous materials are networks of solid material, which contain void spaces. These materials can be further classified depending on the size of the pores present in the material. This chapter work brings fresh insights and improved understanding of nanoporous materials through introducing pore structure and different methods, which are used for porosity measuring. To achieve this purpose improved analysis methods for nanoporous materials are reviewed.

18.1 INTRODUCTION TO MORPHOLOGY AND POROSITY

Union of Pure and Applied Chemists (IUPAC) defines morphology as the "shape, optical appearance, or form of phase domains in substances, such as high polymers, polymer blends, composites, and crystals." As it can be seen, this is a very broad and diffuse definition so two classes of morphology are set apart in this work. Shape and bulk morphology are distinguished, because both are very applicable in the description of the porous networks. The former concerns the particle size, shape and pore structure, the latter classifies the polymers by the molecular architecture of the pore walls. Polymers have the advantage that they can be prepared in almost any micro or macroscopic size and shape. This allows extensive tuning of the shape morphology to the desired application, while keeping the bulk morphological parameters unchanged and vice versa.

TABLE 18.1 Examples of Size-Dependent Shape Morphology

Size range	Shape morphology
Nanometer	Polymer brush
	Micelle
	Microgel
	Pores
Micrometer	Powders
	Poly HIPE (high internal phase emulsion) pores
Macroscopic	Beads
	Discs
	Membranes

TABLE 18.2 Examples of Cross-Link-Dependent Bulk Morphology

Content of cross-links	Bulk morphology
None	Soluble polymer
	Supported polymer brush
Low	Swellable polymer gels
High	Polymer networks
Extra-high	Hyper cross-linked
	Polymers

In a classic view, porous matter is seen as a material that has voids through and through. The voids show a translational repetition in 3-D space, while no regularity is necessary for a material to be termed "porous." A common and relatively simple porous system is one type of dispersion classically described in colloid science, namely foam or, better, solid foam. In correlation with this, the most typical way to think about a porous material is as a material with gas-solid interfaces as the most dominant characteristic. This already presents that classical colloid and interface science as the creation of interfaces due to nucleation phenomena (in this case nucleation of wholes), decreasing interface energy, and stabilization of interfaces is of elemental importance in the formation process of nanoporous materials [1–4].

These factors are often crossed off due to the final products are stable (they are metastable). This metastability is due to the rigid character of the void-surrounding network, which is covalently cross-linked in most cases. However, it must be noticed that most of the porous materials are not stable by thermodynamic means. As soon as kinetic energy boundaries are overcome, materials start to breakdown [1].

Porous materials have been broadly exploited for use in a broad range of applications: for example, as membranes for separation and purification [5], as high surface-area adsorbants, as solid supports for sensors [6] and catalysts, as materials with low dielectric constants in the fabrication of microelectronic devices [7], and as scaffolds to guide the growth of tissues in bioengineering [8].

Porous materials occur widely and have many important usages. They can, for instance, offer a proper method of imposing fine structure on adsorbed materials. They can be applied as substrates to support catalysts and can act as highly selective sieves or cages that only allow access to molecules up to a certain size.

Food is often finely structured. Many biologically active materials are porous, as are many construction and engineering materials. Porous geological materials are of great interest; high porosity rock may contain water, oil or gas; low porosity rock may act as a cap to porous rock, and is of importance for active waste sealing.

18.2 POROSITY

Porosity φ is the fraction of the total soil volume that is taken up by the pore space. Thus it is a single-value quantification of the amount of space available to fluid within a specific body of soil. Being simply a fraction of total volume, φ can range between 0 and 1, typically falling between 0.3 and 0.7 for soils. With the assumption that soil is a continuum, adopted here as in much of soil science literature, porosity can be considered a function of position.

18.2.1 POROSITY IN NATURAL SOILS

The porosity of a soil depends on several factors, including (i) packing density, (ii) the breadth of the particle size distribution (polydisperse vs. monodisperse), (iii) the shape of particles, and (iv) cementing. Mathematically considering an idealized soil of packed uniform spheres, φ must fall between 0.26 and 0.48, depending on the packing. Spheres randomly thrown together will have φ near the middle of this range, typically 0.30 to 0.35. Sand with grains nearly uniform in size (monodisperse) packs to about the same porosity as spheres. In polydisperse sand, the fitting of small grains within the pores between large ones can reduce φ, conceivably below the 0.26 uniform-sphere minimum. Figure 18.1 illustrates this concept. The particular sort of arrangement required to reduce φ to 0.26 or less is highly improbable, however, so φ also typically falls within the 0.30–0.35 for polydisperse sands. Particles more irregular in shape tend to have larger gaps between their nontouching surfaces, thus forming media of greater porosity. In porous rock such as sandstone, cementation or welding of particles not only creates pores that are different in shape from those of particulate media, but also reduces the porosity as solid material takes up space that would otherwise be pore space. Porosity in such a case can easily be less than 0.3, even approaching 0. Cementing material can also have the opposite effect. In many soils, clay and organic substances cement particles together into aggregates. An individual aggregate might have 0.35 porosity within it but the medium as a whole has additional pore space in the form of gaps between aggregates, so that φ can be 0.5 or

greater. Observed porosities can be as great as 0.8 to 0.9 in a peat (extremely high organic matter) soil.

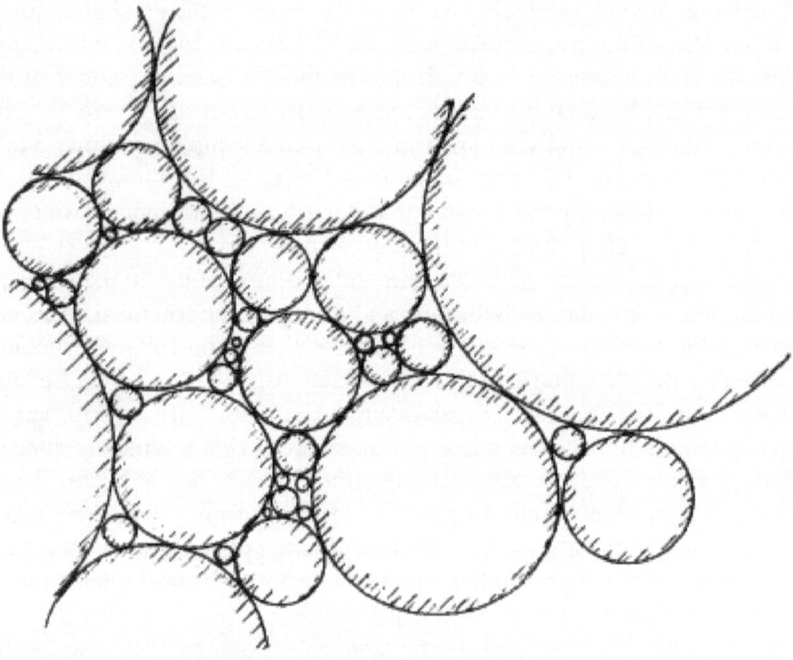

FIGURE 18.1 Dense packing of polydisperse spheres.

Porosity is often conceptually partitioned into two components, most commonly called textural and structural porosity. The textural component is the value the porosity would have if the arrangement of the particles were random, as described above for granular material without cementing. That is, the textural porosity might be about 0.3 in a granular medium. The structural component represents nonrandom structural influences, including macropores and is arithmetically defined as the difference between the textural porosity and the total porosity.

The texture of the medium relates in a general way to the pore-size distribution, as large particles give rise to large pores between them, and therefore is a major influence on the soil water retention curve. Additionally, the structure of the medium, especially the pervasiveness of aggregation, shrinkage cracks, worm-holes, etc. substantially influences water retention.

18.2.2 MEASUREMENT OF POROSITY

The technology of thin sections or of tomographic imaging can produce a visualization of pore space and solid material in a cross-sectional plane. The summed area of pore space divided by total area gives the areal porosity over that plane. An analogous procedure can be followed along a line through the sample, to yield a linear porosity. If the medium is isotropic, either of these would numerically equal the volumetric porosity as defined above.

The volume of water contained in a saturated sample of known volume can indicate porosity. The mass of saturated material less the oven-dry mass of the solids, divided by the density of water, gives the volume of water. This divided by the original sample volume gives porosity.

An analogous method is to determine the volume of gas in the pore space of a completely dry sample. Sampling and drying of the soil must be conducted so as not to com-press the soil or otherwise alter its porosity. A pycnometer can measure the air volume in the pore space. A gas-tight chamber encloses the sample so that the internal gas-occupied volume can be perturbed by a known amount while the gas pressure is measured. This is typically done with a small piston attached by a tube connection. Boyle's law indicates the total gas volume from the change in pressure resulting from the volume change. This total gas volume minus the volume within the piston, connectors, gaps at the chamber walls, and any other space not occupied by soil, yields the total pore volume to be divided by the sample volume.

To avoid having to saturate with water or air, one can calculate porosity from measurements of particle density and bulk density. From the definitions of bulk density as the solid mass per total volume of soil and particle density as the solid mass per solid volume, their ratio ρ_b / ρ_p is the complement of φ, so that:

$$\varnothing = 1 - \rho_b / \rho_p \tag{1}$$

Often the critical source of error is in the determination of total soil volume, which is harder to measure than the mass. This measurement can be based on the dimensions of a minimally disturbed sample in a regular geometric shape, usually a cylinder. Significant error can result from irregularities in the actual shape and from unavoidable compaction. Alternatively, the measured volume can be that of the excavation from which the soil sample originated. This can be done using measurements of a regular geometric shape, with the same problems as with measurements on an extracted sample. Additional methods, such as the balloon or sand-fill methods, have other sources of error.

18.2.3 PORES AND PORE-SIZE DISTRIBUTION: THE NATURE OF A PORE

As soil does not contain discrete objects with obvious boundaries, it could be called individual pores and the precise delineation of a pore unavoidably requires artificial, subjectively established distinctions. This contrasts with soil particles, which are easily defined, being discrete material objects with obvious boundaries. The arbitrary criterion required to partition pore space into individual pores is often not explicitly stated when pores or their sizes are discussed. Because of this inherent arbitrariness, some scientists argue that the concepts of pore and pore size should be avoided. Much valuable theory of the behavior of the soil-water-air system, however, has been built on these concepts, defined using widely, if not universally, accepted criteria.

18.3 POROUS MATERIALS

Porous materials are solid forms of matter permeated by interconnected or noninterconnected pores (voids) of different kinds: channels, cavities or interstices; that can be divided into several classes.

According to the nomenclature suggested by the IUPAC, porous materials are usually classified into three different categories depending on the lateral dimensions of their pores: microporous (<2 nm), mesoporous (between 2 and 50 nm) and macroporous (>50 nm) [9].

Liquid and gaseous molecules have been known to bring forward characteristic transport behaviors in each type of porous material. For example, mass transport can be obtained via viscous flow and molecular diffusion in a macroporous material, through surface diffusion and capillary flow in a mesoporous material and by activated diffusion in a microporous material.

Pores from the nanoscopic to the macroscopic scale are generated depending on the method. A summary of selected methods that can be applied to styrene-codivinylbenzene polymers is given in Table 18.3 [10].

The internal structural architecture of the void space potentially controls the physical and chemical properties, such as reactivity, thermal and electric conductivity, as well as the kinetics of numerous transport processes. The characterization of porous materials, therefore, has been of great practical interest in numerous areas including catalysis, adsorption, purification, separation, etc., where the essential aspects for such applications are pore accessibility, narrow pore size distribution (PSD), relatively high specific surface area and easily tunable pore sizes.

TABLE 18.3 Overview of Methods of Generating Porosity during Polymer Synthesis

Method	Porogene	Porosity	
		Accessibility	Typical size
Foaming	Gas, solvent, supercritical solvent	open/closed	100 nm – 1 mm
Phase separation	Solvent	Open	1 μm – 1 mm
High internal phase emulsion polym.	Emulsion	Open	10 μm – 100 μm
Soft templating	Molecules (solvent)	Micelles	Bicontinuous microemulsion
Hard templating	Colloidal crystals	Porous solids	Open

Ordered porous materials are recognized to be much more interesting because of the control over pore sizes and pore shapes. Their disordered counterparts exhibit high poly dispersity in pore sizes, and the shapes of the pores are irregular. Ordered porous materials seem to be much more homogeneous. In many cases a material possesses more than one kind of porosity. This could be for microporous materials: an additional meso- or macroporosity caused by random grain packing.

For mesoporous materials: an additional macroporosity caused by random grain packing, or an additional microporosity in the continuous network. For macroporous materials: an additional meso- and microporosity, these factors should be taken into consideration when materials are classified according to their homogeneity. A material possessing just one type of pore, even when the pores are disordered, might be more homogenous than one having just a fraction of nicely ordered pores.

Ordered porous solids contain a regularly arranged pore system and it is desired to design materials with different cylindrical, window-like, spherical or slit-like pore shapes and sizes [11].

18.3.1 PROPERTIES OF POROUS MATERIALS

Effective properties of porous materials with randomly dispersed pores represent various mechanical, physical, elastic, thermal and acoustic properties. A number of important properties of porous materials are:

- Porosity
- Specific surface area

- Permeability
- Breakthrough capillary pressure
- Diffusion properties of liquids in pores
- Pore size distribution
- Radial density function

18.3.2 MACROPOROUS MATERIALS AND THEIR USES

Macroporous materials are formed from the packing of monodisperse spheres (polystyrene or silica) into a three-dimensional ordered arrangement, to form face-centered cubic (FCC) or hexagonal close-packed (HCP) structures. The spaces between the packed spheres create a macroporous structure.

Glass or rubbery polymer includes a large number of macropores (50 nm–1 μm in diameter), which persist when the polymer is immersed in solvents or in the dry state.

Macroporous polymers are often network polymers produced in bead form. However, linear polymers can also be prepared in the form of macroporous polymer beads. They swell only slightly in solvents.

Macroporous polymers can be used, for instance, as precursors for ion-exchange polymers, as adsorbents, as supports for catalysts or reagents, and as stationary phases in size-exclusion chromatography columns.

Macroporous materials have many applications in the field of engineering due to their large effective surface area, and can be used for purposes such as filters, catalysts, supports, heat exchangers, and fuel cells. Although microporous and mesoporous materials also possess large surface areas, their small pore diameters (less than 10 nm) do not allow larger molecules to pass through them. Hence, macroporous materials with larger pore diameters are preferred and are of more practical use. Through colloidal crystal templating techniques, three-dimensionally ordered macroporous materials with uniform pore size can be successfully synthesized, thus improving the efficiency of transport of materials through the pores. Furthermore, photonic crystals possessing optical band-gaps can be synthesized from these macroporous structures by infiltrating the macroporous material with a precursor fluid, followed by removal of the original spheres through calcinations.

Photonic crystals are materials in which the dielectric constants vary periodically in space. Due to the alternating dielectric properties, photonic crystals are hence able to control the propagation of photons, by creating a frequency (band-gap) in which light is not able to propagate. Photonic crystals themselves have great potential use in the engineering field. Due to their ability to localize photons, photonic crystals can be used as wave guides in optical

fibers, which would prove very valuable in optical communications for the transfer of information. With the advent of information technology and the need for faster, quicker and more efficient data transmission, the importance and potential of photonic crystals are ever more apparent.

18.3.3 MESOPOROUS MATERIALS

Mesoporous solids consist of inorganic or inorganic/organic hybrid units of long-range order with amorphous walls, tunable textural and structural properties with highly controllable pore geometry and narrow pore size distribution in the 2–50 nm range [12].

The pores can have different shapes such as spherical or cylindrical and be arranged in varying structures, (see Fig. 18.2). Some structures have pores that are larger than 50 nm in one dimension, see e.g. the two first structures in (Fig. 18.2), but there the width of the pore is in the mesorange and the material is still considered to be mesoporous.

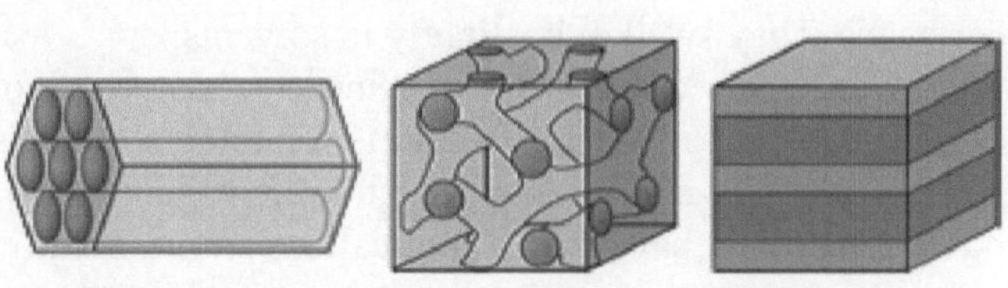

FIGURE 18.2 Different pore structures of mesoporous materials.

Mesoporous materials can have a wide range of compositions but mainly consists of oxides such as SiO_2, TiO_2, ZnO_2, Fe_2O or combinations of metal oxides, but also mesoporous carbon can be synthesized[13–16]. Most commonly is to use a micellar solution and grow oxide walls around the micelles. Both organic metal precursors such as alkoxides [17, 18] as well as inorganic salts such as metal chloride salts [14] can be used. Alternatively a mesoporous template can be used to grow another type of mesoporous material inside it. This is often used for synthesizing e.g. mesoporous carbon [16, 19, 20].

18.3.4 MICROPOROUS MATERIALS

Porous materials are networks of solid material, which contain void spaces. These materials can be further classified depending on the size of the pores present in the material. Microporous solids are materials that contain permanent cavities with diameters of less than 2 nm. Mesoporous materials contain

pores ranging from 2 nm to 50 nm and macroporous materials contain pores of greater than 50 nm [21]. The field of microporous materials contains several classes which are well known [22], including naturally occurring zeolites, activated carbons and silica. Synthetic microporous solids have recently emerged as a potentially important class of materials.

These include Metal Organic Frameworks (MOFs), Microporous Organic Polymers (MOPs) including Covalent Organic Frameworks (COFs) and Polymers of Intrinsic Microporosity (PIMs)[23]. It is the very large surface areas and very small pore sizes of these materials, which make them of specific interest. These two factors permit microporous materials to be useful in applications such as heterogeneous catalysis, separation chemistry, and potential uses in hydrogen or other gas storage [24, 25]. Most synthetic strategies to prepare microporous materials consist of linking together smaller units with di-topic or poly topic functionalities in order to form extended networks much like is displayed in the general diagram of Fig. 18.3.

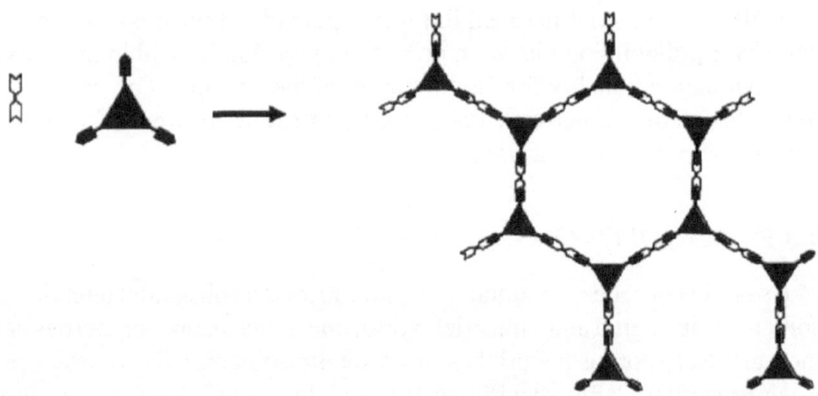

FIGURE 18.3 General schematic showing the linking of polytopic building blocks to form synthetic networks.

Whether a network formed is crystalline or amorphous is generally governed by whether the covalent bonds being formed involve reversible chemistry or not. Crystalline networks are typically formed under reversible reaction conditions that allow error corrections during network formation and produce thermodynamically stable networks. These types of reactions are commonly condensation reactions [26, 27]. On the other hand, when irreversible reactions are employed, such as cross couplings, the networks tend to form in

a disordered manner [28, 29] resulting in amorphous materials. This is because a carbon-carbon bond is formed irreversibly under conditions such as a Sonagashira or Yamamoto couplings resulting in amorphous networks. This is of course unless some other templating measure is taking into account while considering the reaction conditions [30].

Specified factors such as temperature, solvent and solvent-to-head-space ratio play an important role in the formation of a crystalline framework [27]. Apparent solvents can be employed in order to form ordered networks by means of their ability to dissolve the monomer building blocks. If the concentration of a monomer in solution is controlled by a solvent in which it is slightly soluble, then a network is more likely to form under thermodynamic, instead of kinetic control [29]. Solvents could also be used based on their molecular size to act as templates for pores to form around [30]. While this idea of MOP/COF templating is generally understood in qualitative terms (which solvents produce crystalline networks) there has been no research into the quantitative effects (what solvent ratio is required to produce a well-structured network).

Finally, if a material is to exhibit micro porosity, it must be composed of somewhat rigid building blocks in order to impart rigidity within the network and provide directionality for the formation of the network. This rigidity prevents collapse of the network upon itself and results in free volume which becomes the pores within a framework.

18.3.5 NANOPOROUS POLYMERS

In the past few decades, nanomaterials have received substantial attention and efforts from academic and industrial world, due to the distinct properties at the nanoscale. Nanoporous materials as a subset of nanomaterials possess a set of unique properties: large specific surface volume ratio, high interior surface area, exclusive size sieving and shape selectivity, nanoscale space confinement, and specific gas/fluid permeability. Moreover, pore-filled nanoporous materials can offer synergistic properties that can never be reached by pure compounds. As a result, nanoporous materials are of scientific and technological importance and also considerable interest in a broad range of applications that include templating, sorting, sensing, isolating and releasing.

Nanoporous materials can be classified by pore geometry (size, shape, and order) or distinguished by type of bulk materials. Nanoporous materials are considered uniform if the pore size distribution is relatively narrow and the pore shape is relatively homogenous. The pores can be cylindrical, conical, slit-like, or irregular in shape. They can be well ordered with an alignment as

opposed to a random network of tortuous pores. Nanoporous materials cover a wide variety of materials, which can be generally divided into inorganic, organic and composite materials. The majority of investigated nanoporous materials have been inorganic, including oxides, carbon, silicon, silicate, and metal. On the other side, polymers have been identified as materials that offer low cost, less toxicity, easy fabrication process, diverse chemical functionality, and extensive mechanical properties [31, 32].

Naturally, the success of inorganic materials to form nanoporous materials has promoted the development of analogous polymers. More importantly, advances in polymer synthesis and novel processing techniques have led to various nanoporous polymers.

18.3.5.1 NANOPOROUS MATERIALS CONNECTION

A whole variety of nanoporous materials in nature can be found in many different functions. The most common task for nanoporous materials in nature is to make inorganic material much lighter while preserving or improving the high structural stability of these compounds. Often, by filling the voids between inorganic matters the desired properties of the hybrid materials exceed the performances of the pure compounds by several orders of magnitude. Nanoporous materials in nature are organic-inorganic hybrids. Naturally occurring materials exhibit synergistic properties. Neither the organic material filling the void nor the inorganic network materials are able to achieve comparable performances by themselves [33–35].

It is seen that complex mechanisms are involved in the formation of these hierachical materials. Similar to the structure motives on different length-scales cells, vesicles, supra-molecular structures, and biomolecules are involved in the structuring process of inorganic matter occurring in nature. This process is commonly known as biomineralization [34].

It is often not seen in this relation, but it will be shown later that ordered porous materials, and therefore artificial materials, are constructed according to very similar principles. A completely different area where nanoporous materials are highly important is in the lungs, where foam with a high surface area permits sufficient transfer of oxygen to the blood. Even the most recent developments in nanoporous materials, such as their application as photonic materials are already present in nature, the color of butterfly wings, for instance, originates from photonic effects [36, 37].

It can be concluded that nature applies the concept of nanoporous materials (either filled or unfilled) as a powerful tool for constructing all kinds of materials with advanced properties. So it is not surprising how much research

has recently been devoted to porous materials in different areas such as chemistry, physics, and engineering. The current interests in nanoporous materials are now far behind their size-sieving properties.

18.3.5.2 CLASSIFICATION BY NETWORK MATERIAL

One of the most important goals in the field of nanoporous materials is to achieve any possible chemical composition in the network materials "hosting" the pores. It makes sense to divide the materials into two categories: (a) inorganic materials and (b) organic materials. Among the in organic materials, the larger group, it could be found:

(i) Inorganic oxide-type materials. This is the field of the most commonly known porous silica, porous titania, and porous zirconia materials.
(ii) A category of its own is given for nanoporous carbon materials. In this category are the highly important active carbons and some examples for ordered mesoporous carbon materials.
(iii) Other binary compounds such as sulfides, nitrides. Into this category also fall the famous AlPO materials.
(iv) There are already some examples in addition to carbon where just one element (for instance, a metal) could be prepared in a nanoporous state. The most appropriate member of this class of materials is likely to be nanoporous silicon, with its luminescent properties [38, 39]. There are far fewer examples of nanoporous organic materials, such as polymers [40].

18.3.5.3 SUMMARY OF CLASSIFICATIONS

Three main criteria could be defined:

1. size of pores;
2. type of network material;
3. state of order: ordered or disordered materials.

18.4 ORIGIN AND CLASSIFICATION OF PORES IN SOLIDMATERIALS

Solid materials have a cohesive structure, which depends on the interaction between the primary particles. The cohesive structure leads indispensably to a void space, which is not occupied by the composite particles such as atoms, ions, and line particles. Consequently, the state and population of such

voids strongly depends on the interparticle forces. The interparticle forces are different from one system to another; chemical bonding, Vander Waals force, electrostatic force, magnetic force, surface tension of adsorbed films on the primary particles, and so on. Even the single crystalline solid, which is composed of atoms or ions has intrinsic voids and defects. Therefore, pores in solids are classified into intraparticle pores and interparticle pores (Table 18.4) [41].

TABLE 18.4 Classification of Pores from Origin, Pore Width, and Accessibility

Origin and structure		
Intraparticle pore	Intrinsic intraparticle pore	Structurally intrinsic type
	Extrinsic intraparticle pore	injected intrinsic type
		Pure type
		Pillared type
Interparticle pore	Rigid interparticle *pore* (Agglomerated)	
	Flexible interparticle pore (Aggregated)	
Pore width		
Macropore	50 nm	
Mesopore	2 nm<<50 nm	
Micropore	W< 2 nm	
	Supermicropare, 0.7 < W 2 nm	
	Ultramicropore, W nm	
	(Ultrapore, w0.35 nm in this review)	
Accessibility to surroundings		
Open pore	Communicating with external surface	
Closed pore	No communicating with surroundings	
Latent pore	Ultrapore and closed pore	

18.4.1 TRAPARTICLE PORES

18.4.1.1 INTRINSIC IN TRAPARTICLE PORE

Zeolites are the most representative porous solids whose pores arise from the intrinsic crystalline structure. Zeolites have a general composition of Al, Si, and O, where Al-O and Si-O tetrahedral units cannot occupy the space perfectly and therefore produce cavities. Zeolites have intrinsic pores of different connectivities according to their crystal structures [42]. These pores may be named intrinsic crystalize pores. The carbon nanotube has also an intrinsic crystalline pore [43].

Although all crystalline solids other than zeolites have more or less intrinsic crystalline pores, these are not so available for adsorption or diffusion due to their isolated state and extremely small size.

There are other types of pores in a single solid particle. -FeOOH is a precursor material formagnetic tapes, a main component of surface depositsand atmospheric corrosion products of iron-based alloys, and a mineral. The -FeOOH microcrystal is of thin elongated plate [44].

These new created intrinsic intraparticle pores should have their own name different from the intrinsic intraparticle pore. The latter is called a structurally intrinsic intraparticle pore, while the former is called an injected intrinsic intraparticle pore.

18.4.1.2 EXTRINSIC INTRA-PARTICLE PORE

When a foreign substance is impregnated in the parent material in advance this is removed by a modification procedure [45]. This type of pores should be called extrinsic intraparticle pores. Strictly speaking, as the material does not contain other components, extrinsic pure intraparticle pore is recommended. However, the extrinsic intrapore can be regarded as the interparticle pore in some cases.

It can be introduced a pore-forming agent into the structure of solids to produce voids or fissures which work as pores. This concept has been applied to layered compounds in which the interlayer bonding is very weak; some inserting substances wells the interlayer space. Graphite is a representative layered compound; the graphitic layers are weakly bound to each other by the Vander Waals force [46]. If it heated in the presence of intercalants such as K atoms, the intercalants are inserted between the interlayer spaces to form a long periodic structure. K-intercalated graphite can adsorb a great amount of H_2 gas, while the original graphite cannot [47]. The interlayer space opened

by intercalation is generally too narrow to be accessed by larger molecules. Intercalation produces not only pores, but also changes the electronic properties. Montmorillonite is a representative layered clay compound, which swells in solution to intricate hydrated ions or even surfactant molecules [48, 49]. Then some pillar materials such as metal hydroxides are intricate in the swollen interlayer space under wet conditions. As the pillar compound is not removed upon drying, the swollen structure can be preserved even under dry conditions.

The size of pillars can be more than several nm, being different from the above intercalants. As the graphite intercalation compounds and pillar ones need the help of foreign substances, they should be distinguished from the intrinsic intraparticle pore system. Their pores belong to extrinsic intraparticle pores. As the intercalation can be included in the pillar formation, it is better to say that both the pillared and intercalated compounds have pillared in traparticle pores [50].

18.4.2 INTER-PARTICLE PORES

Primary particles stick together to form a secondary particle, depending on their chemical composition, shape and size. In colloid chemistry, there are two gathering types of primary particles. One is aggregation and the other is agglomeration.

The aggregated particles are loosely bound to each other and the assemblage can be readily broken down. Heating or compressing the assemblage of primary particles brings about the more tightly bound agglomerate [51].

There are various interaction forces among primary particles, such as chemical bonding, Vander Waals force, magnetic force, electrostatic force, and surface tension of the thick adsorbed layer on the particle surface. Sintering at the neck part of primary particles produces stable agglomerates having pores. The aggregate bound by the surface tension of adsorbed water film has flexible pores. Thus, interparticle pores have wide varieties in stability, capacity, shape, and size, which depend on the packing of primary particles. They play an important role in nature and technology regardless of insufficient understanding. The interparticle pores can be divided into rigid and flexible pores. The stability depends on the surroundings. Almost all interparticle pores in agglomerates are rigid, whereas those in aggregates are flexible. Almost all sintered porous solids have rigid pores due to strong chemical bonding among the particles. The rigid interparticle porous solids have been widely used and have been investigated as adsorbents or catalysts. Silica gel

is a representative of interparticle porous solids. Ultrafine spherical silica particles form the secondary particles, leading to porous solids [52, 53].

18.5 STRUCTURE OF PORES

The pore state and structure greatly depend on the origin. The pores communicating with the external surface are named open pores, which are accessible for molecules or ions in the surroundings. When the porous solids are insufficiently heated, some parts of pores near the outer shell are collapsed inducing closed pores without communication to the surroundings. Closed pores also remain by insufficient evolution of gaseous substance. The closed pore is not associated with adsorption and permeability of molecules, but it influences the mechanical properties of solid materials, the new concept of latent pores is necessary for the best description of the pore system. This is because the communication to the surroundings often depends on the probe size, in particular, in the case of molecular resolution porosimetry. The open pore with a pore width smaller than the probe molecular size must be regarded as a closed pore. Such effectively closed pores and chemically closed pores should be designated the latent pores [54]. The combined analysis of molecular resolution porosimetry and small angle X-ray scattering (SAXS) offers an effective method for separated determination of open and closed (or latent) pores, which will be described later. The porosity is defined as the ratio of the pore volume to the total solid volume[55].

The geometrical structure of pores is of great concern, but the three-dimensional description of pores is not established in less-crystalline porous solids. Only intrinsic crystalline intraparticle pores offer a good description of the structure. The hysteresis analysis of molecular adsorption isotherms and electron microscopic observation estimate the pore geometry such as cylinder (cylinder closed at one end or cylinder open at both ends), cone shape, slit shape, interstice between closed-packing spheres and inkbottle. However, these models concern with only the unit structures. The higher order structure of these unit pores such as the network structure should be taken into account. The simplest classification of the higher order structures is one-, two- and three-dimensional pores. Some zeolites and alumino phosphates have one-dimensional pores and activated carbons have basically two-dimensional slit-shaped pores with complicated network structures [56].

The IUPAC has tried to establish a classification of pores according to the pore width (the shortest pore diameter), because the geometry determination of pores is still very difficult and molecular adsorption can lead to the reliable parameter of the pore width. The pores are divided into three categories:

macropores, mesopores, and micropores, as mentioned above. The fact that nanopores are often used instead of micropores should be noted.

These sizes can be determined from the aspect of N, adsorption at 77 K, and hence N_2 molecules are adsorbed by different mechanisms multilayer adsorption, capillary condensation, and micropore filling for macropores, mesopores, and micropores, respectively. The critical widths of 50 and 2 nm are chosen from empirical and physical reasons. The pore width of 50 nm corresponds to the relative pressure of 0.96 for the N_2 adsorption isotherm. Adsorption experiments above that are considerably difficult and applicability of the capillary condensation theory is not sufficiently examined. The smaller critical width of 2 nm corresponds to the relative pressure of 0.39 through the Kelvin equation, where an unstable behavior of the N, adsorbed layer (tensile strength effect) is observed. The capillary condensation theory cannot be applied to pores having a smaller width than 2 nm. The micropores have two subgroups, namely ultra-micropores (0.7 nm) and super-micropores ($0.7 \text{ nm} < w < 2 \text{ nm}$). The statistical thickness of the adsorbed N2 layer on solid surfaces is 0.354 nm. The maximum size of ultra-micropores corresponds to the bi-layer thickness of nitrogen molecules, and the adsorbed N_2 molecules near the entrance of the pores often block further adsorption. The ultra-micropore assessment by N_2 adsorption has an inevitable and serious problem. The micropores are divided into two groups.

Recently the molecular statistical theory has been used to examine the limitation of the Kelvin equation and predicts that the critical width between the micropore and the mesopore is 1.3–1.7 nm (corresponding to 4–5 layers of adsorbed N_2), which is smaller than 2 nm [57].

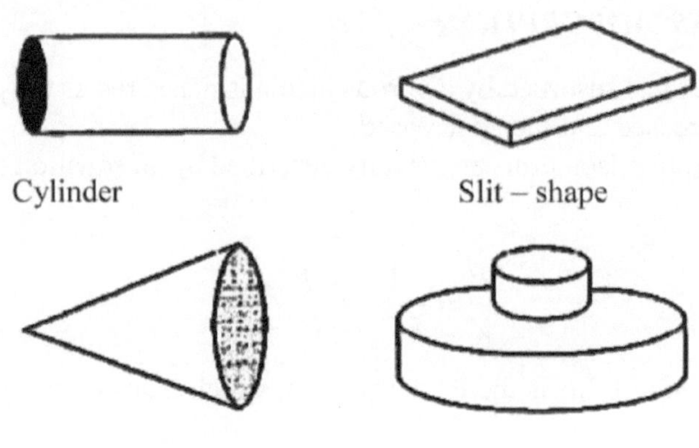

FIGURE 18.4 Different pore shapes

18.6 POROSITY AND PORE SIZE MEASUREMENT TECHNIQUES ON POROUS MEDIA

Crushing measure the volume of the porous material, crush it to remove the void space, and remeasure the volume.

- Optically this may involve filling the pores with a material such as black wax or Wood's metal, sectioning and inspecting with a microscope or scanning electron microscope.
- Imbibition weighing before and after filling the pores with a liquid.
- Gas Adsorption measure the change in pressure as a gas is adsorbed by the sample.
- Mercury Intrusion Measure the volume of mercury forced into the sample as a function of pressure.
- Thermo porosimetry fill the pores with a liquid, freeze it, then measure the heat evolved as the sample is warmed, until all the liquid is melted.
- NMR Cryoporometry fill the pores with a liquid, freeze it, then measure the amplitude of the NMR signal from the liquid component as the sample is warmed, until all the liquid is melted.
- Small Angle Neutron Scattering (SANS) scatter neutrons from the pores, then the smaller the dimensions of the variations in density distribution, the larger the angle through which the neutrons will be scattered.

Many of these methods give results that quite frequently differ from one another. This is often because they are in fact measuring different things some measurements are directly on the pores themselves. Others (such as mercury intrusion) are in effect measuring the necks that give access to the pores [58].

18.6.1 GAS ADSORPTION

The volume of gas adsorbed by a porous media is measured as the pressure of the gas is increased and then decreased.

Gas adsorption isotherms are usually described by an equation of the form [59]:

$$\frac{P}{V} = \frac{1}{bV_m} + \frac{P}{V_m} \tag{2}$$

An extension of this is the BET equation, which treats multilayer adsorption [60]:

$$\frac{x}{V.(1-x)} = \frac{1}{c.V_m} + \frac{(c-1).x}{c.V_m} \tag{3}$$

Where $x = \frac{p}{p_0}$.

One may go further and calculate the specific surface area of the solid:

$$S = \frac{N.\sigma^0.V_M}{M_v} \tag{4}$$

The gram molecular volume is 22.41. In this chapter, pores diameters determined by gas adsorption to establish a relationship between the melting point depression (as measured by NMR cryoporometry) and the pore diameter in the porous medium.

Gas adsorption commonly uses the BJH method based on the Kelvin equation to calculate pore size distributions from the desorption P(v) curve, using a pore model of right cylinders [61–63]:

$$RT\,Ln\frac{P}{P_0} = -2\gamma.\frac{V_M}{R_k}.Cos(\phi) \tag{5}$$

where R = gas constant = 8.314 J° K^{-1} mol^{-1}; T = boiling point of nitrogen = 77.3 K; = angle of contact between liquid and walls = 0°.

For Nitrogen: V_M = 34.6 mL. mol^{-1}; = $8.855.10^{-3}$ Nm^{-1}

This is for cylindrical geometry, and may be written in the form [63]:

$$\frac{R_k}{2} = -\frac{\gamma V_M}{RT\,Ln\frac{P}{P_0}}.Cos(\phi) \tag{6}$$

This may be recast into a more general form for other pore geometries:

$$\frac{v_p}{a_p} = -\frac{\gamma V_M}{RT\,Ln\frac{P}{P_n}}.Cos(\phi) \tag{7}$$

Thus for nonintersecting spherical pore geometry one obtains:

$$\frac{v_p}{a_p} = \frac{R_k}{3} \tag{8}$$

For the spherical pore geometry with multiple throats with 50% of the pore area intact, one obtains:

$$\frac{v_p}{a_p} = \frac{2R_k}{3} \tag{9}$$

18.6.2 MERCURY INTRUSION

In mercury intrusion the volume of mercury forced into the sample is measured as a function of the applied pressure. Since this capillary pressure is related to the diameter of the neck leading to the pores, rather than the pore diameters themselves, the results are not strictly comparable with gas adsorption.

If dV is the volume of pores having entry diameters between D_e and D_e + dD_e [64]:

$$dV = \alpha(D_e).dD_e \tag{10}$$

It can be written as:

Mercury intrusion calibration has not been used, with the exception of providing nominal pore diameters for large pore diameter.

18.6.3 THERMOPOROSIMETRY (DTA/DSC)

Differential Thermal Analysis or Differential Scanning Calorimetry is a technique that can be used to study the behavior of liquids in pores, in a similar manner to NMR cryoporometry. When applied to porous materials, the names Thermoporometry, Thermoporosimetry are frequently used [65–67].

The sample and liquid () is placed in a small sealed capsule, and a thermocouple used to monitor the difference in temperature to an empty capsule, as both are smoothly first reduced in temperature and then increased in temperature.

The signal obtained is not directly proportional to the volume of liquid melted, as in the case of NMR cryoporometry, but is a measure of the heat absorbed/emitted by the sample. It is divided by the thermal mass of the sample (i.e., different porous substrates will have different scale factors) and is multiplied by the temperature-scanning rate.

Further, in the case of very small pore sizes with deep super coolings, the amplitude of the transition becomes so small and spread-out that it is often not

seen, where NMR cryoporometry obtains clear transitions of very good signal-to-noise, as one goes from: no melted liquid to all liquid in pores melted' over a wide temperature range.

18.7 MODERN ANALYSIS OF NANOPOROUS MATERIALS

18.7.1 GAS ADSORPTION METHODS

Gas sorption represents a widely used technique for characterizing micro and mesoporous materials and provides porosity parameters such as pore size distributions, surface areas, and pore volumes. In the following, a brief description of classical methods of analyzing sorption data is followed by an overview of recent advances in the interpretation of sorption experiments.

In a typical sorption experiment the uptake of gasses such as nitrogen, krypton, and CO_2 is measured as a function of relative pressures $p/p_0<1$ at constant temperature. p and p0 are the equilibrium vapor pressures of the liquid in the pores and that of the bulk liquid, respectively. The interaction between the pore walls and the adsorbate is based on physical sorption (Vander Waals interaction) and leads to the formation of adsorbate layers at low p/p_0.

Simplistically, the macroscopic laws of classical thermodynamics predict that the confinement of pores with radii on the nanometer scale leads to the condensation of gas inside the pores at a pressure smaller than p. In a typical sorption experiment the adsorbed volume is plotted versus p/p_0, and this "sorption isotherm" is the superposition of different uptake mechanisms. At low p/p_0adsorption in micropores takes place, which is supposed to be a process of volume filling rather than capillary condensation [68, 69].

The shape of an isotherm itself distinguishes between representative types of nanoporous materials, based on the classifications by IUPA Cor de Boer [12, 70].

The isotherms of microporous materials are characterized by a steep increase of the isotherm at low p/p_0, ending up in a plateau at larger p/p_0. In a mesoporous substrate, with increasing values of p/p_0, a liquid-like adsorbate film of statistical thickness $t(p/p_0)$ is formed on the pore walls. At a certain pressure, capillary condensation takes place, filling the mesopores with liquid, which is apparent in isotherms as a pronounced increase in the adsorbed amount. The total pore volume ("porosity") is given by the overall uptake of adsorbate.

The first explanation of the capillary condensation in a single infinite cylindrical mesopore based on the macroscopic Kelvin equation was introduced as:

$$ln(p / p_0) = 2\sigma V_L / RTr_m \tag{12}$$

Based on the classical treatment of Cohan and the Kelvin equation, the condensation of a liquid in a nanoporous material at a certain p/p_0 can be related to the corresponding mesopore size, thus also providing a pore size distribution (PSD). Since the condensation starts at a relative pressure p/p_0, where the walls are covered by a film of thickness $t(p/p_0)$, in the so-called modified Kelvin equation $2/r_m$ is replaced by

$$f / (r - t(\frac{p}{p_o}) \tag{13}$$

where r is the "true" mesopore radius and f is the meniscus shape factor, which is 1 or 2 for the filling or emptying of the mesopore, respectively. This procedure represents the basis of the well-known "Barrett-Joyner-Halenda (BJH)"method, currently the procedure most frequently used to determine PSDs.

The dependence $t(p/p_0)$ can be described by the approach of Frenkel-Halsey-Hill describing the sorption on nonporous silica [71].

An improved treatment takes into account the influence of surface forces on adsorbed film equilibrium and stability, which leads to predictions for capillary condensation and desorption pressures that are substantially different from those of Cohan' s theory [72]. In addition, the pore geometry significantly affects thermodynamic properties of confined fluids and their adsorption behavior [73–75].

The confinement is stronger in spherical pores compared with cylindrical pores of the same diameter, leading to a shift of the capillary condensation to lower p/p_0.

Broekh off and de Boer described the condensation in spherical ink-bottle pores, connected by narrow cylindrical windows, based on the Kelvin-Cohan approach, also predicting the appearance of hysteresis [75].

The progress in synthesizing mesoporous materials with well-defined pore morphologies in the terms of uniform mesopore sizes and pore shapes allowed the testing, optimization, and further development these classical approaches. Based on the BJH method, the mesopore size cold be determined by using the "KJS" approach [76, 77], which corrects the Kelvin equation by an empirical additive constant (0.3) for the mesopore radius, which, however, does not have a theoretical foundation:

$$r(p / p_0)[nm] = 2\sigma V_L / RT \ln(p / p_0) + t(p / p_0) + 0.3 \tag{14}$$

A further approach to calculating mesopore sizes is based on the Gurvich approach [69]:

$$D = 4V_{0.4} / S_{BET} \tag{15}$$

The specific surface area can be obtained from the method introduced by Brunauer, Emmet, and Teller (BET).

The BET method is based on the assumption that multiplayer formation takes place prior to capillary condensation and that the equilibrium state is characterized by different rate constants for adsorption and desorption for the mono-and multilayer. The adsorbed amount, p/p_0, and the monolayer capacity are related by the BET equation, from which the specific surface area is calculated as:.

Further widely used methods for determining structural parameters are comparative plots such as the t-plot anda-plot methods [69, 78, 79].

Typically, the amount adsorbed on the porous solid under study is plotted as a function of the amount adsorbed on an ideally nonporous reference solid with similar surface characteristics, providing parameters such as the overall pore volume, specific surface area, and micropore volumes. This procedure has been used to determine microporosity in mesoporous material to show nonnegative intercepts in the comparative plot [71, 80, 81].

The procedures described above are not appropriate for the characterization of microporous materials, particularly microporous carbons and zeolites. Among the most frequently used evaluation procedures are the phenomenological models based on Dubinin's theory of volume filling of micropores, such as the Dubinin-Radushkevich (DR), Dubinin-Astakhov, and Dubinin-Stoeckli equations. Another approach is the Saito-Foley method, which is an extension of the Horvath-Kawazoe method [82–85].

It has turned out that the macroscopic thermodynamics of the classical methods described above do not provide reliable descriptions of materials with mesopore sizes below about 4 nm for oxidic materials [86].

One of the main shortcomings of these approaches lies in the non-consideration of fluid-wall interactions. Recent progress in under-standing capillary condensation deals with molecular level models. The methods of the grand canonical Monte Carlo (GCMC) simulations [87], molecular dynamics [88], and density functional theory (DFT) allow direct modeling of capillary condensation/desorption phase transitions and are capable of generating hysteresis loops of simple fluids sorbed in model pores. Neimark and Ravikovitch have shown that the nonlocal density functional theory (NLDFT), with properly

chosen parameters of fluid-fluid and fluid-solid intermolecular interactions, quantitatively predicts desorption branches of hysteretic isotherms [89–91].

This method was tested against Monte Carlo simulations and was shown to provide reliable pore sizes and wall thicknesses in porous materials.

Moreover, the NLDFT predictions of equilibrium and spontaneous capillary condensation transitions for pores wider than 6 nm were well approximated by the macroscopic equations of the Derjaguin-Broekhoff-de Boer theory, while the results of the traditional Cohan equation (BJH method)were shown to be significantly in error [75].

In spite of the recent progress in the theoretical understanding of sorption phenomena, certain issues of sorption are still unclear:

1. The nature of the hysteresis in mesoporous materials is still subject to intensive theoretical and experimental research and is not yet fully understood. While certain materials with small mesopores exhibit equilibrium capillary condensation and show pronounced hysteresis loops with parallel adsorption and desorption branches [71, 80, 92].
   In contrast, other types of mesoporous materials show a steep decrease in the desorption curve [93]. It was concluded that hysteresis is a function of the pore size and the temperature [94].
2. The simultaneous presence of mesopores and irregular micropores in the walls greatly prevents the evaluation of surface areas and the quantification of the micropore volumes. Although it was shown that the use of comparative plots underestimates the microporosity, t-plots and a-plots are still frequently used for these materials [95].
3. The dependence $t(p/p_0)$ is still a matter of discussion because of the lack of independent techniques for determining film thicknesses, and usually reference data from nonporous materials are used. An exact knowledge of $t(p/p_0)$ is needed for the determination of mesopore sizes by classical methods and for testing DFT models. t(p/p0) can be also determined by a combination of nitrogen sorption and small-angle neutron scattering [95, 96].

In mercury porosimetry (MP), gas is evacuated from the sample, which is then immersed in mercury, and an external pressure is applied to gradually force then on wetting mercury into the sample. By monitoring the incremental volume of mercury intruded for each applied pressure, the pore size distribution of the sample can be estimated in terms of the volume of the pores intruded for a given diameter. The evaluation of pore sizes from MP is based on the Washburn equation, quantifying the pressure p required to force a non wetting fluid into a circular cross-sectional capillary of diameter [97].

MP allows the determination of PSD between 3 nm and200 nm and is therefore inappropriate for microporous materials but more suitable for pore sizes above ca. 30 nm compared with nitrogen sorption. As further parameters, the total pore volume V_{tot} is accessible from the total intruded volume of mercury at the highest pressure deter-mined, and the total pore surface S is calculated from:

$$S = \left(1/\gamma |cos\theta| \right) \int_0^{V_{tot}} pdV \quad (16)$$

MP has intrinsic shortcuts in determining mesopore sizes, especially below 20 nm. During the measurement, high pressures used to force mercury into small pores may compress the sample.

Damage or compression of highly porous silica has been reported previously [98–100].

In addition, mercury porosimetry overestimates the volume of the smallest pores in the case of ink-bottle-shaped pores by the small openings[100].

18.7.2 ELECTRON MICROSCOPY

Electron microscopy (transmission electron microscopy (TEM) for micro and mesoporous samples and scanning electron microscopy (SEM) for macroporous samples) is an indispensable tool for the investigation of porous materials. The biggest advantage of these techniques is that they deliver an optical image of the samples.

18.7.3 DIFFRACTION TECHNIQUES

Experiments using elastic X-ray and neutron scattering have turned out to be an invaluable tool for the characterization of various types of porous materials, providing quantitative parameters such as the pore size, surface area, and pore volume. In addition, diffraction techniques allow the determination of the shape and, in particular, the spatial distribution of the pores, for both highly ordered arrays such as inMCM-41 and a more disordered arrangement of pores, for instance in activated carbons. In the early 1900 s, Max von Laue, W. L. Bragg, W. H. Bragg, and others laid the groundwork for X-ray crystallography, which has become a powerful method of visualizing complex inorganic and organic crystalline materials [101].

In spite of the variety of different diffraction techniques currently available, they all are based on the same physical phenomenon, namely the scat-

tering of X-rays and neutrons by the atoms through their electrons or nuclei, respectively. In the Fraunhofer approximation, the interaction of X-rays with electrons leads to the superposition of the coherent scattering of the basic scattering centers (atoms, molecules, or pores), without changing the energy of the incoming X-rays ("elastic scattering"). The resulting coherent scattering pattern is directly related to the mutual position, size, and scattering power of these scattering units. For almost all types of nanoporous materials, no single crystal diffraction datum is obtainable. Therefore, most of the most prominent types of nanoporous materials (zeolites, highly ordered mesoporous materials such as MCM-41or SBA-type materials, and porous carbons) are studied in diffraction experiments as polycrystalline powders. In this case, the powder diffraction raw data are obtained as 1D plots of the coherent scattering intensity versus the scattering angle 2q. Only in the case of thin porous films with an oriented alignment of the pores relative to the substrate 2D diffraction can patterns be obtained [102, 103].

The main problem in analyzing scattering patterns of any kind of polycrystalline nanoporous materials lies in the extraction of a maximum of structural information: the highly advanced crystallographic strategies for the interpretation of 3D diffraction patterns of single crystals are not applicable, requiring different evaluation approaches, which will be briefly described in this section. Depending on the length scale of the pore size, two methods can be distinguished regarding both the experimental realization and the theoretical treatment. Basically, the characteristic length scaled describing the pore system and the corresponding diffraction angle 2q are related by the Bragg equation:

$$s = \frac{1}{d} = (2 sin\theta) / \lambda \tag{17}$$

The crystalline porous materials with pore sizes below the nanometer scale can be investigated by wide-angle scattering (WAS) techniques. The WAS scattering from a polycrystalline porous material is given by:

$$I\left(\left|\vec{S}\right|\right) = \left|Z(\vec{S})\right|^2 \left|F(\vec{S})\right|_w^2 \tag{18}$$

The main difficulty in the interpretation of diffraction patterns of polycrystalline material arises from the spatial average, which leads to a loss of information compared with single crystals. Once the crystal structure has been solved, the pore size and shape are obtained from the atom coordinates.

Nanoporous materials with pores sizes above 1 nm are characterized by a more disordered spatial distribution of the pores and are studied by so-called

small-angle scattering (SAS), with the use of either X-rays (SAXS) or neutrons (SANS). The evaluation of SAS data for porous materials is usually based on the approximation that the material can be regarded as also-called two-phase system: in the case of X-ray scattering, basic theoretical considerations show that the SAS of such materials arises from the scattering at the void-solid interface and is related to the scattering contrast $(\delta_1 - \delta_2)^2$, where δ_1 is the average electron density of the voids (δ_1=0) and δ_2 represents the average electron density of the solid. The SAS of a sufficiently ordered nanoporous material with pores of a distinct shape is also given by Eq. (18), where the lattice factor has the same meaning as in wide-angle scattering, and the form factor here corresponds to the shape of the mesopore/solid rather than the electron density of single atoms [104, 105].

Polycrystalline mesoporous materials with a distinct pore shape and a well-defined 3D or 2D alignment give rise to SAS patterns with a characteristic sequence of reflections ("peaks"). The pore structures of even the most highly ordered mesoporous materials show a considerably lower order and symmetry, which lead to a much smaller number of possible meso structures and corresponding SAS patterns. Therefore, the SAS patterns of this limited number of possible regular mesopore structures and space groups can serve as "finger prints," allowing an almost unambiguous assignment of a certain mesopore lattice structure, if a sufficient number of reflection peaks are obtained. In combination with TEM, SAXS experiments turned out to be a powerful technique for determining the alignment and structure in a variety of mesoporous MCM- and SBA-type materials and other structures [13].

As the main information, SAXS provides the crystallographic space group of the mesopore arrangement and the corresponding lattice parameter. While the 3D alignment of the mesopores can be obtained from the bare SAS peak positions, the determination of mesopore sizes from single-peak analyzes involves substantial uncertainties, because both the peak profiles and intensities can be substantially super imposed by various factors such as smearing, incoherent background scattering, the arrangement of the mesopores and background fluctuations, and the presence of additional intra wall micropores. In particular, so far no satisfactory approach has been developed to quantitatively simulate the influence of 2D and 3D disorder on SAS. Even highly ordered mesopore systems show a certain 2D displacement of the cylindrical mesopores on the hexagonal lattice, which together with the polydispersity of the pores may result in a non-negligible overlap of the peaks, thus severely aggravating a meaningful pore size analysis [106].

Aside from the mesoporous materials with a high degree of order in terms of a regular spatial distribution and a high uniformity of pore size and shape,

SAS is also applied to mesopores showing a certain disorder in the mesopore alignment and a pronounced poly dispersity of pore sizes, which leads to SAS patterns with only few broad reflections [95].

The SAS of this type of material, possessing only a "liquid-like" local ordering of a system of polydisperse mesopores can be reasonably approximated with the use of hard-sphere structure factors such as the Percus-Yevick approach [95].

It is noteworthy that the SAXS data obey the Porod law,

$$\lim_{s \to \infty} s^4 I(s) = \frac{k}{2\pi^3 l_p} \tag{19}$$

which is theoretically predicted for a two-phase system [104].

A further challenge in the interpretation of SAS data from mesoporous materials is the presence of additional micropores, located in the mesopore walls and created by the chains after template removal. However, this approach suffers from the general problems of single-peak analyzes.

A different approach for the evaluation of SAS data, without assuming a specific model, was recently pursued by using the concept of the so-called chord-length distribution (CLD) g(r). g(r) is a statistical function describing the probability of finding a chord of length r that is a connector of two points on the solid-void interface[107].

The only preassumption of this concept is the formal description of the pore system as a two-phase system, but no assumptions about the pore shape and distribution are needed. Therefore the CLD approach is most appropriate for disordered pore systems. The CLD evaluation method provides various useful structural parameters, such as the Porod length (which is the first moment of g(r)), from which an average pore size and wall thickness can be calculated by

$$\begin{aligned} l_{Pore} &= l_P / (1-\phi) \\ l_{wall} &= l_P / \phi \end{aligned} \tag{20}$$

Furthermore, the inner surface area per unit volume S/V is related to l_p as:

$$l_P = 4\phi(1-\phi)V / S \tag{21}$$

Various evaluation methods have been used to determine the micropore size from SAXS such as:

1. An experimental SAS technique for the investigation of mesoporous silicas is: In experimental setup, SANS is combined with in situ nitrogen sorption at T =77 K. In essence, complete SANS curves are obtained during nitrogen sorption at arbitrary points on the isotherm, thus allowing the investigation of the subsequent steps of pore filling as a function of the pore size.
   The experiment takes advantage of the fortunate situation that the scattering contrast between amorphous material and condensed nitrogen (at T =77 K) is almost zero [95, 108].
2. SAS techniques are also applied intensely to elucidate microporous carbons. In this case, the pore structure is characterized by a significant degree of polydispersity in pore size and shape and an irregular spatial distribution of the micropores. Whereas in certain studies the evaluation is based on a dilute system of polydisperse spheres, the application of the CLD concept has shown that the micropores in carbons are more consistent with an acute, needle-like shape. In addition, recent detailed SAXS studies indicate that the apparent fractal structures reported for certain microporous carbons are due to a misinterpretation of SAS data [109].

It has been shown that the CLD concept provides an appropriate method for elucidating the microporous structure of carbons both qualitatively and quantitatively. Based on this method, the changes in the porosity that occur with typical physicochemical processing such as thermal treatment could be described by various structural parameters such as the average pore size, wall thickness, and angularity of the micropores.

In a recent study, the changes in the void-solid microstructure due to these treatments were studied by a combination of SAXS and WAXS.

18.7.4 POSITRON ANNIHILATION

In the past 20 years, positron annihilation lifetime spectroscopy (PALS) has been developed into a powerful tool for the detection and quantification of defects on the atomic scale in various types of solids. PALS is sensitive to different kinds of defects, such as dislocations and vacancies in metals or crystals, grain boundaries, and voids and pores. Similar to scattering techniques, PALS is a noninvasive technique and thereby allows the detection of inaccessible pores. In the area of micro and mesoporous materials, PALS is predominantly applied to porous polymers and thin porous films. PALS is based on the decay of positrons into twoγphotons ("annihilation") and has been described in various publications [110, 111].

Naas the radioactive source, the formation of positrons (b^+) by radioactive decay is accompanied by the simultaneous emergence of ag-quantum of 1.273 MeV, which defines the starting signal of the positron lifetime measurement. Entering the sample, the positrons lose their high energy by inelastic collisions with electrons. These "thermalized" positrons have energies on the order of a few meV, form positroniums (Ps, the electron-positron bound state), and diffuse through the solid until annihilation after their specific lifetime in the solid, which is measured as the time difference between the creation of the 1.273-MeVg-quantum and the annihilation radiation (two 511-keVg rays). The natural lifetime of Ps of 142 ns is reduced by annihilation with electrons during collisions. The lifetimes, the inverse of the annihilation rates, become longer when a positron or positronium is localized at spaces with lower electron density such as voids.

Thus, positrons can be used as a probe to investigate average sizes of the free volume, size distribution, and the free volume concentration by measuring their lifetimes.

The raw data of PALS are plots of the annihilation radiation signal as a function of time. In the case of defect concentration porosities that are not too high, there will be at least two or three lifetime components in the spectra, which are usually analyzed as a sum of exponentials after background subtraction. Furthermore, the spectra can be deconvoluted with CONTIN software [112, 113].

Basically, PALS allows the calculation of both the porosity and the pore size distribution. Since the determination of pore size distributions is based on the trapping of positrons in voids of varying size, PALS was reported to lead to a single, average lifetime in the case of interconnected pore systems.

The fundamental problem in the evaluation of PALS data lies with the exact relationship between the intensity of the long-lived components of a PALS spectrum and the concentration of the voids of a certain size, which is still a matter of intense research [114–116].

The extension of PALS to voids larger than 1 nm causes further uncertainties and is subject to ongoing research [116, 117].

PALS has been successfully applied to the investigation of porosity in various types of porous materials, such as thin siliceous films, cement, porous silicon, zeolites, and mesoporous silicas such as vycor glass [117, 118].

18.7.5 MERCURY POROSIMETRY

In mercury porosimetry (MP), gas is evacuated at elevated temperatures and low pressures from the sample placed inside a pressure chamber. Subsequent-

ly the sample is immersed in mercury and an external pressure is applied to force mercury into the sample. By gradually increasing the pressure and monitoring the incremental volume of mercury intruded for each new applied pressure, the pore size distribution of the sample can be estimated in terms of the volume of the pores intruded for a given diameter. Mercury is a non wetting fluid at room temperature for most porous materials of technological interest. The evaluation of pore sizes from MP is based on the Wash burn equation, quantifying the pressure required to force an on wetting fluid into a circular cross-sectional capillary of diameter by [97]:

$$P = (4\gamma COS\,\theta) / D \tag{22}$$

MP allows the determination of average pore sizes (and their distribution) between 3 nm and 200 nm. So it is improper for microporous materials and suitable for pore sizes above ca. 30 nm than nitrogen sorption. As further parameters, the total pore volume is accessible from the total intruded volume of mercury at the highest pressure determined, and the total pore surface is calculated from:

$$S = 1/\gamma\left|COS\theta\right|\int_{0}^{V_{tot}} pdV \tag{23}$$

Because of inherent shortcuts MP is not used as frequently as gas sorption138 techniques for samples with mesopore sizes below 20 nm. During the measurement, high pressures used to force mercury into small pores may compress the sample [98].

In addition, mercury porosimetry overestimates the volume of the smallest pores in the case of ink-bottle-shaped pores, because the intrusion of mercury into the larger pores is determined by the small openings.

Moreover, it has to be pointed out that poresize distributions by the Washburn equation are not a geo-metrical relationship, but a physical characteristic of a porous medium, because MP is based on transport and relaxation phenomena [100].

18.8 CONCLUDING REMARKS

This chapter attempted to cover the field of nanoporous materials. In addition, it was tried to give an introduction to state-of-the-art overview for microporous, mesoporous and macroporous materials, while the preference lies on ordered pore structures. Different techniques for measuring pore size distri-

bution were investigated such as: optically technique, imbibitions weighing, gas adsorption, mercury intrusion and etc. Many of these methods give results that quite frequently differ from one another. This is often because they are in fact measuring different things some measurements are directly on the pores themselves. Others (such as mercury intrusion) are in effect measuring the necks that give access to the pores. For obtaining better understanding of modern analysis methods for nanoporous analysis, gas adsorption, electron microscopy, positron annihilation and mercury porosimetry were investigated in detail for different situation. Among them, gas adsorption can be successfully used in many samples and diffraction techniques can analysis as 3D networks.

KEYWORDS

- **Analysis methods**
- **Porosity measurement**
- **Porous material**

REFERENCES

1. Polarz, S. & Smarsly, B., *Nanoporous materials.* Journal of nanoscience and nanotechnology, 2002, *2(6),* 581–612.
2. Hiemenz, P. C. & Rajagopalan, R., *Principles of Colloid and Surface Chemistry, revised and expanded.* Vol. 14. 1997, CRC Press.
3. McDowell-Boyer, L. M., Hunt, J. R., & Sitar, N., *Particle transport through porous media.* Water Resources Research, 1986, *22(13),* 1901–1921.
4. Auset, M. & A. Keller, A., *Pore-scale processes that control dispersion of colloids in saturated porous media.* Water Resources Research, 2004, *40(3).*
5. Bhave, R. R., *Inorganic membranes synthesis, characteristics, and applications.* Vol. 312. 1991, Springer.
6. Lin, V. S.-Y., et al., *A porous silicon-based optical interferometric biosensor.* Science, 1997, *278(5339),* 840–843.
7. Hedrick, J., et al. *Templating nanoporosity in organosilicates using well-defined branched macromolecules.* in *Materials Research Society Symposium Proceedings.* 1998, Cambridge Univ Press.
8. Hubbell, J. A. & Langer, R. *Tissue engineering* Chem. Eng. News 1995, 13, 42–45.
9. Schaefer , D. W., *Engineered porous materials* MRS Bulletin 1994, 19, 14–17.
10. Hentze, H. P., & Antonietti, M. *Porous Polymers and Resins.* Handbook of Porous Solids: 1964–2013.
11. Endo, A., et al., *Synthesis of ordered microporous silica by the solvent evaporation method.* Journal of materials science, 2004, *39(3),* 1117–1119.
12. Sing, K., et al., *Physical and biophysical chemistry division commission on colloid and surface chemistry including catalysis.* Pure and Applied Chemistry, 1985, *57(4),* 603–619.

13. Kresge, C., et al., *Ordered mesoporous molecular sieves synthesized by a liquid-crystal template mechanism.* nature, 1992, 359(6397), 710–712.
14. Yang, P., et al., *Generalized syntheses of large-pore mesoporous metal oxides with semi-crystalline frameworks.* nature, 1998, 396(6707), 152–155.
15. Jiao, F., K. Shaju, M., & P. Bruce, G., *Synthesis of Nanowire and Mesoporous Low-Temperature $LiCoO_2$ by a Post-Templating Reaction.* Angewandte Chemie International Edition, 2005, *44(40),* 6550–6553.
16. Ryoo, R., et al., *Ordered mesoporous carbons.* Advanced Materials, 2001, *13(9),* 677–681.
17. Beck, J., et al., *Chu, DH Olson, EW Sheppard, SB McCullen, JB Higgins and JL Schlenker.* J. Am. Chem. Soc, 1992, *114(10),* 834.
18. Zhao, D., et al., *Triblock copolymer syntheses of mesoporous silica with periodic 50 to 300 angstrom pores.* Science, 1998, 279(5350), 548–552.
19. Joo, S. H., et al., *Ordered nanoporous arrays of carbon supporting high dispersions of platinum nanoparticles.* nature, 2001, 412(6843), 169–172.
20. Kruk, M., et al., *Synthesis and characterization of hexagonally ordered carbon nanopipes.* Chemistry of materials, 2003, *15(14),* 2815–2823.
21. Rouquerol, J., et al., *Recommendations for the characterization of porous solids (Technical Report).* Pure and Applied Chemistry, 1994, *66(8),* 1739–1758.
22. Schüth, F., Sing, K. S.W., & Weitkamp, J., *Handbook of porous solids*. 2002, Wiley-Vch.
23. Maly, K. E., *Assembly of nanoporous organic materials from molecular building blocks.* Journal of Materials Chemistry, 2009, *19(13),* 1781–1787.
24. Davis, M. E., *Ordered porous materials for emerging applications.* nature, 2002, 417(6891), 813–821.
25. Morris, R. E. & Wheatley, P. S., *Gas storage in nanoporous materials.* Angewandte Chemie International Edition, 2008, *47(27),* 4966–4981.
26. Cote, A. P., et al., *Porous, crystalline, covalent organic frameworks.* Science, 2005, 310(5751), 1166–1170.
27. El-Kaderi, H. M., et al., *Designed synthesis of 3D covalent organic frameworks.* Science, 2007, 316(5822), 268–272.
28. Jiang, J. X., et al., *Conjugated microporous poly (aryleneethynylene) networks.* Angewandte Chemie International Edition, 2007, *46(45),* 8574–8578.
29. Ben, T., et al., *Targeted synthesis of a porous aromatic framework with high stability and exceptionally high surface area.* Angewandte Chemie, 2009, *121(50),* 9621–9624.
30. Eddaoudi, M., et al., *Modular chemistry: secondary building units as a basis for the design of highly porous and robust metal-organic carboxylate frameworks.* Accounts of Chemical Research, 2001, *34(4),* 319–330.
31. Lu, G. Q., Zhao, X.S., & T. Wei, K., *Nanoporous materials: science and engineering.* Vol. 4. 2004, Imperial College Press.
32. Holister, P., Vas, C.R., & Harper, T., *Nanocrystalline materials.* Technologie White Papers, 2003(4).
33. Smith, B. L., et al., *Molecular mechanistic origin of the toughness of natural adhesives, fibres and composites.* nature, 1999, 399(6738), 761–763.
34. Mann, S. & Ozin, G. A., *Synthesis of inorganic materials with complex form.* nature, 1996, 382(6589), 313–318.
35. Mann, S., *Molecular tectonics in biomineralization and biomimetic materials chemistry.* nature, 1993, *365(6446),* 499–505.
36. Busch, K. & John, S., *Photonic band gap formation in certain self-organizing systems.* Physical Review E, 1998, *58(3),* p. 3896.

37. Argyros, A., et al., *Electron tomography and computer visualisation of a three-dimensional 'photonic'crystal in a butterfly wing-scale.* Micron, 2002, *33(5),* 483–487.
38. Sailor, M. J. & Kavanagh, K.L., *Porous silicon–what is responsible for the visible luminescence?* Advanced Materials, 1992, *4(6),* 432–434.
39. Koshida, N. & Gelloz, B., *Wet and dry porous silicon.* Current opinion in colloid & interface science, 1999, *4(4),* 309–313.
40. Hentze, H.-P. & Antonietti, M., *Template synthesis of porous organic polymers.* Current Opinion in Solid State and Materials Science, 2001, *5(4),* 343–353.
41. Nakao, S.-I., *Determination of pore size and pore size distribution: 3. Filtration membranes.* Journal of Membrane Science, 1994, *96(1),* 131–165.
42. Barrer, R. M., *Zeolites and their synthesis.* Zeolites, 1981, *1(3),* 130–140.
43. Iijima, S., *Helical microtubules of graphitic carbon.* nature, 1991, 354(6348), 56–58.
44. Kaneko, K. & Inouye, K., *Adsorption of water on FeOOH as studied by electrical conductivity measurements.* Bulletin of the Chemical Society of Japan, 1979, *52(2),* 315–320.
45. Maeda, K., et al., *Control with polyethers of pore distribution of alumina by the sol-gel method.* Chem. Ind., 1*989(23),* 807.
46. Matsuzaki, S., Taniguchi, M., & Sano, M., *Polymerization of benzene occluded in graphite-alkali metal intercalation compounds.* Synthetic metals, 1986, *16(3),* 343–348.
47. Enoki, T., Inokuchi, H., & Sano, M., ESR study of the hydrogen-potassium-graphite ternary intercalation compounds. *Physical Review B*, 1988, *37(16),* p. 9163.
48. Pinnavaia, T. J., *Intercalated clay catalysts.* Science, 1983, 220(4595), 365–371.
49. Yamanaka, S., et al., *High surface area solids obtained by intercalation of iron oxide pillars in montmorillonite.* Materials research bulletin, 1984, *19(2),* 161–168.
50. Inagaki, S., Fukushima, Y., & Kuroda, K., Synthesis of highly ordered mesoporous materials from a layered polysilicate. *J. Chem. Soc., Chem. Commun.,* 1*993(8),* 680–682.
51. Vallano, P. T. & Remcho, V. T., Modeling Interparticle and Intraparticle (perfusive) Electroosmotic Flow in Capillary Electrochromatography. *Analytical Chemistry*, 2000, *72(18),* 4255–4265.
52. Levenspiel, O., *Chemical Reaction Engineering.* Vol. 2. 1972, Wiley New York etc.
53. Li, Q., et al., *Interparticle and Intraparticle Mass Transfer in Chromatographic Separation.* Bioseparation, 1995, *5(4),* 189–202.
54. Setoyama, N., et al., *Surface Characterization of Microporous Solids with Helium Adsorption and Small Angle X-ray Scattering.* Langmuir, 1993, *9(10),* 2612–2617.
55. Marsh, H., *Introduction to Carbon Science.* 1989.
56. Kaneko, K., *Determination of pore size and pore size distribution: 1. Adsorbents and catalysts.* Journal of Membrane Science, 1994, *96(1),* 59–89.
57. Seaton, N. & Walton, J., *A new analysis method for the determination of the pore size distribution of porous carbons from nitrogen adsorption measurements.* Carbon, 1989, *27(6),* 853–861.
58. Dullien, F. A., *Porous media: fluid transport and pore structure*. 1991, Academic press.
59. Langmuir, I., *The adsorption of gases on plane surfaces of glass, mica and platinum.* Journal of the American Chemical society, 1918, *40(9),* 1361–1403.
60. Brunauer, S., P. Emmett, H., & Teller, E., *Adsorption of gases in multimolecular layers.* Journal of the American Chemical society, 1938, *60(2),* 309–319.
61. Barrett, E. P., L. Joyner, G., & Halenda, P. P., *The determination of pore volume and area distributions in porous substances. I. Computations from nitrogen isotherms.* Journal of the American Chemical society, 1951, *73(1),* 373–380.

62. Thomson, W., *LX. On the equilibrium of vapour at a curved surface of liquid.* The London, Edinburgh, and Dublin Philosophical Magazine and Journal of Science, 1871, 42(282), 448–452.
63. Gregg, S. J., Sing, K. S.W., & Salzberg, H., *Adsorption surface area and porosity.* Journal of The Electrochemical Society, 1967, *114(11)*, 279C-279C.
64. Ritter, H. & Drake, L., *Pressure porosimeter and determination of complete macropore-size distributions. Pressure porosimeter and determination of complete macropore-size distributions.* Industrial & Engineering Chemistry Analytical Edition, 1945, *17(12)*, 782–786.
65. Brun, M., et al., *A new method for the simultaneous determination of the size and shape of pores: the thermoporometry.* Thermochimica Acta, 1977, *21(1)*, 59–88.
66. Jallut, C., et al., *Thermoporometry.: Modelling and simulation of a mesoporous solid.* Journal of Membrane Science, 1992, *68(3)*, 271–282.
67. Ishikiriyama, K., Todoki, M., & Motomura, K., *Pore size distribution (PSD) measurements of silica gels by means of differential scanning calorimetry, I. Optimization for determination of PSD.* Journal of colloid and interface science, 1995, *171(1)*, 92–102.
68. Leofanti, G., et al., *Surface area and pore texture of catalysts.* Catalysis Today, 1998, *41(1)*, 207–219.
69. Sing, K. & Gregg, S., *Adsorption, surface area and porosity.* Adsorption, Surface Area and Porosity, 1982.
70. De Boer, J., Everett, D., & Stone, F., *The structure and properties of porous materials.* Butterworths, London, 1958, 10, 68.
71. Ravikovitch, P. I. & Neimark, A. V., *Characterization of micro-and mesoporosity in SBA-15 materials from adsorption data by the NLDFT method.* The Journal of Physical Chemistry B, 2001, *105(29)*, 6817–6823.
72. Cole, M. W. & Saam, W., *Excitation spectrum and thermodynamic properties of liquid films in cylindrical pores.* Physical Review Letters, 1974, *32(18)*, 985.
73. Evans, R., Marconi, U. M. B., & Tarazona, P., *Capillary condensation and adsorption in cylindrical and slit-like pores.* Journal of the Chemical Society, Faraday Transactions 2, Molecular and Chemical Physics, 1986, *82(10)*, 1763–1787.
74. Jiang, S., Gubbins, K. E., & Balbuena, P. B., *Theory of adsorption of trace components.* The Journal of Physical Chemistry, 1994, *98(9)*, 2403–2411.
75. Jc, B. & Deboer, J., Studies on Pore Systems in Catalysts. 11. Pore distribution calculations from adsorption branch of a nitrogen adsorption isotherm in case of ink-bottle type pores. *Journal of Catalysis*, 1968, *10(2)*, 153.
76. Kruk, M., Jaroniec, M., & Sayari, A., *Application of large pore MCM-41 molecular sieves to improve pore size analysis using nitrogen adsorption measurements.* Langmuir, 1997, *13(23)*, 6267–6273.
77. Kruk, M., et al., *New approach to evaluate pore size distributions and surface areas for hydrophobic mesoporous solids.* The Journal of Physical Chemistry B, 1999, *103(48)*, 10670–10678.
78. Jaroniec, M. & Kaneko, K., *Physicochemical foundations for characterization of adsorbents by using high-resolution comparative plots.* Langmuir, 1997, *13(24)*, 6589–6596.
79. Kruk, M., et al., *Characterization of high-quality MCM-48 and SBA-1 mesoporous silicas. Chemistry of Materials*, 1999, *11(9)*, 2568–2572.
80. Kruk, M., et al., *Characterization of the porous structure of SBA-15. Chemistry of Materials*, 2000, *12(7)*, 1961–1968.
81. Kruk, M., et al., *Determination of pore size and pore wall structure of MCM-41 by using nitrogen adsorption, transmission electron microscopy, and X-ray diffraction.* The Journal of Physical Chemistry B, 2000, *104(2)*, 292–301.

82. Dubinin, M. & Astakhov, V., *Development of the concepts of volume filling of micropores in the adsorption of gases and vapors by microporous adsorbents.* Bulletin of the Academy of Sciences of the USSR, Division of chemical science, 1971, *20(1),* 3–7.
83. Dubinin, M. & Stoeckli, H., *Homogeneous and heterogeneous micropore structures in carbonaceous adsorbents.* Journal of colloid and interface science, 1980, *75(1),* 34–42.
84. Saito, A. & H. Foley, C., *Argon porosimetry of selected molecular sieves: experiments and examination of the adapted Horvath-Kawazoe model.* Microporous Materials, 1995, *3(4),* 531–542.
85. Kawazoe, H., *Method for the calculation of effective pore size distribution in molecular sieve carbon.* Journal of Chemical Engineering of Japan, 1983, *16(6),* 470–475.
86. Fraissard, J. P., *Physical Adsorption: Experiment, Theory, and Applications*. Vol. 491. 1997, Springer.
87. Maddox, M., Olivier, J., & Gubbins, K., *Characterization of MCM-41 using molecular simulation: heterogeneity effects.* Langmuir, 1997, *13(6),* 1737–1745.
88. Tarazona, P., Marconi, U. M. B., & Evans, R., *Phase equilibria of fluid interfaces and confined fluids: non-local versus local density functionals.* Molecular Physics, 1987, *60(3),* 573–595.
89. de Keizer, A., Michalski, T., & Findenegg, G.H., *Fluids in pores: experimental and computer simulation studies of multilayer adsorption, pore condensation and critical-point shifts.* Pure Appl. Chem, 1991, *63(7),* p. 1495.
90. Tarazona, P., *Erratum: Free-energy density functional for hard spheres [Phys. Rev. A 31, 2672 (1985)].* Physical Review A, 1985, *32,* p. 3148.
91. Neimark, A. V., et al., *Pore size analysis of MCM-41 type adsorbents by means of nitrogen and argon adsorption.* Journal of colloid and interface science, 1998, *207(1),* 159–169.
92. Köhn, R. & Fröba, M., *Nanoparticles of 3d transition metal oxides in mesoporous MCM-48 silica host structures: Synthesis and characterization.* Catalysis Today, 2001, *68(1),* 227–236.
93. Göltner, C. G., et al., *On the microporous nature of mesoporous molecular sieves.* Chemistry of materials, 2001, *13(5),* 1617–1624.
94. Thommes, M., Köhn, R., & Fröba, M., *Sorption and pore condensation behavior of nitrogen, argon, and krypton in mesoporous MCM-48 silica materials.* The Journal of Physical Chemistry B, 2000, *104(33),* 7932–7943.
95. Smarsly, B., et al., *SANS investigation of nitrogen sorption in porous silica.* The Journal of Physical Chemistry B, 2001, *105(4),* 831–840.
96. Jaroniec, M., Kruk, M., & Olivier, J. P., *Standard nitrogen adsorption data for characterization of nanoporous silicas.* Langmuir, 1999, *15(16),* 5410–5413.
97. Van Brakel, J., Modrý, S., & Svata, M., *Mercury porosimetry: state of the art.* Powder Technology, 1981, *29(1),* 1–12.
98. Johnston, G., et al., *Compression effects in mercury porosimetry.* Powder Technology, 1990, *61(3),* 289–294.
99. Brown, S. M. & Lard, E. W., *A comparison of nitrogen and mercury pore size distributions of silicas of varying pore volume.* Powder Technology, 1974, *9(4),* 187–190.
100. Allen, T., *Particle Size Measurement, Vol. 1.* Powder Technology Series, Powder Sampling and Particle Size Measurement. Chapman & Hall, 1997.
101. Bragg, W. & Bragg, W., *The reflection of X-rays by crystals.* Proceedings of the Royal Society of London. Series A, 1913, *88(605),* 428–438.
102. Inagaki, S., et al., *An ordered mesoporous organosilica hybrid material with a crystal-like wall structure.* nature, 2002, *416(6878)*, 304–307.

103. Grosso, D., et al., *Highly Organized Mesoporous Titania Thin Films Showing Mono-Oriented 2D Hexagonal Channels.* Advanced Materials, 2001, *13(14),* 1085–1090.
104. Guinier, A. & Fournet, G., *Small-angle scattering of X-rays.* 1955, Wiley.
105. Feigin, L., D. Svergun, I., & G. Taylor, W., *Structure analysis by small-angle X-ray and neutron scattering.* 1987, Springer.
106. Sauer, J., Marlow, F., & Schüth, F., *Simulation of powder diffraction patterns of modified ordered mesoporous materials.* Physical Chemistry Chemical Physics, 2001, *3(24),* 5579–5584.
107. Dubey, P. A., Schönfeld, B., & Kostorz, G., *Shape and internal structure of Guinier-Preston zones in AlAg.* Acta metallurgica et materialia, 1991, *39(6),* 1161–1170.
108. Hoinkis, E., *Small angle neutron scattering study of C6D6 condensation in a mesoporous glass.* Langmuir, 1996, *12(17),* 4299–4302.
109. Albouy, P.-A. & Ayral, A., *Coupling X-ray scattering and nitrogen adsorption: an interesting approach for the characterization of ordered mesoporous materials. Application to hexagonal silica.* Chemistry of materials, 2002, *14(8),* 3391–3397.
110. Shantarovich, V., et al., *Positron annihilation lifetime study of high and low free volume glassy polymers: effects of free volume sizes on the permeability and permselectivity.* Macromolecules, 2000, *33(20),* 7453–7466.
111. Staab, T., Krause-Rehberg, R., & Kieback, B., *Review Positron annihilation in fine-grained materials and fine powders—an application to the sintering of metal powders.* Journal of materials science, 1999, *34(16),* 3833–3851.
112. Dlubek, G., et al., *Water in local free volumes of polyimides: A positron lifetime study.* Macromolecules, 1999, *32(7),* 2348–2355.
113. Gregory, R. B., *Free-volume and pore size distributions determined by numerical Laplace inversion of positron annihilation lifetime data.* Journal of *Applied Physics*, 1991, *70(9),* 4665–4670.
114. Eldrup, M., Lightbody, D., & Sherwood, J., *The temperature dependence of positron lifetimes in solid pivalic acid.* Chemical Physics, 1981, *63(1),* 51–58.
115. Shantarovich, V., *Some Aspects of Positronium Interaction with Elementary Free Volumes in Polymers.* Acta Physica Polonica A, 2001, *99(3–4),* 487–495.
116. Ito, K., Nakanishi, H., & Ujihira, Y., *Extension of the equation for the annihilation lifetime of ortho-positronium at a cavity larger than 1 nm in radius.* The Journal of Physical Chemistry B, 1999, *103(21),* 4555–4558.
117. Nakanishi, H. & Ujihira, Y., *Application of positron annihilation to the characterization of zeolites.* The Journal of Physical Chemistry, 1982, *86(22),* 4446–4450.
118. Goworek, T., et al., *Temperature variations of average o-Ps lifetime in porous media.* Radiation Physics and Chemistry, 2000, *58(5),* 719–722.

INDEX

C

F

G

Q

R

S